Microwave Integrated Circuits

Microwave Technology Series

The *Microwave Technology Series* publishes authoritative works for professional engineers, researchers and advanced students across the entire range of microwave devices, sub-systems, systems and applications. The series aims to meet the reader's needs for relevant information useful in practical applications. Engineers involved in microwave devices and circuits, antennas, broadcasting communications, radar, infra-red and avionics will find the series an invaluable source of design and reference information.

Series editors:
Michel-Henri Carpentier
Professor in 'Grandes Écoles', France,
Fellow of the IEEE, and President of the French SEE

Bradford L. Smith
International Patents Consultant and Engineer
with the Alcatel group in Paris, France,
and a Senior Member of the IEEE and French SEE

Titles available

1. **The Microwave Engineering Handbook Volume 1**
   Microwave components
   Edited by Bradford L. Smith and Michel-Henri Carpentier
2. **The Microwave Engineering Handbook Volume 2**
   Microwave circuits, antennas and propagation
   Edited by Bradford L. Smith and Michel-Henri Carpentier
3. **The Microwave Engineering Handbook Volume 3**
   Microwave systems and applications
   Edited by Bradford L. Smith and Michel-Henri Carpentier
4. **Solid-state Microwave Generation**
   J. Anastassiades, D. Kaminsky, E. Perea and A. Poezevara
5. **Infrared Thermography**
   G. Gaussorgues
6. **Phase Locked Loops**
   J.B. Encinas
7. **Frequency Measurement and Control**
   Chronos Group
8. **Microwave Integrated Circuits**
   Edited by I. Kneppo and J. Fabian

Microwave Integrated Circuits

I. KNEPPO

Slovak Institute of Metrology
Bratislava, Slovak Republic

J. FABIAN

Electrotechnical Institute of
the Slovak Academy of Sciences,
Bratislava, Slovak Republic

P. BEZOUŠEK
P. HRNÍČKO
M. PAVEL

Research Institute of
Radiotechnology
Pardubice, Czech Republic

CHAPMAN & HALL
London · Glasgow · New York · Tokyo · Melbourne · Madras

Published by Chapman & Hall, 2-6 Boundary Row, London SE1 8HN, UK

Chapman & Hall, 2–6 Boundary Row, London SE1 8HN, UK

Blackie Academic & Professional, Wester Cleddens Road, Bishopbriggs, Glasgow G64 2NZ, UK

Chapman & Hall Inc., One Penn Plaza, 41st Floor, New York NY 10119, USA

Chapman & Hall Japan, Thomson Publishing Japan, Hirakawacho Nemoto Building, 6F, 1-7-11 Hirakawa-cho, Chiyoda-ku, Tokyo 102, Japan

Chapman & Hall Australia, Thomas Nelson Australia, 102 Dodds Street, South Melbourne, Victoria 3205, Australia

Chapman & Hall India, R. Seshadri, 32 Second Main Road, CIT East, Madras 600 035, India

Published in co-edition with Ister Science Press Limited
Ister Science Press Limited, Staromestska 6, Bratislava, Slovak Republic

First edition 1994

Printed in Great Britain by Clays Ltd, St Ives Plc, Bungay, Suffolk

ISBN 0 412 54700 7

A catalogue record for this book is available from the British Library

∞ Printed on permanent acid-free text paper, manufactured in accordance with ANSI/NISO Z39.48-1992 (Permanence of Paper).

Contents

Contributors

P. Bezoušek

Research Institute of Radiotechnology, Pardubice, Czech Republic

J. Fabian

Electrotechnical Institute of the Slovak Academy of Sciences, Bratislava, Slovak Republic

F. Hrníčko

Research Institute of Radiotechnology, Pardubice, Czech Republic

I. Kneppo

Slovak Institute of Metrology, Bratislava, Slovak Republic

M. Pavel

Research Institute of Radiotechnology, Pardubice, Czech Republic

List of symbols

A) Physical quantities

a	waveguide width; parameter of the Richards' transformation; parameter of the coplanar line
b	waveguide height
B	parameter of the Richards' transformation
B_a, B_b	susceptance
c	light velocity in vacuum
C, C_c	capacitance, collector-base capacitance
d	distance of the lateral wall from the line
e	distance of the slot centre from the waveguide centre
$\mathbf{E}$	electric field intensity
$\mathbf{E}_0$	electric intensity before introducing losses
f	frequency
f_c	waveguide cut-off frequency
f_{cm}	m-th mode cut-off frequency
$f_{\max}$	maximum oscillation frequency of the transistor
f_n	normalization frequency
f_{stat}	limit frequency of using results of quasistatic methods
f_T	transit time cut-off frequency of a transistor
$F_{\min}$	minimum noise figure
g, g_m	normalized slot width, transconductans for $U_S = 0\,\mathrm{V}$, respectively
g_i	values of prototype elements

G	conductivity
G_a	available power gain
$G(x, y; x', y')$	Green's function
h	substrate thickness
h_1	height of the dielectric layer under substrate
h_2	height of dielectric layer above substrate
h_{21e}	current gain of the transistor in common emitter configuration
$\mathbf{H}$	magnetic field intensity
$\mathbf{H}_0$	magnetic field intensity before introducing losses
I	longitudinal current in the strips; isolation attenuation
I_{dss}	drain current of transistor
I_L	transition attenuation (loss)
J	surface current density
$J_{xin}(\beta)$	Fourier image of the base function of the current density
k	wave number
k_c	equivalent dielectric constant at the limit frequency
k_e	equivalent dielectric constant
k, k_i, k', k'_i	moduli of elliptic integral
$K(k), K(k_i)$	complete elliptic integral of first type
$K'(k), K'(k_i)$	complementary elliptic integral of first type
l_i	length of the line segment
l_o	length of the capacity of the stub
l_r	resonator length
l_s	active segment length
L	inductance; coupling attenuation of a coupler
n	number of vertices of n-angle; filtre degree
n_j	normal line to the j-th part of the conductor
N	number of conductive strips
$N_n(kr)$	Bessel's function of second type of n-th order

p	approximation coefficient; complex frequency; auxiliary constant
$P, P_{\max}$	power, maximum power
p	dipole moment
$P(f)$	correcting function
q	approximation coefficient, filling factor, number of transfer zeros
q_s	correction for screening
q_t	correction for the microstrip thickness
Q	charge on the strip
r	cylindrical coordinate
r_i	internal diameter of a radial stub
r_L	external diameter of a radial stub
R, R_0	resistance, terminating resistance
R'_b	equivalent substrate resistance
RL	return losses
R_s, R_{sj}	specific high frequency resistance, specific high frequency resistance of the j-th part of the conductor surface
s	slot width; curve enclosing the strip conductor
S	directivity, surface area
S_v	surface area of the conductor cross section
$S_{12}(j)$	transfer function
t	time, conductor thickness
u	normalized conductor width
U	voltage at the strip centre
U_{DS}	drain-source voltage
U_p, U_P	breakdown voltage, pinch-off voltage, respectively
v_f	phase velocity
v	wave velocity
V, V_1, V_2, V_3	potential of conductors
${}_kV_m^i$	input voltage pulse from direction m

${}_{k+1}V_n^r$	reflected voltage pulse in direction n
w	complex variable; conductor width
w'	corrected strip width
w_c, w_L, w_1, w_2	conductor width
w_{ef}	effective strip width
W	wave energy per unit length
x, x', x_b	parameter
x, x_i	Cartesian coordinate
x	charge coordinate
y, y_i	Cartesian coordinate
$\mathbf{y}$	unit vector in direction of y-axis
y'	charge coordinate
Y_0	characteristic admittance
Y_{0p}	characteristic admittance of a capacitive sub
z	Cartesian coordinate; complex variable
Z_i	input wave impedance
Z_T	input terminal impedance
Z_0, Z_{0A}	characteristic impedance
$Z_{0\infty}$	characteristic impedance for $f \to \infty$
$Z_0(kr)$	wave impedance of radial line
Z_{0e}, Z_{0o}	characteristic impedance of even and odd modes respectively
α	attenuation constant, current gain, relaxation constant
α_0	current gain at zero frequency
α_C	conductive losses of unit length line
α_d	attenuation constant of the TEM mode resulting from dielectric losses
α_D	dielectric losses of unit length line
α_i	vertex angle
α_r	centre angle of radial stub
α_l	temperature coefficient of the length expansivity

α_z	transformation parameter
α_{ε_r}	temperature coefficient of relative permitivity
β	phase constant, coefficient of voltage coupling
γ	propagation constant
Γ	reflection coefficient
Γ_e, Γ_o	reflection coefficient of even and odd modes, respectively
δ, δ_{ef}	loss angle, effective loss angle
δ_n	penetration depth
$\delta(x - x')$	δ-function
Δ	increment
$\varepsilon, \varepsilon(\omega)$	permittivity
ε_r	relative permittivity
ε_{ef}	effective permittivity
η_{add}	adding efficiency
Θ	cylindrical coordinate
Θ_i, Θ_L	angle
λ_g, λ_L	wave length in line
λ_c	cut-off wave length
μ, μ_0	permeability, permeability of vacuum
Π^e, Π^m	Hertz vector
ρ	charge density; triangular polar coordinate
τ_b, τ_c, τ_e	time of flight through the base, collector, emittor
τ_{ec}	total transistor delay
τ_{rc}	time constant of the collector
φ_d	differential phase shift
Φ, Φ_k	potential, potential on the surface of the k-th conductor
Φ_D	scalar dipole potential
ψ	angle
ω	angular frequency
ω'	cut-off angular frequency

B) Abbreviations

BFL	Buffered FET Logic
DCFL	Direct Coupled FET Logic
SDFL	Schottky Diode FET Logic
HEMT	High Electron Mobility Transistor
MODFET	MOdulation Doped FET
TEGFET	Two-dimensional Electron Gas FET
SDHT	Selective Doped Heterostructure Transistor
SISFET	Semiconductor Insulator Semiconductor FET
MESFET	MEtal Semiconductor FET
FET	Field Effect Transistor
MIC	Microwave Integrated Circuit
MMIC	Monolitic Microwave Integrated Circuit
SWR	Standing Wave Ratio
VSWR	Voltage Standing Wave Ratio
TEM	Transverse ElectroMagnetic (wave)

1

Introduction

In general, the origin of the development of microwave circuits dates back to 1930, when microwave systems were introduced, advantageously implemented by waveguide parts. In attempts to provide the efficient transfer of the microwave power from the supply into the waveguide transmission line, considerable effort was developed in the field of the analysis of microwave circuits. Works by P. H. Smith (Smith, 1969) as a designer of graphic means for solving problems of transmission lines are well known. World War II accelerated the progress of the development of microwave procedures, which have found their application in the radar technique. This was a period of the origination of the hybrid T-element and of the first directional coupler and these parts immediately found their use in practice. The classic methods of solving the circuits led to a change from relationships for the voltage and current and impedance and admittance matrices to terms as transmitted and reflected waves, which resulted to a concept of scattering matrices. This opened possibilities of solving multiport networks. In this period, the waveguide and coaxial technique represented elements of microwave circuits and they complemented each other by their characteristics. The waveguide technique was used for its capability to transmit higher powers at low losses. Resonators based on cavity waveguides achieved high quality coefficients Q. In contrast to this, the coaxial technique offered a wide frequency range due to the absence of dispersion effects. The 1950s were characterized by an origin of special applications of two-conductor lines, as for example the strip line (Barret and Barnes, 1951). The advantage was taken of the feature that the characteristic impedance of the line can be easily controlled by the width of the middle strip, usually produced by the photo-etching technology on the copperclad dielectric substrate. This technology also found a wider application in the production of directional couplers and power dividers.

The microstrip is a further type of line occurring after that, however, due to larger loss per unit length resulting from radiation and due to its low dielectric constant ($\varepsilon = 2.5$) of the originally considered substrates, it did not find wide use in the microwave technique at that time. This, however, simultaneously initiated the development for a further period.

In the 1960s, a revival of microstrip lines actually occurred. This was also facilitated by the discovery of the elegant analysis of microstrip structures by Wheeler, 1964, which was based on the conformal representation. The new methods and technologies initiated a rapid development in the use of microstrip lines. The abilities of planar microwave transmission structures, rapid development of microwave semiconductor elements, technique of thin layers and photolithography resulted in a development of a new field of the technology — microwave integrated circuits (MIC). We can tell that these circuits mean an extension of the technology of hybrid integrated circuits into microwave frequencies.

The integrated technology plays an important role in the current microwave technique and it represents a key factor in the successful development of the microwave technique and in their penetration into different regions of the science and technology, into electronics for investment as well as consumption purposes. This technology presents a basic assumption of the economically efficient implementation of modern high-frequency systems with excellent technical and operational parameters, high reliability, resistance and stability, which are compact and in comparison with the preceding waveguide systems also characterized by smaller size and weight.

The high-productivity production of microwave integrated circuits and microwave systems on the integrated technological basis with low losses corresponding to the rejection of products and exceeding of allowances desired, is tightly related to a high level of theoretical as well as experimental methods of the microwave integrated circuit design. A wide base of the knowledge, method of the analysis and synthesis and of checked procedures is an ultimate requirement here, since the integrated structure allows at most for minimum repair interventions into the physically accomplished circuits in order that it could be unnecessary to adjust or alter unsuitable and, due to an imprecise design, deflected values of electric parameters of the integrated circuit. This factor is even more remarkably manifested in the series and mass production of a certain type of the microwave integrated circuit, where the lowest losses considerably affect the economy of the whole production.

The purpose of the authors of the present book is to yield to an expert reader as much information as possible on theoretical methods of analysis and synthesis of microwave elements and circuits produced by the integrated technology. The material arranged in the book is to a considerable extent a result of the research and development works of the authors, and it was also experimentally checked by the authors. The book presents material sufficient for providing its use as a practical handbook for experts in the research, designers of integrated circuits in industry and also as an auxiliary text for students at technical universities. However, the authors are aware that the current rapid development of the microwave integrated technology cannot be completely described in the monograph.

Chapter 2 deals with the analysis of passive circuit elements starting from transmission lines of different types and their MIC applications. The second part considers discontinuities, their characteristics and applications in the MIC. In the third part, the problems of lumped elements are analysed.

Chapter 3 concerns semiconductor circuit elements such as Schottky diodes, varactor diodes, PIN diodes and transistors (bipolar, MESFET, HEMT).

Chapter 4 considers the basic circuits. The first part is aimed at methods of MIC synthesis with the use of the matrix approach. In the second part, basic linear circuits are analysed (non-reflecting load, resonators, attenuators, etc.) and the last part describes basic active and non-linear circuit (detectors, mixers, oscillators and amplifiers).

Chapter 5 includes the problems of measuring and testing. The first part considers measuring methods from the metrological standpoint. In the second part, the problems of the measuring set (test fixture and probes) are analysed and in the third part, calibration approaches are discussed. The last part considers the noise measurments.

REFERENCES

Barret R. M. and M. H. Barnes, Microwave printed circuits, *Radio and T. V. News, Vol. 46*, p. 16, Sept. 1951.

Smith P. H., *Electronic Applications of the Smith Chart in Waveguide, Circuit and Component Analysis*, McGraw Hill, New York, 1969.

Wheeler H. A., Transmission Line Properties of Parallel Wide Strips by Conformal Mapping Approximation, *IEEE Trans. Microwave Theory Tech., Vol. MTT-12*, pp. 280–289, 1964.

2

Analysis of passive circuit elements

F. Hrníčko and M. Pavel

2.1 TRANSMISSION LINES FOR MICROWAVE INTEGRATED CIRCUITS

In the microwave integrated circuits (MICs) special types of lines are used that, in general, are formed by means of diverse configurations of conducting strips on different substrates. The latter are represented by various types of lines such as microstrip, coplanar, slotlines or transmission lines on suspended or multi-layer substrates as well as coupled lines. For some important types of lines accurate solutions for quasi-static approximations were obtained, whereby the line's dispersion was determined with the use of a series of numerical methods that for the purpose of computer aided design can be approximated by closed relationships and vice versa the relationships for the synthesis can also be obtained.

2.1.1 Basic characteristics of homogeneous transmission lines

Basic component of the MIC is formed by homogeneous transmission lines enabling transmission of the electromagnetic field energy from the generator to the load or they stand as a part of a distributed circuit design. Usually, strip conductors located in definite places at dielectric sheets interfaces are used. Further, we shall confine ourselves just to the structures showing longitudinal uniformity. Transverse dimensions of such a line are independent of the coordinate along the direction of propagation. Basic parameters of the line are found using CAD-approximations or more tedious analytical methods from its dimensions in the transverse plane.

In the transmission lines used in the MICs, various types of waves can be excited. Apart from the waves propagating along the interface formed by the dielectric substrates waves can also be induced that are radiated into surrounding space. We try to suppress such unwanted waves by a suitable choice of transverse line dimensions and in the way the circuit is fed.

Transmission lines structure can be classified as *open*, *side-open* or *closed*. This classification says in what type of space the electromagnetic energy is propagating. By computation, various numerical solutions of the field are obtained differing both by their character and formal description. Open MIC lines are not resistant against perturbations and they, themselves can become a source of disturbation if the transverse dimensions of the structure are chosen improperly. On the other side, the improper transmission of information on the lines in closed structures is not affected by external fields. Most popular transmission lines in open structures are microstrip, coplanar waveguide, coplanar strips, slotline, suspended microstrip line, and some other lines. These types of lines, as a rule are enclosed in conductive boxes whose effect can either be neglected under the assumption of the properly selected size. In the other case the closed transmission structure should be investigated. To the closed structures the lines belong such as stripline, suspended stripline, finline etc. In many cases the closed structure transmission line problem can be analysed as a side-open one advantageously.

Transverse arrangement of some transmission lines for MICs are depicted in Fig. 2.1. In the following chapters the brief description of their properties, the way and limits of their application and accurate closed-form equations are given. They are most often achieved by many parameter curve fitting of precise analytical results. Some of the analytical methods are presented in Chapter 2.1.8 (or in the Appendix). More precise elaborating of those methods lies beyond the scope of this book.

Practically speaking, what we are interested in is the propagation along the transmission lines mentioned in the above. It is necessary to determine the distribution of the intensity of both the electric and magnetic field for the propagation mode under study. It can be achieved by solving the wave equations under definite boundary conditions.

Multiconductor transmission lines enable purely TEM or non-

TEM modes propagation in comparison with waveguides and dielectric waveguides.

The dominating mode of a two-conductor line with homogeneous filling is represented by the TEM wave with the only non zero compo-

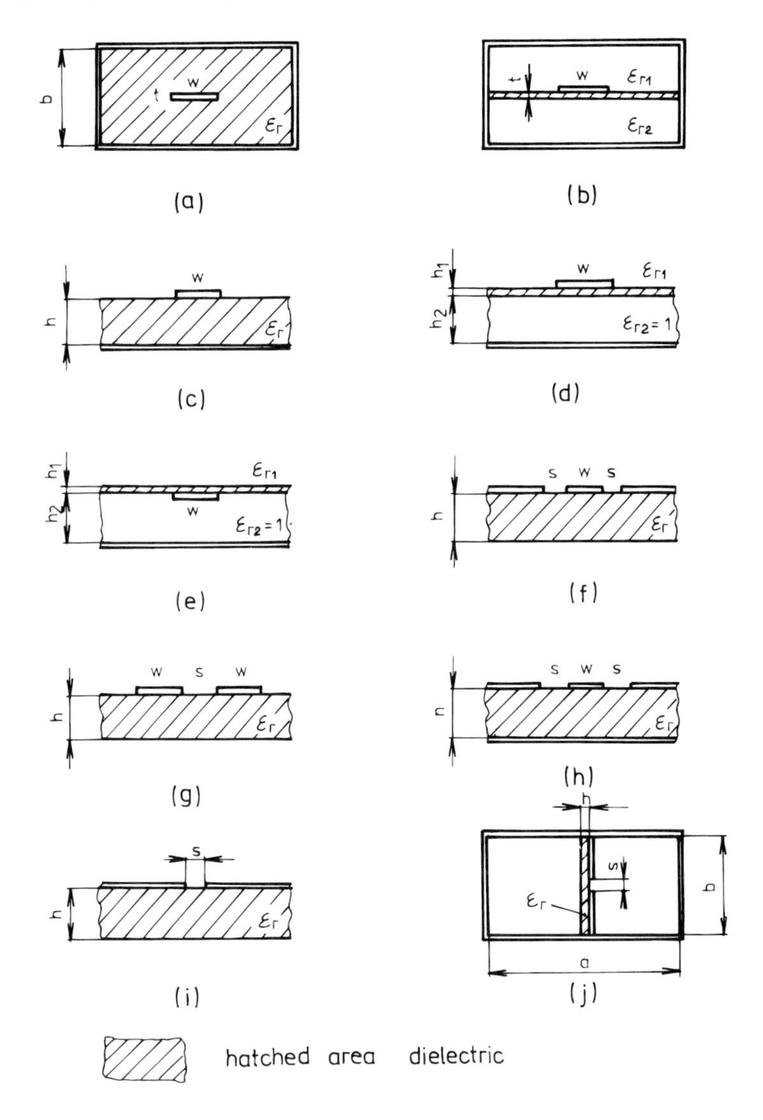

Fig. 2.1 Transmission lines for MICs (a) strip line (b) suspended stripline (c) microstrip (d) suspended microstrip (e) inverted microstrip (f) coplanar waveguide (g) coplanar strips (h) back side grounded CPW (i) slotline (j) fin line.

nents in a plane perpendicular to the propagation direction without dispersion. In a medium with an inhomogeneous filling the dominant mode is representated by a hybrid wave (HEM) with non zero longitudinal field components E_z and H_z. It can be demonstrated on a microstrip transmission line using the train of thoughts as follows. The relationship between tangential components of electric field at the boundary of the media 1, 2 (Fig. 2.2) can be expressed by

$$E_{x_1} = E_{x_2} \tag{2.1}$$

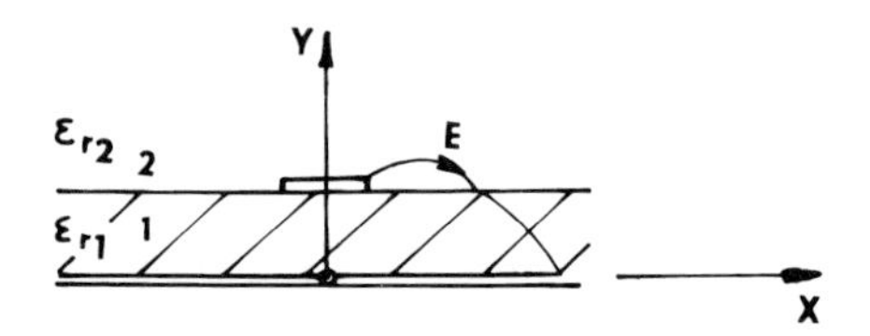

Fig. 2.2 Open transmission line structure.

Introducing $\varepsilon_r = \varepsilon_{r1} > 1$, $\varepsilon_{r2} = 1$, and using Maxwell equations we can write

$$(\nabla \times H)_{x_1} = \varepsilon_r (\nabla \times H)_{x_2} \tag{2.2}$$

By transcribing the equations for individual components and using the continuity equation for the normal component of magnetic flow we arrive at

$$\varepsilon_r \frac{\partial H_{z1}}{\partial y} - \frac{\partial H_{z2}}{\partial y} = (\varepsilon_r - 1) \frac{\partial H_y}{\partial z} \tag{2.3}$$

Since $\varepsilon_{r_1} > 1$ and $H_y \neq 0$, the left side of (2.3) must be non-zero — what can hold true just under the condition that $H_z \neq 0$. Similarly, it can be proved that the longitudinal component of electric field must be non-zero, as well. The field distribution is now defined by the wave equation whose solution yields the dispersion characteristic of the transmission line. Under the assumption that the maximum cross-sectional parameter of a transmission line or the distance between the lines is much less than half wavelength in the medium surrounding transmission lines, the longitudinal components can be neglected in comparison with transverse ones, approximating the hybrid wave by the quasi-TEM one. Here several artificial quantities are being used (such as effective permittivity, filling factor, effective transmission line width) in a definite models for a hybrid transmission line dispersion description.

The basic parameters describing a transmission line are as follows: characteristic impedance Z_0, propagation constant γ, dispersion, cut-off propagation frequency, and maximum power handling.

The definition of the characteristic impedance Z_0 in a TEM line is based on the quantities such as the voltage U between the transmission conductors and longitudinal current I through the conductor

$$Z_0 = \frac{U}{I} = \frac{\left|\int_C E\,\mathrm{d}s\right|}{\left|\int_{S_v} I_z\,\mathrm{d}S_v\right|} \tag{2.4}$$

where C is one of the characteristic curves connecting the conductors of the transmission line, S_v is the cross-section area of the conductor, or alternatively, using the quantities of the substitutional diagram of the transmission line (Fig. 2.3)

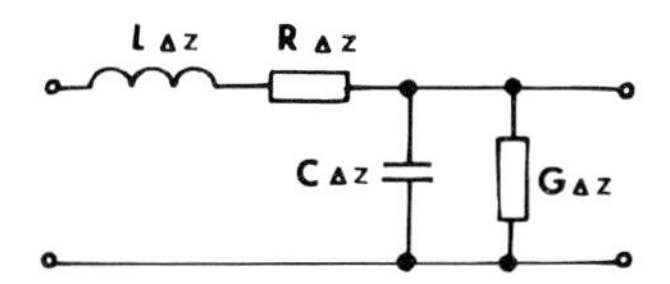

Fig. 2.3 Equivalent circuit of transmission line with length Δz.

$$Z_0 = \sqrt{\frac{R + j\omega L}{G + j\omega C}} \tag{2.5a}$$

where R, L, G, C are the series resistance and inductance, and the parallel conductivity and the capacity per a unit line length. Thus for a low-loss transmission line we can write

$$Z_0 \simeq \sqrt{\frac{L}{C}} \tag{2.5b}$$

In case of a hybrid wave propagation analysis one of the following definitions in accordance with the structure nature

$$Z_0 = \frac{2P}{|I|^2} \tag{2.6a}$$

and

$$Z_0 = \frac{|U|^2}{2P} \tag{2.6b}$$

or the definition according to (2.4) is used. In the above formulations P is the time average of the power propagating by the line and it can be determined by integrating the Poynting vector over the cross section S of the lines

$$P = \frac{1}{2}\mathrm{Re}\int_S (E \times H^*)\,\mathrm{d}S \tag{2.7}$$

It can be shown that the formulation $(P \to I)$ fits the best to the microstrip problems while the dual slot problems are best approximated using the formulation $(P \to U)$.

For every real transmission line propagation constant γ is a complex number $\gamma = \alpha + j\beta$, where α is the attenuation constant of the line and β is the phase constant of the line. All the field components of a progressive wave, i.e. the wave propagating from the source are proportional to $\exp(-\gamma)$. The propagation constant can be calculated from the dispersion equation.

Losses

Every transmission structure shows a loss. The losses in a homogeneous transmission line can be divided into dielectric, conductivity and radiation losses. The dielectric loss is caused by losing the RF energy in the material of a dielectric substrate whose origin is connected with the mechanism of polarization of a dielectric depending on a series of factors such as the purity, homogeneity of a dielectric, temperature, etc. The input quantity for the computation is represented by the value of $\tan\delta$ defined for dielectric materials as follows

$$\varepsilon(\omega) = \mathrm{Re}\varepsilon(\omega) + j\mathrm{Im}\varepsilon(\omega) = \mathrm{Re}\varepsilon(\omega)(1 + \tan\delta) \tag{2.8}$$

Conductivity losses are caused by finite conductivity of the materials used in transmission lines. They increase with the frequency since the penetration depth δ_n is decreasing due to the skin-effect

$$\delta_n = \left(\frac{\rho}{\pi\mu f}\right)^{0.5} \tag{2.9}$$

where ρ is the resistivity of the conducting material, μ the magnetic permeability of the material ($\mu = \mu_0$), f — the operating frequency. Due to the above, the efficient cross-section of the conductor is decreased for RF currents with corresponding increase of the RF resistance of the conductor. If the surface roughness of the conductor is comparable with the penetration depth, the path of the high-frequency currents is increased and again, the efficient resistance of the conductor increases. This is of importance in MIC lines, since the conductors are deposited on the dielectric substrate surface and they follow its shape. Thus, the smooth substrate surface is one of the conditions for low conductivity losses.

The constant of attenuation α_{d} resulting from dielectric losses in material can be determined for a quasi-TEM mode and small losses ($\tan\delta \ll 1$) by modifying the relationship for the attenuation constant of the TEM mode in a structure with homogeneous filling in the form

$$\alpha_d = \frac{\omega}{2}\sqrt{\mu\varepsilon_r\varepsilon_0}\tan\delta \qquad [\mathrm{Np\,m^{-1}}] \tag{2.10}$$

When the filling is not homogeneous and the top dielectric (air) is loss-free, then it is possible to write

$$\alpha_d = \frac{\omega}{2}\sqrt{\mu\varepsilon_{ef}\varepsilon_0}\tan\delta_{ef} \tag{2.10a}$$

With the help of filling factor q and effective permittivity ε_{ef} the effective loss factor $\tan\delta_{ef}$ is defined as follows (Schneider, 1969):

$$\tan\delta_{\mathrm{ef}} = q\frac{\varepsilon_r}{\varepsilon_{ef}}\tan\delta \tag{2.11}$$

where $\varepsilon_{ef} = 1 + q(\varepsilon_r - 1)$ is the effective permittivity in the loss-free case. For the determination of α_d it is inserted from (2.11) into (2.10a). The attenuation of the line of a unit length α_D [$\mathrm{dB\,m^{-1}}$] can be then expressed as

$$\alpha_D \doteq 27.3q\frac{\varepsilon_r}{\varepsilon_{ef}}\frac{\tan\delta}{\lambda_g} \tag{2.12}$$

where $\lambda_g = c/(f\sqrt{\varepsilon_{ef}})$ is the wavelength in the lossless line; c is the velocity of light in vacuum.

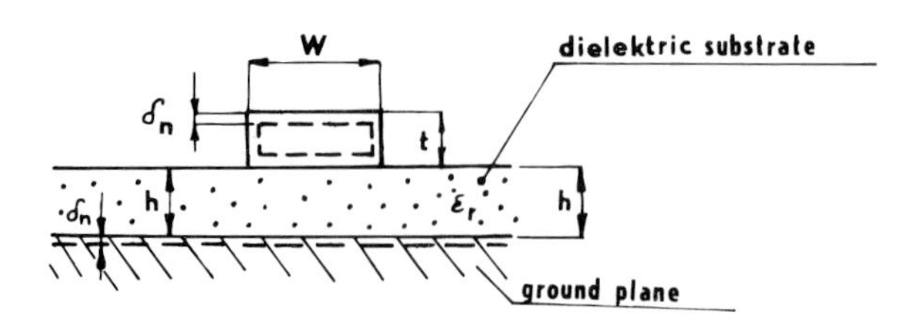

Fig. 2.4 Model of microstrip transmission line for conductive loss determination.

The determination of conductivity losses with the help of the field distribution in the line space is very complicated. For their calculation it is advantageous to employ a model proposed for the microstrip line (Pucel et al., 1968). The concept of this procedure is based on the relationship between the value of the RF resistance of a conductor and change of its inductance resulting from a surface process. The following consideration will concern a microstrip line with a cross-section shown in Fig. 2.4. As a result of the finite depth of the penetration, the effective distance of conductors h will be enlarged and the effective width of conductors above the earthing plane will be reduced. In this way, the inductance L of the line of a unit length will be increased by a value ΔL_j. The increase of the effective resistance ΔR_j, when assuming $w \gg \delta_n$ and $t \gg \delta_n$, is associated with the inductance change as follows:

$$\Delta R_j = \omega \Delta L_j = 2\pi f \mu_0 \frac{\delta_n}{2} \left(\frac{\partial L}{\partial n_j} \right) = \frac{R_{sj}}{\mu_0} \left(\frac{\partial L}{\partial n_j} \right) \tag{2.13}$$

where R_{sj} is the resistance of the j-th part of the surface of conductors and $\partial L / \partial n_j$ is the inductance change depending on the normal line to the j-th part of the conductor surface. The conduction losses are defined by

$$\alpha_c \cong \frac{P_c}{2P(z)} \qquad [\mathrm{Np\,m^{-1}}] \tag{2.14}$$

where P_c is the power loss in the conductors and $P(z)$ is the power transferred. After arranging and expressing in $\mathrm{dB\,m^{-1}}$ we can write for the microstrip line

$$\alpha_c = \frac{4.343}{Z_0 \mu_0} \left[R_{s_1} \left(-2\frac{\partial L}{\partial w} - 2\frac{\partial L}{\partial t} - 2\frac{\partial L}{\partial h} \right) + R_{s_2} \frac{\partial L}{\partial h} \right] \tag{2.15}$$

Losses for four of $50\,\Omega$ lines with the quasi-TEM mode was compared by (Spielman, 1977).

The coplanar waveguide has considerably larger losses than the microstrip line, whereas the inversion microstrip and suspended lines exert losses reduced by factors exceeding two. This results from the fact that the width of strips for the same impedance is larger by a factor of two to three and thus, a major part of the field energy is concentrated in the air. For the slotted line it is impossible to determine the losses in this way. They were determined experimentally and they are about twice as high as those in the microstrip line.

Maximum power handling is limited by electric strength of dielectric and conductors warming-up due to the conductivity loss. The maximum electric strength gives us the peak power value while the temperature increase due to conductivity and dielectric loss limits the average power. At a normal atmospheric pressure the breakdown the electric field is 2.9×10^6 V/m for dry air. Calculating the maximum electric field intensity the peak power handling capacity is given immediately. The average power on the line is influenced by the attenuation constant of a propagated signal, conductor surface size, maximum conductor temperature, and heat conductivity of dielectric material that is in a contact with conductors.

Coupled lines. The coupled MIC lines are systems of n conductors ($n \geq 3$) deposited on a simple or multilayer substrate. The simplest case is two parallel microstrip lines on a simple substrate (Fig. 2.5). This system includes three conductors. Two basic quasi-TEM modes (in general $n - l$ basic modes) will propagate through it with different arrangements of the electromagnetic field. Whereas for the TEM structure each superimposition of eigen modes is an eigen mode of the line, again for structures with unhomogeneous filling, the intrin-

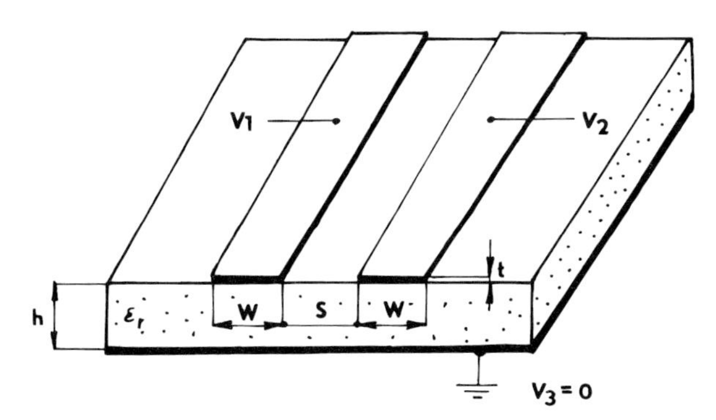

Fig. 2.5 Coupled microstrip line.

sic modes exert different velocities of the propagation and thus, their superimposition is no eigen mode of the line. The quasi-TEM mode of the line will be assumed and the static field problem will be solved in the transverse cross-section of the line. It is first necessary to define the potential of conductors 1, 2, 3. From the condition of a zero potential at an infinite distance it follows that $V_3 = 0$. The potential of one of the conductors can be chosen and it determines the excitation amplitude, which are the line characteristics independent due to the linearity of Maxwell's equations. The ratio of potential values of both conductors, $k = V_2/V_1$, is, however, of an essential importance for solving the problem, since it affects the static capacitance of the line $C_1(k)$, $C_2(k)$ and thus also the effective permittivity $\varepsilon_{ef_1}(k)$ and $\varepsilon_{ef_2}(k)$ for both strips. For the constant k of the line intrinsic mode the equation

$$\varepsilon_{ef_1}(k) = \varepsilon_{ef_2}(k) \tag{2.16}$$

must be valid, which has two solutions k_1, k_2 corresponding to two basic modes of coupled lines. In certain cases, the problem can be essentially simplified. This is the case of a line through which it is possible to situate a longitudinal plane of the mirror symmetry ($w_1 = w_2$). The basic modes in this system are the even ($k = 1$) and odd ($k = -1$) modes. This follows from the fact the above mentioned plane of the symmetry is a plane of symmetry or antisymmetry of the electromagnetic field for the even or odd mode, respectively. Both lines are identical from the standpoint of the field distribution and the condition expressed by (2.16) is adhered to. For the effective permittivity of the even as well as odd mode ε_{ef_e}, ε_{ef_o} it holds that

$$\varepsilon_{ef_e} > \varepsilon_{ef_o} \tag{2.17}$$

since in the case of the first even mode a prevalent portion of the energy is concentrated in the dielectric.

Different values of the velocities of basic modes in the line result in problems in the synthesis of circuits and deteriorate optimum parameters of circuits. Certain methods have been elaborated making possible a compensation for this difference. This will be detailed in a Section 4.3 dealing with couplers.

Transmission line overview (Table 2.1). Planar transmission lines differ not only by the basic electrical parameters that together with frequency determine the range of applications but also by mechanical parameters and technological possibilities that enable parallel or series component mounting and transmission line compatibility.

For a practical choice of the most convenient transmission line for a required MIC the possibilities of various types of lines are sumarized in Table 2.1.

Table 2.1 Parameter comparison for various MIC transmission lines

Parameter	Microstrip	Suspended Microstrip	Slotline	CPW	CPS	Fin-line
Impedance range [Ω]	15–120	25-180	50–300	25–155	45–280	
Frequency range [GHz]	0–60	0–90	3–60	0–60	0–60	6–100
ϵ_r range	2–20	2–10	10–20	2–20	2–20	2–4
Dispersion	small	very small	high	medium	medium	low
Attenuation loss	low	very low	high	medium	medium	low
Q unloaded	high	very high	low	medium	medium	high
$P_{\max}$	high	medium	low	medium	medium	medium
Component mounting Series Shunt	 easy via holes edge plating	 easy difficult	 difficult easy	 easy easy	 easy easy	 difficult easy
Coupled lines	easy	easy	possible, but unwanted modes should be considered			

a) Microstrip line — lines on glass or ceramica substrates using the thin or thick film technology as well as on plated laminate materials such as RT-duroid, CuClad, etc. Required line shapes are etched photo-chemically.

b) Suspended microstrip line — 2 to 3 times lower conductivity loss compared with microstrip due to wider strip widths for the same impedance, higher phase velocity. Quartz substrates for millimetre wave frequencies are frequently used.

c) Coplanar waveguide (CPW) — ground conductor appears on the same side of a substrate as a signal one, higher design flexibility and higher MIC integration is achieved. Special probes enable easy on-wafer measurement in MMICs.

d) Coplanar strips (CPS) — symmetrical line, higher impedance are available, higher flexibility and MIC integration is achieved.

e) Slotline — is advantageous for high impedance lines in MICs and for hybrid circuits in combination with microstrip line.

f) Fin-line — Various types of fin-line are used for higher microwave and millimetre wave circuits especially (unilateral, bilateral, ...) in combination with waveguide technique.

For design and development of MICs very precise CAD models are necessary to provide us with the first attempt solution of the problem. Advanced software packages from Compact Software, EEsof, etc. are available for linear frequency domain MIC simulation at present. For easy and immediate solutions of particular problems closed-form formulae for precise transmission line characteristics are of great interest. In the following chapters Precise CAD models are submitted for wide design parameter range.

2.1.2 Microstrip lines

The microstrip line (Fig. 2.6) is the most frequently used type of line in the MIC. It is formed by a strip conductor of a width w and thickness t, situated on the top side of a planar dielectric substrate and grounding conductor, which covers the bottom side. The dielectric

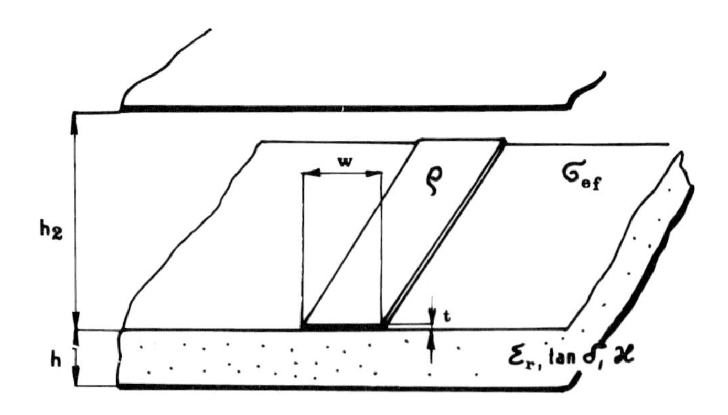

Fig. 2.6 Characteristic parameters of microstrip transmission line (σ_{ef}-mean value of the surface roughness, $\varkappa$- mean value of the thermal conductivity, ϱ-specific resistance).

substrate is characterized by its relative permittivity ε_r, thickness h, loss angle $\tan\delta$, mean value of the surface roughness σ_{ef} and thermal conductivity $\varkappa$. The conductor material and thus also the specific resistance of the conductors are different.

The hybrid wave is a prevalent phenomenon in the lines. The propagation of this wave is described by the equivalent quasi-TEM wave. The distribution of the wave electric and magnetic field intensity is depicted in Fig. 2.7. At low frequencies it is of advantage

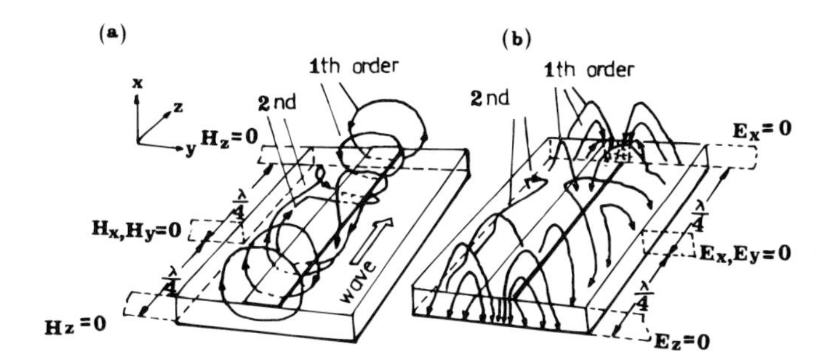

Fig. 2.7 Scattering of (a) magnetic and (b) electric fields of hybrid wave (quasi-TEM) for microstrip transmission line at high frequencies.

to analyse the line with the help of quasistatic methods. The upper limit of the use of these results is about

$$f_{stat} \doteq 0.04 \frac{Z_0\,(0)}{h} \qquad [\mathrm{GHz}, \Omega, \mathrm{mm}] \qquad (2.18)$$

where $Z_0(0)$ is the characteristic impedance of the line for $f = 0$. Above this limit, it is necessary to correct the results of quasistatic methods with the help of a suitable dispersion model, which is usually determined by an approximation of dispersion dependences obtained by solving the wave equation or experimentally. The limit of the use of the microstrip line is the cut-off frequency f_{cl} of the first mode of a higher order, which is obtained from the following relationship (Hoffman, 1983)

$$f_c m = \frac{cm}{2w_{ef}\,(0)\sqrt{\epsilon_{ef}\,(0)}} \doteq \frac{0.4 m Z_0\,(0)}{h} \qquad [\mathrm{GHz}, \Omega, \mathrm{mm}] \qquad (2.19)$$

where m is the number of the mode, c is the light velocity in the vacuum, $w_{ef}(0)$ is the effective width of the strip and $\varepsilon_{ef}(0)$ is the

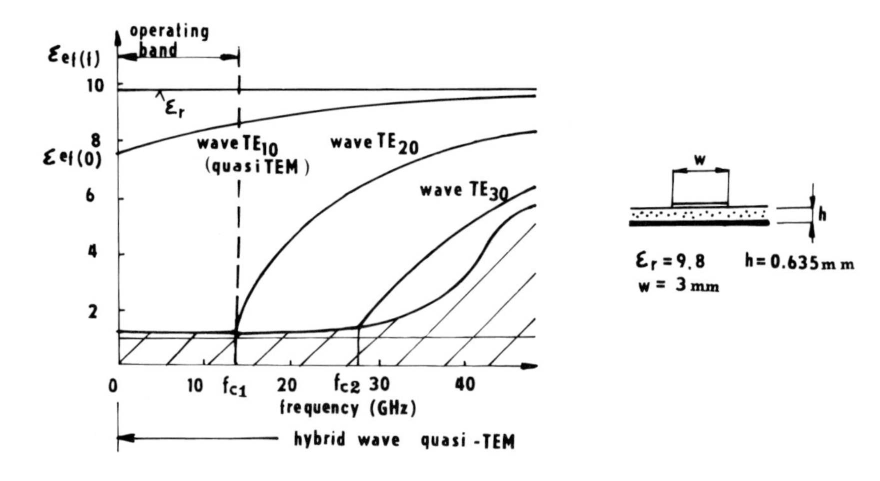

Fig. 2.8 Dispersion diagram of the basic wave and of waves of higher order.

effective relative permittivity for the static case. The dispersion diagram of the basic wave and of waves of higher order is in Fig. 2.8. The effective width w_{ef} of the microstrip line is determined as a width of a waveguide with a height h, which is filled with material with a permittivity ε_{ef}, the longer wave of the waveguide being electric walls and shorter sides being magnetic walls of the waveguide, by the relationship

$$w_{ef}(f) = \frac{120\pi h}{Z_0(f)\sqrt{\varepsilon_{ef}(f)}} \tag{2.20}$$

For a computerized design and optimization of microstrip circuits, complete design relationships are required for the characteristic impedance Z_0 and effective dielectric constant ε_{ef} including effects of the geometric size of the line parameters of the substrate and frequency. Very precise and simple equations for the quasistatic solution are presented in Hammerstad and Jensen (1980). The model is based on an equation for the impedance Z_{0_A} of a microstrip in a homogeneous medium ($\varepsilon_r = 1$) and on an equation for ε_{ef}

$$Z_{0_A}(u) = \frac{\eta_0}{2\pi}\ln\left[\frac{f(u)}{u} + \sqrt{1+\left(\frac{2}{u}\right)^2}\right] \tag{2.21}$$

$$f(u) = 6 + (2\pi - 6)\,e^{\left[-\left(\frac{30.666}{u}\right)^{0.7528}\right]} \tag{2.22}$$

where η_0 is the wave impedance of the medium ($376.73\,\Omega$ in a vacuum), u is the standardized strip width ($u = w/h$). The effective permittivity ε_{ef} determines the phase velocity v_f of the wave in the line and it can be determined with the help of the filling factor q

$$\varepsilon_{ef}\left(u, \varepsilon_r\right) = \frac{\varepsilon_r + 1}{2} + q\frac{\varepsilon_r - 1}{2} \tag{2.23}$$

where the filling factor for open line is

$$q = \left(1 + \frac{10}{u}\right)^{-a(u)b(\varepsilon_r)} \tag{2.24}$$

$$a\left(u\right) = 1 + \frac{1}{49}ln\frac{u^4 + \left(\frac{u}{52}\right)^2}{u^4 + 0.432} + \frac{1}{18.7}\ln\left[1 + \left(\frac{1}{18.1}\right)^3\right] \tag{2.25}$$

$$b\left(\varepsilon_r\right) = 0.564\left(\frac{\varepsilon_r - 0.9}{\varepsilon_r + 3}\right)^{0.053} \tag{2.26}$$

The characteristic impedance Z_0 can be determined from the relationship

$$Z_0\left(u, \varepsilon_r\right) = \frac{Z_{0_A}\left(u\right)}{\sqrt{\varepsilon_{ef}\left(u, \varepsilon_r\right)}} \tag{2.27}$$

The precision of these relationships is higher than 0.2% for $\varepsilon_r \leq 128$, $0.01 \leq u \leq 100$.

The correction for the metal coating thickness can be introduced for example according to (Wheeler, 1977) with the help of the corrected width $w' = w + \Delta w$ where

$$\Delta w = \frac{t}{\pi}\left\{1 + \ln 4 - 0.5\ln\left[\left(\frac{t}{h}\right)^2 + \left(\frac{t}{\pi w}\right)^2\right]\right\} \tag{2.28}$$

In equations (2.21) to (2.26) the parameter u is substituted by the standardized corrected width of this strip w'/h and thus, the quantities $\varepsilon_{ef}(u, \varepsilon_r, \tau)$ and $Z_0(u, \varepsilon_r, \tau)$ are obtained, where τ is the standardized thickness of the metal coating, $\tau = t/h$.

The microstrip line is frequently situated in a shielded structure with a screening lid at a finite distance h_2 above the top plane of the dielectric substrate. The effect of the lid on the line parameters cannot be neglected for $h_2 \leq 10h$ however, it should be expressed by a suitable modification of the proposal relationships (March, 1981). The characteristic impedance Z_{0A} in the air corrected for distance h_2 is of

$$Z_{0_A}\left(u,\tau,\frac{h_2}{h}\right) = Z_{0A}\left(u,\tau\right) - \Delta Z_{0_A}\left(\frac{h_2}{h}\right) \tag{2.29}$$

where

$$\Delta Z_{0_A}\left(\frac{h_2}{h}\right) = PQ \tag{2.30}$$

$$P = 270\left\{1 - \tanh\left[1.192 + 0.706\sqrt{1+\frac{h_2}{h}} - \frac{1.389}{1+\frac{h_2}{h}}\right]\right\} \tag{2.31}$$

$$Q = 1.0109 - \arctan\frac{0.012u + 0.177u^2 - 0.027u^3}{1+\left(\frac{h_2}{h}\right)^2} \tag{2.32}$$

For the determination of $\varepsilon_{ef}(u,\varepsilon_r,\tau,h_2/h)$ the filling factor is divided in a similar manner into three components

$$q\left(u,\varepsilon_r,\tau,\frac{h_2}{h}\right) = \left(q - q_\tau\right)q_s \tag{2.33}$$

where q is the filling factor for $h_2 \Rightarrow \infty$ given by the relationship (2.24), q_t is the correction for the microstrip thickness,

$$q_\tau = \frac{2}{\pi}\tau\frac{\ln 2}{\sqrt{u}} \tag{2.34}$$

and q_s is the correction for the finite distance of the screening plane h_2

$$q_s = \tanh\left(1.043 + 0.121\frac{h_2}{h} - 1.164\frac{h}{h_2}\right) \qquad \text{for } \frac{h_2}{h} \geq 1 \tag{2.35}$$

The value $\varepsilon_{ef}(u,\varepsilon_r,\tau,h_2/h)$ is determined by substituting $q(u,\varepsilon_r,\tau,$ $h_2/h)$ from (2.33) into (2.23) and $Z_0(u,\varepsilon_r,\tau,h_2/h)$ by substituting the value $Z_{0_A}(u,\tau,h_2/h)$ from relationship (2.29) and $\varepsilon_{ef}(u,\varepsilon_r,\tau,$

h_2/h) into (2.27). These dependences for $\varepsilon_r = 10$ and $t = 0$ are shown in Fig. 2.9.

The effect of the side wall (Fig. 2.10a) at a distance d from the line can be determined from an analogy with a parallel coupled line (see Section 2.1.8), where the slot between the strip is of $s = 2d$. The parameters Z_0 and ε_{ef} can then be determined as parameters of an odd mode Z_{Oo}, ε_{efo}. This effect is shown in Fig. 2.10b.

A very precise dispersion model for the effective dielectric constant was developed in Kirschning and Jansen (1982). It employs the mathematical structure of the Getsinger's formula describing the dispersion of the microstrip in the form

$$\varepsilon_{ef}(f) = \varepsilon_r - \frac{\varepsilon_r - \varepsilon_{ef}(0)}{1 + P(f)} \tag{2.36}$$

where the term $P(f)$ is simulated as

$$P(f) = P_1 P_2 \left[(0.1844 + P_3 P_4) f_n\right]^{1.5763} \tag{2.37}$$

$$P_1 = 0.27488 + \left[0.6315 + \frac{0.525}{(1 + 0.0157 f_n)^{20}}\right] u - 0.065683 e^{-8.7513} \tag{2.38a}$$

$$P_2 = 0.33622 \left[1 - e^{-0.03442\varepsilon_r}\right] \tag{2.38b}$$

$$P_3 = 0.363 e^{-4.6u} \left[1 - e^{-\left(\frac{f_n}{38.7}\right)^{4.97}}\right] \tag{2.38c}$$

$$P_4 = 1 + 2.751 \left[1 - e^{-\left(\frac{\varepsilon_r}{15.916}\right)^{8}}\right] \tag{2.38d}$$

where f_n [GHz mm] is the standardized frequency ($f_n = f_h \approx h/\lambda_0$) and $\varepsilon_{ef}(0)$ is the static value of the effective dielectric constant given for example by the relationship (2.23) for $t = 0$. The precision of this model with respect to the hybrid numerical solution (Jansen, 1978) is higher than 0.6% in a range of parameters $0.1 \leq u \leq 100, 1 \leq \varepsilon_r \leq 20$ and $0 \leq h/\lambda_0 \leq 0.13$.

For the definition of the characteristic impedance the formulation (2:6a) seems to be the most useful one, which is based on the mean power transferred P and longitudinal current I in the strip conductor. On the basis of a multidimensional computerized approximation of results of the hybrid mode method (Jansen and Kirschning, 1983), a very precise model of the characteristic impedance dispersion was

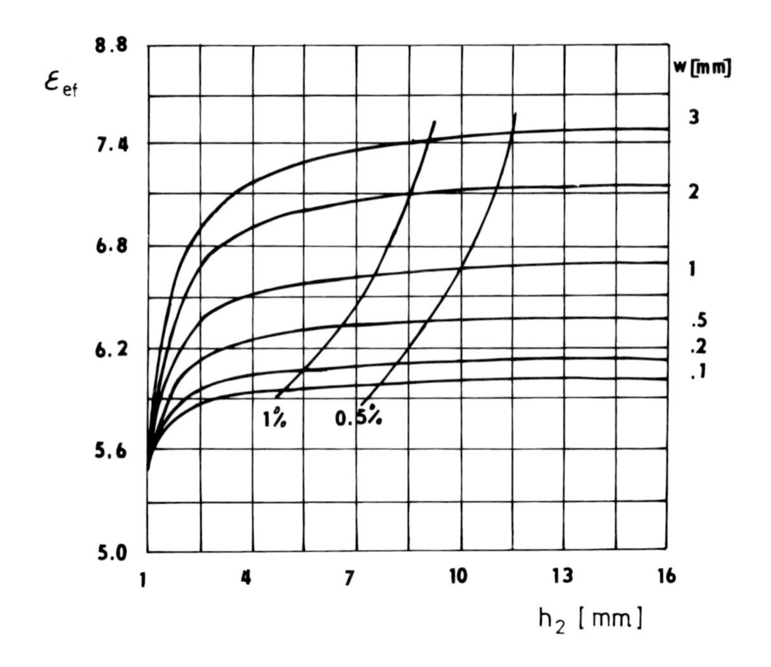

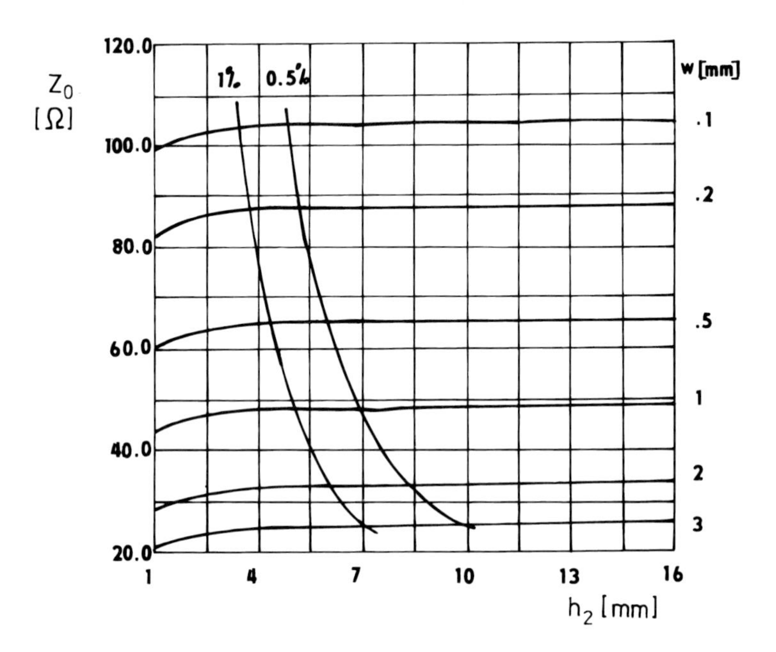

Fig. 2.9 Static parameters of side opened microstrip line on corrund substrate ($\varepsilon_r = 10$, $h = 1\,\text{mm}$, $h_2 = 1 \div 16\,\text{mm}$) with level 0.5% and 1% deviation from line parameters without top cover.

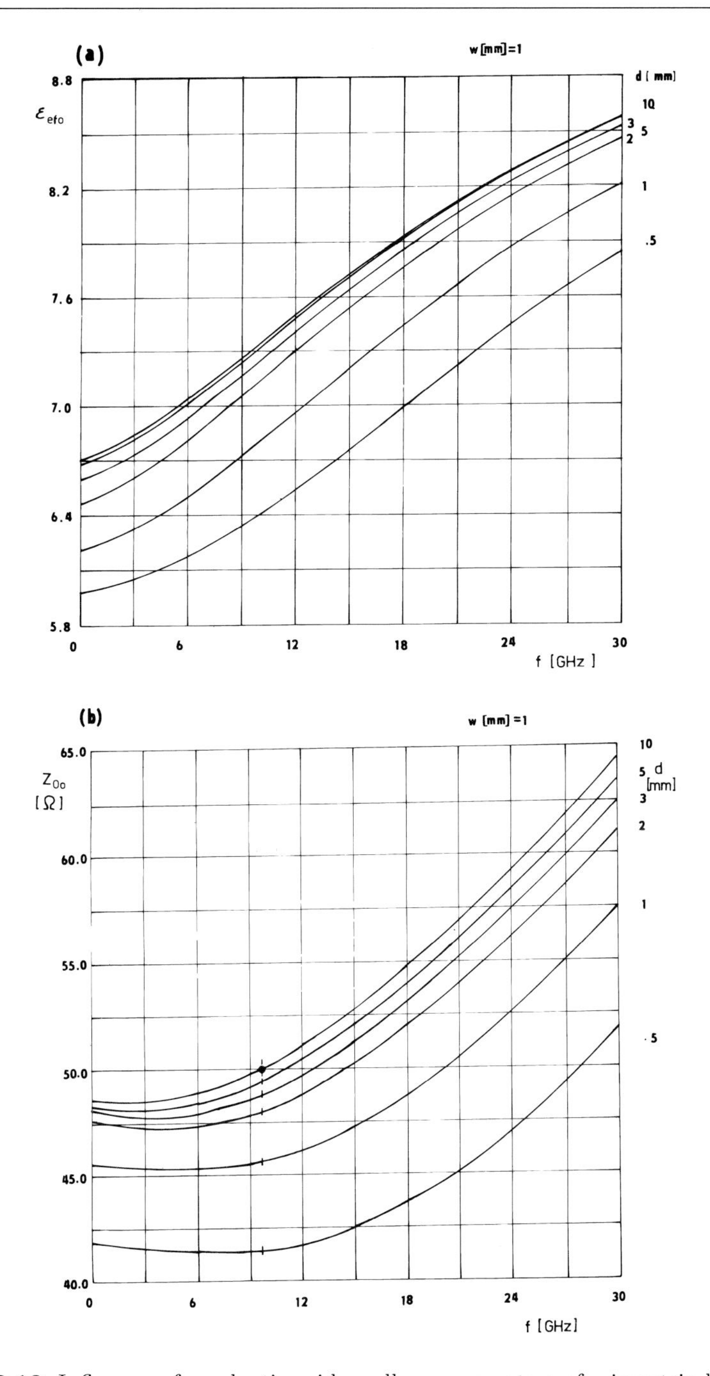

Fig. 2.10 Influence of conductive side wall on parameters of microstrip line on corrund substrate ($\varepsilon_r = 10$, $h = 1\,\text{mm}$, $t = 5\mu\text{m}$, $w = 1\,\text{mm}$) (a) frequency dependence of ε_{efo} (b) frequency dependency of Z_{0o}.

developed for the computer proposal purposes, which is determined by a sequence of the following relationships

$$R_1 = 0.03891\varepsilon_r^{1.4}$$

$$R_2 = 0.267u^7$$

$$R_3 = 4.766e^{-3.228u^{0.641}}$$

$$R_4 = 0.016 + (0.0514\varepsilon_r)^{4.524}$$

$$R_5 = \left(\frac{f_n}{28.842}\right)^{12}$$

$$R_6 = 22.2u^{1.92}$$

$$R_7 = 1.206 - 0.3144e^{-R_1}\left(1 - e^{-R_2}\right)$$

$$R_8 = 1 + 1.275\left[1 - e^{-0.004625R_3\varepsilon_r^{1.674}\left(\frac{f_n}{18.365}\right)^{2.745}}\right]$$

$$R_9 = 5.086R_4\frac{R_5}{0.3838 + 0.386R_4}\frac{e^{-R_6}}{1 + 1.2992R_5}\frac{(\varepsilon_r - 1)^6}{1 + 10(\varepsilon_r - 1)^6}$$

$$R_{10} = 0.00044\varepsilon_r^{2.136} + 0.0184$$

$$R_{11} = \frac{\left(\frac{f_n}{19.47}\right)^6}{1 + 0.0962\left(\frac{f_n}{19.47}\right)^6}$$

$$R_{12} = \frac{1}{1 + 0.00245u^2}$$

$$R_{13} = 0.9408\varepsilon_{ef}(f_n)^{R_8} - 0.9603$$

$$R_{14} = (0.9408 - R_9)\varepsilon_{ef}(0)^{R_8} - 0.9603$$

$$R_{15} = 0.707R_{10}\left(\frac{f_n}{12.3}\right)^{1.097}$$

$$R_{16} = 1 + 0.0503\varepsilon_r^2R_{11}\left(1 - e^{-\left(\frac{u}{15}\right)^6}\right)$$

$$R_{17} = R_7\left[1 - 1.1241\frac{R_{12}}{R_{16}}e^{\left(-0.026f_n^{1.15656} - R_{15}\right)}\right] \quad (2.39)$$

$$Z_0(f_n) = Z_0(0)\left(\frac{R_{13}}{R_{14}}\right)^{R_{17}} \tag{2.40}$$

where $Z_0(0)$ is the static value of the characteristic impedance given for example, by relationship (2.27) for $t = 0$. The precision of the given model is better than 1% in comparison with Jansen (1978) for a range of the design parameters $0.1 \leq u \leq 10, 1 \leq \varepsilon_r \leq 18, 0 \leq h/\lambda_0 \leq 0.1$. The model can also be used with a lower precision beyond the range of these parameters, since the relationships include the proper asymptotic behaviour. The dispersion dependences for open microstrip line on a ceramic substrate ($\varepsilon_r = 10$) are shown in Fig. 2.11.

When neglecting losses by radiation, the total losses can be expressed as a sum of conductivity α_C and dielectric α_D losses from relationship (2.12), so that

$$\alpha = \alpha_C + \alpha_D \qquad \left[\mathrm{dB\,m^{-1}}\right] \tag{2.41}$$

Closed relationships for conductivity losses are taken from Gupta *et al.* (1979)

$$\alpha_C = 1.38A\frac{R_S}{hZ_0}\frac{32-\left(\frac{w'}{h}\right)^2}{32+\left(\frac{w'}{h}\right)^2} \quad \left[\mathrm{dB\,m^{-1}}\right] \quad \text{for } \frac{w}{h} \leq 1 \tag{2.42}$$

$$\alpha_C = 6.1\cdot 10^{-5}A\frac{R_S Z_0 \varepsilon_{ef}}{h}\left[\frac{w'}{h} + \frac{0.667\frac{w'}{h}}{\frac{w'}{h}+1.444}\right] \quad \left[\mathrm{dB\,m^{-1}}\right] \text{ for } \frac{w}{h} \geq 1$$

where

$$A = 1 + \frac{1+\frac{1}{\pi}\ln\left(\frac{2B}{t}\right)}{\frac{w'}{h}} \tag{2.43}$$

$$\begin{aligned} B &= h \qquad \text{for } \frac{w}{h} \geq \frac{1}{2\pi} \\ B &= 2\pi w \qquad \text{for } \frac{w}{h} \leq \frac{1}{2\pi} \end{aligned} \tag{2.44}$$

$$R_S = \sqrt{\pi f \mu_0 \rho}\left\{1 + \frac{2}{\pi}\arctan\left[1.4\left(\frac{\sigma_{ef}}{\delta_n}\right)^2\right]\right\} \tag{2.45}$$

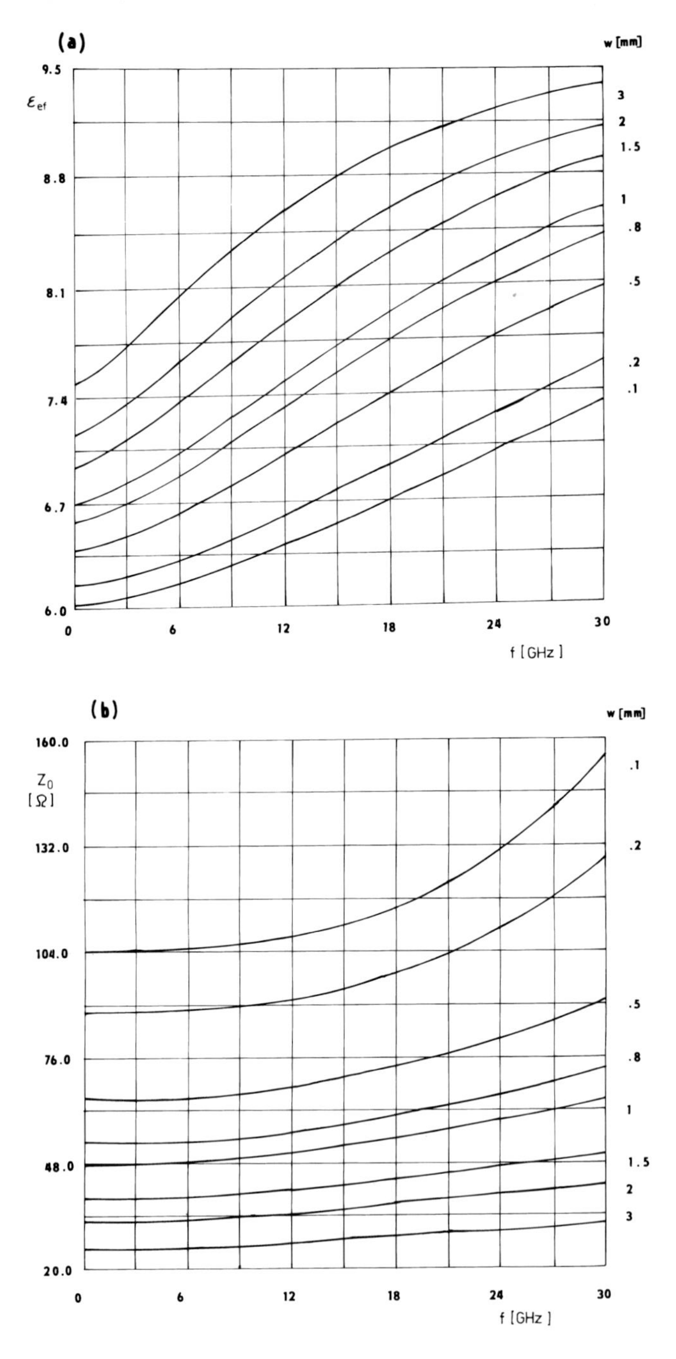

Fig. 2.11 Frequency dependence of parameters of microstrip line on corrund substrate ($\varepsilon_r = 10$, $h = 1\,\text{mm}$, $t = 5\mu\text{m}$) for different width of line (a) $\varepsilon_{ef}(f)$ (b) $Z_0(f)$.

where σ_{ef} is the mean roughness of the surface of conductors, δ_n is given by relationship (2.9), the corrected width of the microstrip $w' = w + \Delta w$ is determined with the help of (2.28) and ρ is the specific resistance of conductors.

2.1.3 Suspended and inverted microstrip line

These are the basic type of lines used in the upper part of the microwave band and lower part of the millimetre wave band. Their transverse cross-sections are shown in Fig. 2.12. The most important fact is that the use of an air gap under the substrate considerably decreases the dispersion of the line parameters and reduces losses in the line. The results of the quasistatic approximation can be used up to higher frequencies.

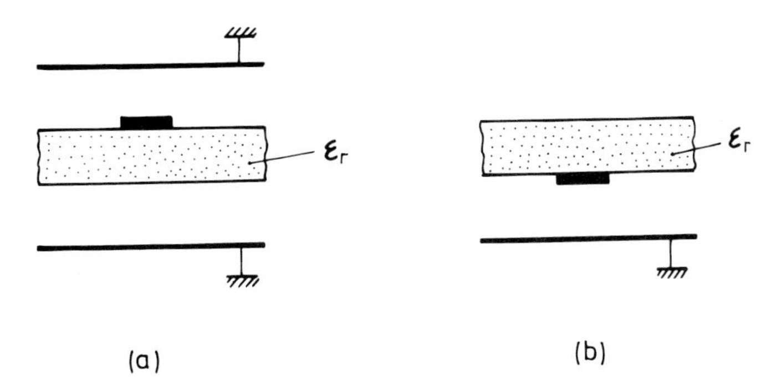

Fig. 2.12 Transverse cross-sections of suspended (a) and inverted (b) microstrip line.

In work (Tomar and Bhartia, 1987) a quasistatic model is presented for $1 \leq \varepsilon_r \leq 20$, $0.5 \leq w/h_1 \leq 10$, $0.05 \leq h/h_1 \leq 1.5$, $t = 5\,\mu\text{m}$ with a precision higher than 0.6% for the analysis and higher than 1% for the synthesis. This is determined by an approximation of a precise solution obtained with the help of the variation method in the Fourier's transformed region. The effective permittivity of the suspended microstrip line is simulated by the relationship

$$\sqrt{\varepsilon_{ef}(0)} = (1 - f_1 f_2)^{-1} \tag{2.46}$$

where

$$f_1 = 1 - \frac{1}{\sqrt{\varepsilon_r}} \tag{2.47}$$

and

$$f_2^{-1} = \sum_{i=0}^{3}\sum_{j=0}^{3}\sum_{k=0}^{3} d_{ijk}\,(\ln \varepsilon_r)^k \left(\frac{h_i}{h}\right)^j \left(\frac{w}{h_i}\right)^i \tag{2.48}$$

The values of coefficients d_{ijk} are presented in Table 2.2. The charac-

Table 2.2 The values of coefficients d_{ijk} for suspended line

k	0	1	2	3
$d_{00k} * 10^2$	176.2576	−43.1240	13.4094	−1.7010
$d_{01k} * 10^4$	4665.2320	−1790.4000	291.5853	−8.0888
$d_{02k} * 10^6$	−3025.5070	−141.9368	−3099.4700	777.6151
$d_{03k} * 10^8$	2481.5690	1430.3860	10085.5500	−2599.1320
$d_{10k} * 10^4$	−1410.2050	149.9293	198.2892	−32.1679
$d_{11k} * 10^4$	2548.7810	1531.8310	−1027.5200	138.4182
$d_{12k} * 10^6$	999.3135	−4036.7910	1762.4120	−298.0241
$d_{13k} * 10^8$	−1983.7890	8523.9290	−5235.4600	1145.7880
$d_{20k} * 10^5$	1954.9720	333.3873	−700.7473	121.3212
$d_{21k} * 10^5$	−3931.0900	−1890.7190	1912.2660	−319.6794
$d_{22k} * 10^7$	−532.1326	7274.7210	−4955.7380	941.4134
$d_{23k} * 10^8$	100.2083	1110.4070	1101.2700	270.0047
$d_{30k} * 10^6$	−983.4028	−255.1229	455.8729	−83.9468
$d_{31k} * 10^6$	1936.3170	779.9975	−995.5454	183.1357
$d_{32k} * 10^8$	62.8550	−3462.5000	2909.9230	−614.7068
$d_{33k} * 10^9$	−35.2531	601.0291	−643.0814	161.2689

teristic impedance $Z_0(0)$ is determined with the help of relationships (2.21), (2.22), (2.27), where

$$u = \frac{\frac{w}{h_1}}{1 + \frac{h}{h_1}} \tag{2.49}$$

For the graphic dependence of parameters of the line on physical dimensions of its structures see Fig. 2.13.

The effective permittivity of an inversed microstrip line is simulated by the relationship

$$\sqrt{\varepsilon_{ef}} = (1 + f_1 f_2) \tag{2.50}$$

where

$$f_1 = \sqrt{\varepsilon_r} - 1 \tag{2.51}$$

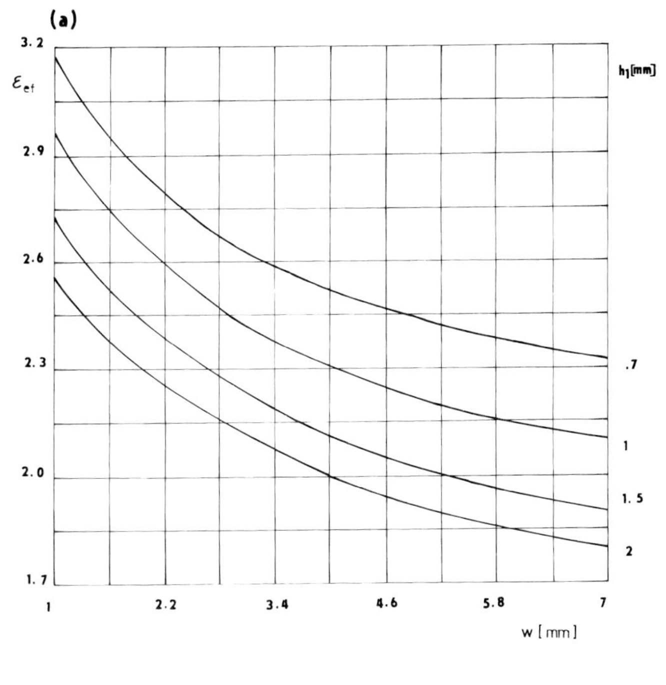

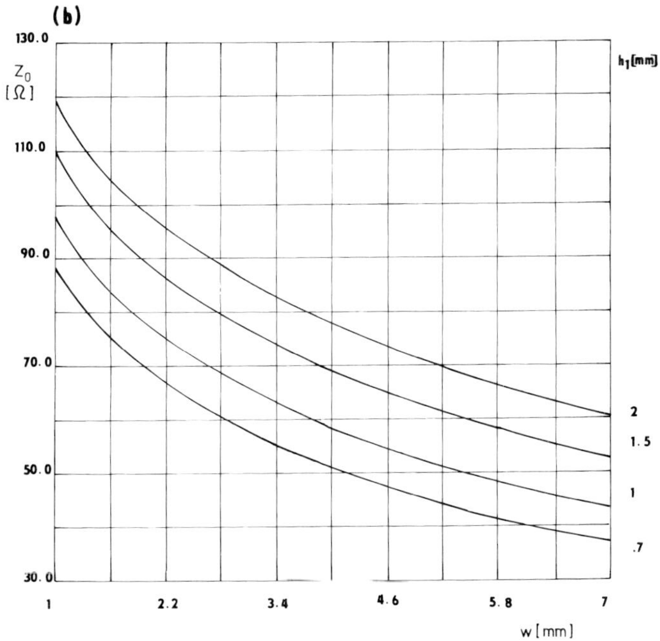

Fig. 2.13 Static parameters of suspended microstrip line on corrund substrate ($\varepsilon_r = 10$, $h = 1\,\text{mm}$) (a) $\varepsilon_{ef}(w)$ (b) $Z_0(w)$.

Table 2.3 The values of coefficients d_{ijk} for inverted line

k	0	1	2	3
$d_{00k} * 10^3$	2359.4010	−97.1644	−5.7706	11.4112
$d_{01k} * 10^5$	4855.9472	−3408.5207	15296.7300	−2418.1785
$d_{02k} * 10^5$	1763.3470	961.0481	−2089.2800	375.8805
$d_{03k} * 10^6$	−556.0909	−268.6165	623.7094	−119.1402
$d_{10k} * 10^3$	219.0660	−253.0864	208.7469	−27.3285
$d_{11k} * 10^3$	915.5589	338.4033	−253.2933	40.4743
$d_{12k} * 10^5$	−1957.3790	−1170.9360	1480.8570	−347.6403
$d_{13k} * 10^6$	486.7425	279.8323	−431.3625	108.8240
$d_{20k} * 10^5$	5602.7670	4403.3560	−4517.0340	743.2717
$d_{21k} * 10^5$	−2823.4810	−1562.7820	3646.1500	−823.4223
$d_{22k} * 10^6$	253.8930	158.5529	−3235.4850	919.3661
$d_{23k} * 10^7$	−147.0235	62.4342	887.5211	−270.7555
$d_{30k} * 10^6$	−3170.2100	−1031.0520	2715.3270	−819.3120
$d_{31k} * 10^6$	596.3251	188.1409	−1741.4770	465.6756
$d_{32k} * 10^7$	124.9655	577.5381	1366.4530	−481.1300
$d_{33k} * 10^9$	−530.2099	−2666.3520	−3220.0960	1324.4990

and f_2 is given by (2.48), where the values of coefficients d_{ijk} are presented in Table 2.3. The characteristic impedance $Z_0(0)$ is also determined from relationships (2.21), (2.22), (2.27), where

$$u = \frac{w}{h_1} \tag{2.52}$$

For the graphic dependence of the line parameters on the structure dimensions see Fig. 2.14.

By solving the wave equation for the suspended microstrip line it was found that the dispersion of the line parameters cannot be always neglected. The dispersion increases with increasing ε_r and with reducing the strip width. The dependence of the dispersion on the thickness of the metal coating can be neglected for typical values of the metallization thicknesses. In Tomar and Bhartia (1987a), closed relationships were simulated for GaAs ($\varepsilon_r = 12.9$). For the frequency dependence of the effective dielectric permittivity $\varepsilon_{ef}(f)$ the Getsinger's relationship was employed

$$\varepsilon_{ef}(f) = \varepsilon_r - \frac{\varepsilon_r - \varepsilon_{ef}(0)}{1 + G\left(\frac{f}{f_p}\right)^2} \tag{2.53}$$

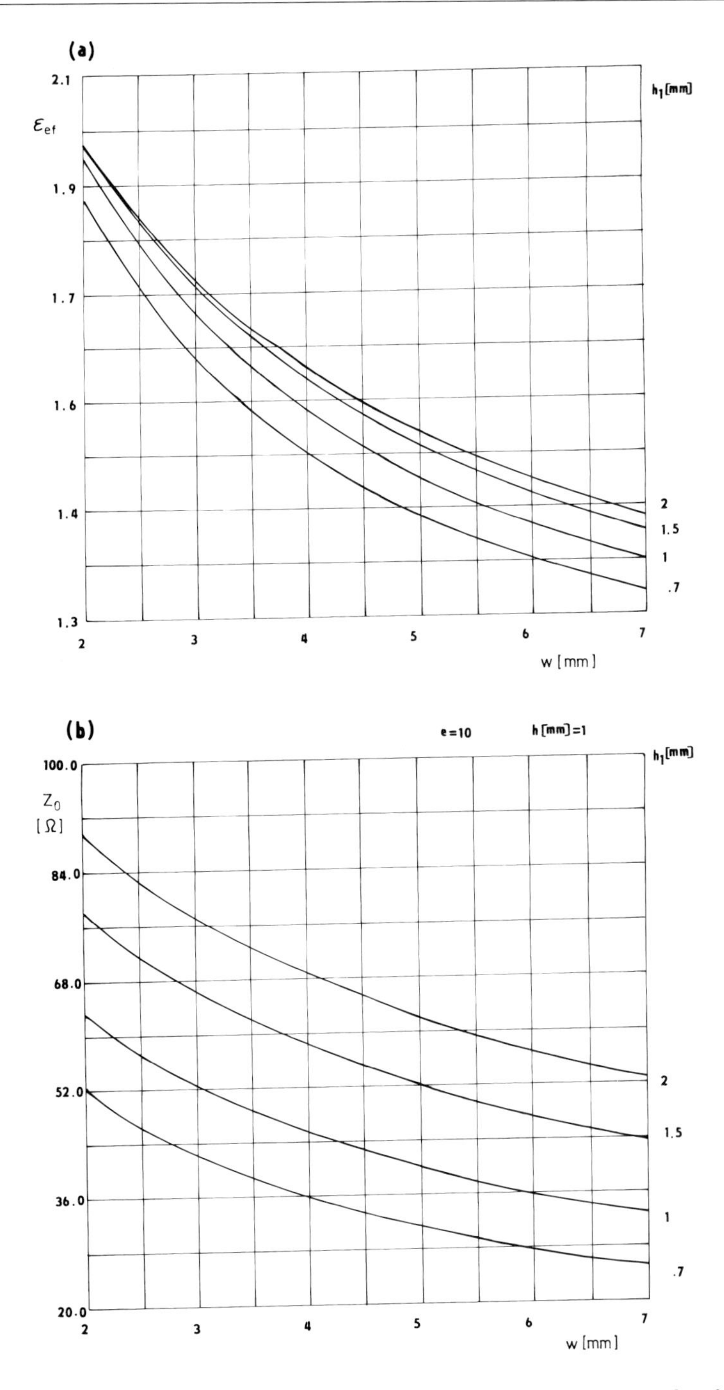

Fig. 2.14 Static parameters of inverse microstrip line on corrund substrate ($\varepsilon_r = 10$, $h = 1\,\text{mm}$) (a) $\varepsilon_{ef}(w)$, (b) $Z_0(w)$.

where the static value $\varepsilon_{ef}(0)$ is given by relationship (2.46) and

$$f_p = \frac{Z_0\,(0)}{0.512\pi^2} \frac{h}{h+h_1} \qquad [\text{GHz}, \Omega] \qquad (2.54)$$

where $Z_0(0)$ is quasistatic value of the characteristic impedance and the parameter G is given by the relationship

$$G = \sum_{i=0}^{2} \sum_{j=0}^{4} d_{ij} \left(\frac{h}{h_1}\right)^j Z^i \qquad (2.55)$$

For values of coefficients d_{ij} see Table 2.4. This model is valid for

Table 2.4 Values of coefficients d_{ij} for GaAs substrate

i \ j	0	1	2	3	4
0	0.0194	−0.2398	0.8977	−0.9924	0.3468
1	−0.0003	0.0096	−0.0346	0.0384	−0.0135
2	0	0	0.0004	−0.0004	0.0001

$0.1 \leq w/h_1 \leq 1$ and $20\,\text{GHz} \leq f \leq 100\text{GHz}$ with a precision of $\pm 2\%$ with respect to precise theoretical data. The dispersion dependences for the other substrate material can be determined by the computerized approximation of precise theoretical data with the use of the same model. The dispersion dependences for RT-duroid ($\varepsilon_r = 2.22$) and quartz ($\varepsilon_r = 3.78$) are graphically represented in Tomar and Bhartia (1986) for parameters $0.1 \leq h/h_1 \leq 0.1$, $0.5 \leq w/h_1 \leq 8$ and $f \leq 100\,\text{GHz}$.

2.1.4 Coplanar line

In certain applications, it is advantageous to use the coplanar line. It is topologically most similar to the coaxial line and it is easy to couple parallel elements to it. In a wider sense of words, the coplanar lines also include strip lines where all the conductors are situated in one plane. These are symmetric lines (coplanar waveguide, coplanar strips, slotline), non-symmetric modifications of these

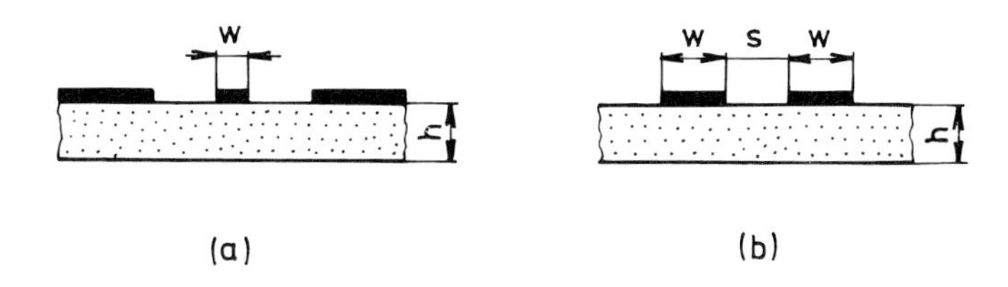

Fig. 2.15 Transverse cross-section through (a) the coplanar line (b) the complementary structure of coplanar strips.

lines and coupled coplanar lines. They can be open, side-open or closed. These lines are separately considered in this monograph in three groups. Slotlines and coupled lines were independently treated. The coplanar lines *sensu stricto* are the coplanar waveguide and the complementary structure of coplanar strips corresponding to it.

A transverse cross-section through the coplanar line is shown in Fig. 2.15. It is formed by a strip conductor of a width w, with external conductors situated along its both sides, which are infinitely wide in the ideal case. All the three conductors are deposited on one side of the dielectric substrate of a thickness h.

The lateral conductors have the same potential. The dominant mode in the line is a hybrid wave, which can be approximated by the quasi-TEM wave at low frequencies. The electric and magnetic field lines are shown in Fig. 2.16. The coplanar lines were studied with

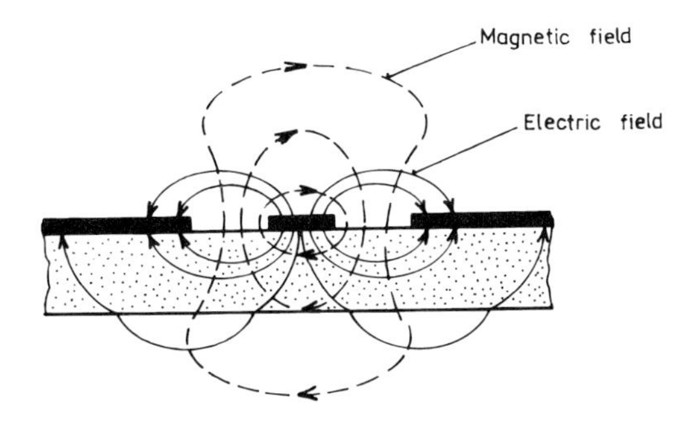

Fig. 2.16 Coplanar waveguide with infinite ground planes — static field lines.

the use of static as well as dynamic methods. The most frequently employed methods were the method of the conformal transformation for the quasi-TEM approximation and the spectral method for the precise analysis including effects of the dispersion, shielding and thickness of metal coating. Since the coplanar waveguide is particularly used in circuits with a higher integration, attention will also be paid to modified structures where the effect of the screening distance h_2 and of the finite width of conductors on the line parameters will be determined (Ghione and Naldi, 1987).

The coplanar waveguide with upper shielding is shown in Fig. 2.17. The analytical expressions for the quasi-TEM approximation are ob-

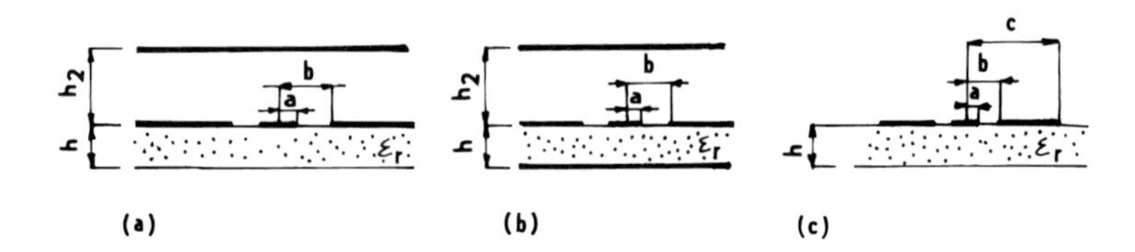

Fig. 2.17 Coplanar waveguide (a) with upper shielding (b) with the top shielding and metal coated bottom side of substrate (c) with external conductors of a finite width.

tained by simulating the slot as a magnetic wall. This assumption can hardly be checked for a large width of slot and small distance of the top shielding plate, however, an excellent agreement with experimental data was achieved for the most frequently used dimensions of the line. The total capacitance per unit length of the line can be calculated as a sum of capacitances of the upper (air) and bottom (air and dielectric base) half-plane. With the help of a sequence of conformal transformations the following relationship is obtained for the total capacitance.

$$C_1(\varepsilon_r) = 2\varepsilon_0 \frac{K(k_2)}{K(k_2')} + 2\varepsilon_0 \frac{K(k)}{K(k')} + 2(\varepsilon - 1)\varepsilon_0 \frac{K(k_1)}{K(k_1')} \tag{2.56}$$

where

$$k = \frac{a}{b} \qquad a = \frac{w}{2} \qquad b = s + \frac{w}{2} \tag{2.57}$$

$$k_1 = \frac{\sinh\left(\frac{\pi a}{2h_2}\right)}{\sinh\left(\frac{\pi b}{2h}\right)} \tag{2.58}$$

$$k_2 = \frac{\tanh\left(\frac{\pi a}{2h_2}\right)}{\tanh\left(\frac{\pi b}{2h_2}\right)} \tag{2.59}$$

$K(k)$ being a complete elliptic integral of the first type and $k_i' = 1 - k_i^2$. The effective permittivity

$$\varepsilon_{ef_1} = \frac{C_1(\varepsilon_r)}{C_1(1)} = 1 + (\varepsilon_r - 1)\frac{\frac{K(k_1)}{K(k_1')}}{\frac{K(k_2)}{K(k_2')} + \frac{K(k)}{K(k')}} \tag{2.60}$$

and

$$Z_{0_1} = \frac{60\pi}{\sqrt{\varepsilon_{ef_1}}}\frac{1}{\frac{K(k_2)}{K(k_2')} + \frac{K(k)}{K(k')}} \tag{2.61}$$

The curves of the constant impedance and effective permittivity depending on the ratios a/b and h_2/b for $h/b = 1$ and $\varepsilon_r = 10$ are plotted in Fig. 2.18. For $h_2 \Rightarrow \infty$ the structure is changed to an unscreened coplanar waveguide.

The complementary structure of coplanar strips is shown as a cross-section in Fig. 2.15b. The line is formed by two strip conductors of a width w, deposited on a substrate of a thickness h width a gap between them of a width s. The potentials on the conductors are different. The effective permittivity can be determined by the conformal mapping method as

$$\varepsilon^*_{ef_1} = 1 + \frac{\varepsilon_r - 1}{2}\frac{K(k)}{K(k')}\frac{K(k_1')}{K(k_1)} \tag{2.62}$$

where

$$k = \frac{s}{2w + s} \tag{2.63}$$

and

$$k_1 = \frac{\tanh\left(\frac{\pi s}{4h}\right)}{\tanh\left[\frac{\pi(2w+s)}{4h}\right]} \tag{2.64}$$

The characteristic impedance Z^*_{01} is determined by the relationship

$$Z^*_{0_1} = \frac{120\pi}{\sqrt{\varepsilon^*_{ef_1}}}\frac{K(k)}{K(k')} \tag{2.65}$$

The coplanar waveguide with the top shielding and metal coated bottom side of the substrate is shown in Fig. 2.17b. By the conformal

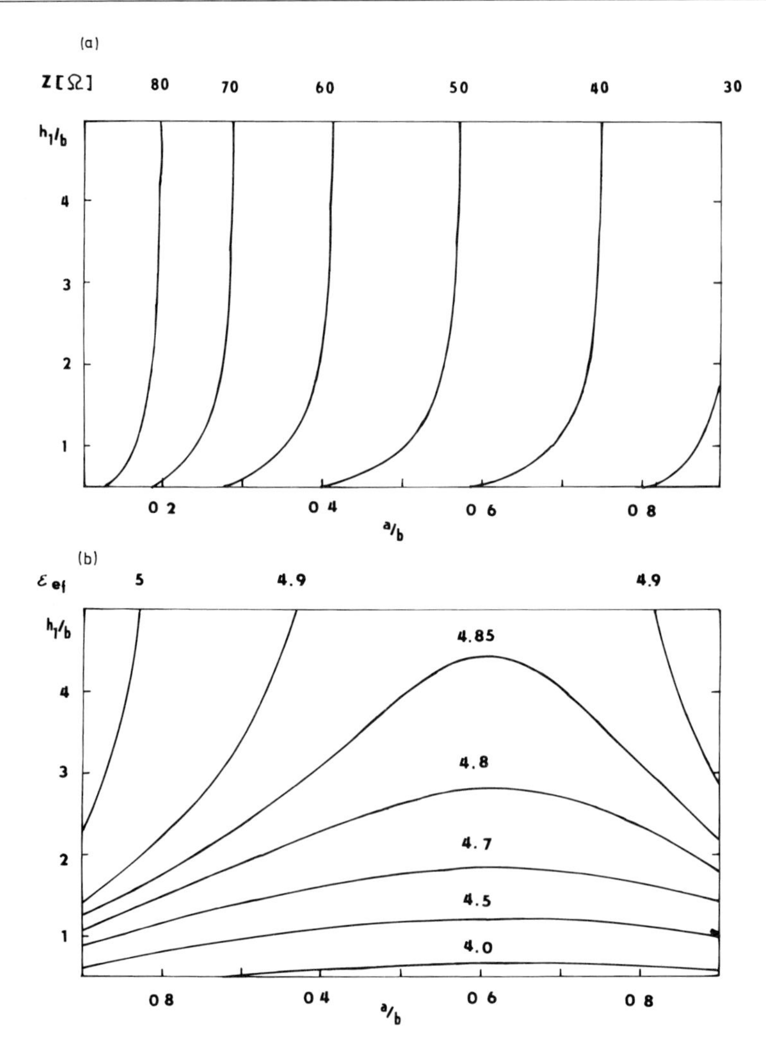

Fig. 2.18 Curves of (a) constant impedance (b) constant ε_{ef} for coplanar waveguide on Fig.2.17a versus relation $\frac{a}{b}$, thickness of substrate and high of shade ($\varepsilon_r = 10$, $\frac{h}{b} = 1$).

representation method the following formula is obtained for the resulting capacitance

$$C_2(\varepsilon_r) = 2\varepsilon_0\varepsilon_r \frac{K(k_3)}{K(k_3')} + 2\varepsilon_0 \frac{K(k_4)}{K(k_4')} \tag{2.66}$$

where

$$k_3 = \frac{\tanh\left(\frac{\pi a}{2h}\right)}{\tanh\left(\frac{\pi b}{2h}\right)} \tag{2.67}$$

$$k_4 = \frac{\tanh\left(\frac{\pi a}{2h_2}\right)}{\tanh\left(\frac{\pi b}{2h_2}\right)} \tag{2.68}$$

The effective permittivity and characteristic impedance

$$\varepsilon_{ef_2} = \frac{C_2\left(\varepsilon_r\right)}{C_2\left(1\right)} = 1 + \left(\varepsilon_r - 1\right)\frac{\frac{K(k_3)}{K(k_3')}}{\frac{K(k_3)}{K(k_3')} + \frac{K(k_4)}{K(k_4')}} \tag{2.69}$$

and

$$Z_{0_2} = \frac{60\pi}{\sqrt{\varepsilon_{ef_2}}}\frac{1}{\frac{K(k_3)}{K(k_3')} + \frac{K(k_4)}{K(k_4')}} \tag{2.70}$$

Curves of the constant impedance and effective permittivity depending on the ratios a/b and h_2/b for $h/b = 1$, $\varepsilon_r = 10$ are depicted in Fig. 2.19. The coplanar waveguide with external conductors of a finite width is shown in Fig. 2.17c. The resulting capacitance of the line per unit length can be determined for the even mode as

$$C_3\left(\varepsilon_r\right) = 4\varepsilon_0\frac{K\left(k_5\right)}{K\left(k_5'\right)} + 2\varepsilon_0\left(\varepsilon_r - 1\right)\frac{K\left(k_6\right)}{K\left(k_6'\right)} \tag{2.71}$$

where

$$k_5 = \frac{a}{b}\sqrt{\frac{1 - \frac{b^2}{c^2}}{1 - \frac{a^2}{c^2}}} \tag{2.72}$$

$$k_6 = \frac{\sinh\left(\frac{\pi a}{2h}\right)}{\sinh\left(\frac{\pi b}{2h}\right)}\sqrt{\frac{1 - \frac{\sinh^2\left(\frac{\pi b}{2h}\right)}{\sinh^2\left(\frac{\pi c}{2h}\right)}}{1 - \frac{\sinh^2\left(\frac{\pi a}{2h}\right)}{\sinh^2\left(\frac{\pi c}{2h}\right)}}} \tag{2.73}$$

The following relationships hold for the effective permittivity and characteristic impedance:

$$\varepsilon_{ef_3} = 1 + \frac{1}{2}\frac{K\left(k_6\right)}{K\left(k_6'\right)} \tag{2.74}$$

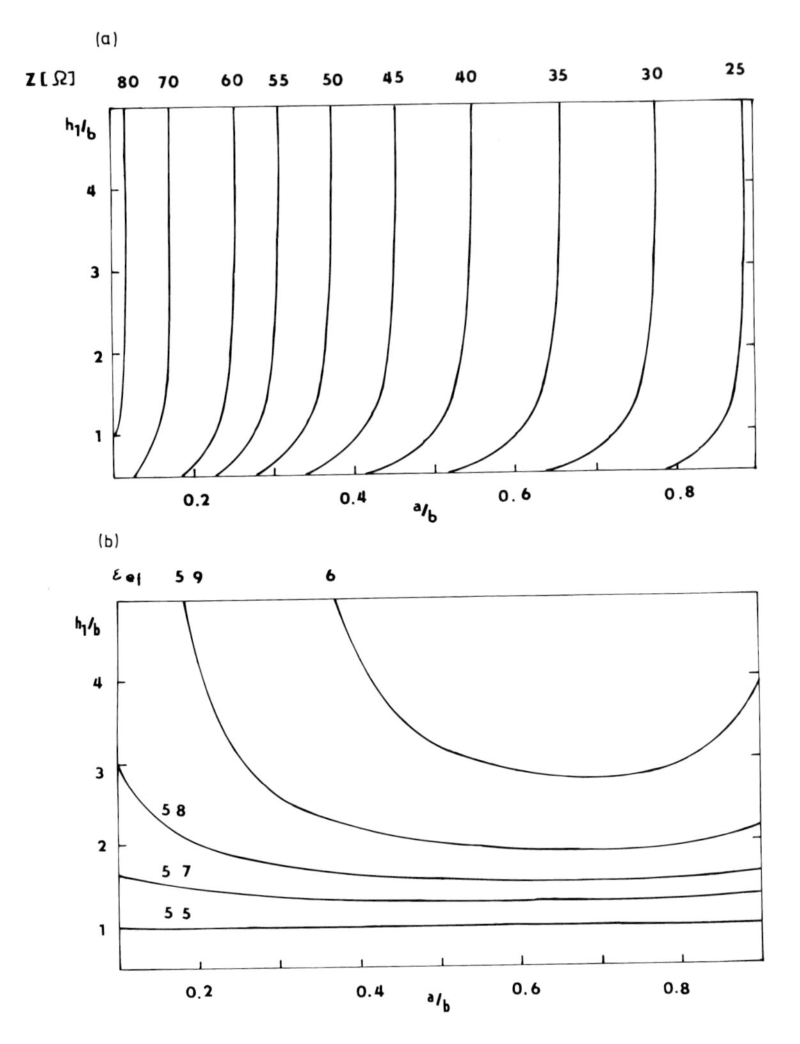

Fig. 2.19 Curves of (a) constant impedance (b) constant ε_{ef} for planar waveguide on Fig.2.17b versus relation $\frac{a}{b}$, thickness of substrate and high of shade ($\varepsilon_r = 10$, $\frac{h}{b} = 1$).

and

$$Z_{0_3} = \frac{30\pi}{\sqrt{\varepsilon_{ef3}}} \frac{K(k_5')}{K(k_5)} \tag{2.75}$$

Curves of the constant impedance and effective permittivity, depending on the ratios a/b and b/c for $h/b = 1$, $\varepsilon_r = 10$ are plotted in Fig. 2.20. A parasitic odd mode can also propagate through the

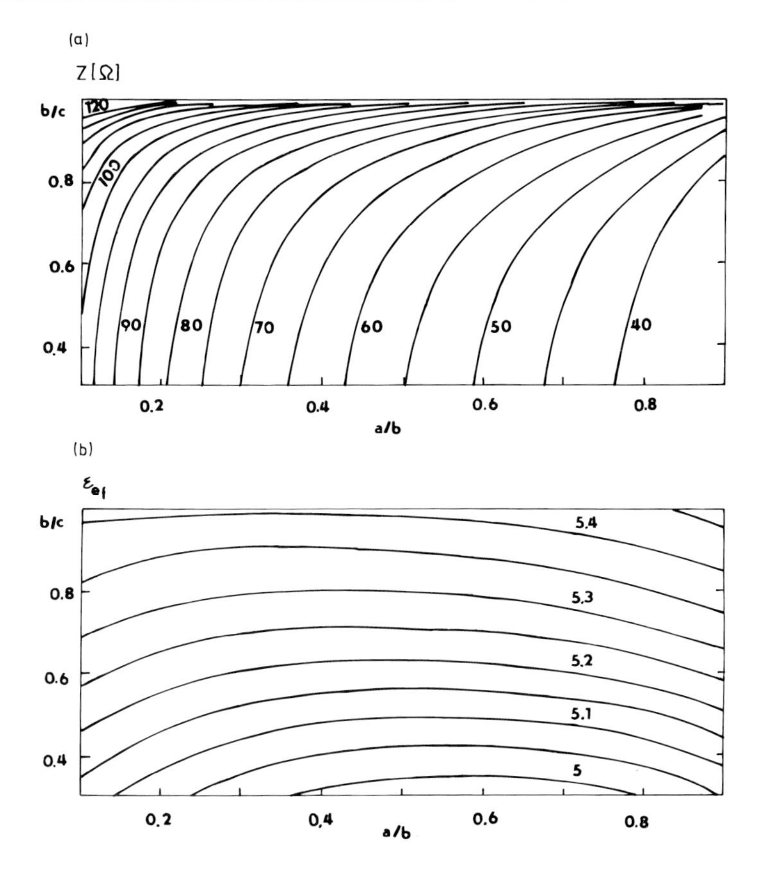

Fig. 2.20 Curves of (a) constant impedance (b) constant ε_{ef} for coplanar waveguide on Fig.2.17c versus relation $\frac{a}{b}$, $\frac{b}{c}$ for $\frac{h}{b} = 1$ and $\varepsilon_r = 10$.

line. In the case of commonly used widths of side conductors, the impedance level of the odd mode is comparable with that of the even mode and thus, it is of importance to provide perfect grounding of external conductors to prevent the odd mode propagation.

The effect of the finite thickness of strip conductors on the coplanar line with the top shielding for $h_2 \Rightarrow \infty$ can be considered similarly as in the case of the microstrip line (Gupta *et al.*, 1979). The effective widths of the strip w' and of slot s' are introduced with the use of equations

$$\begin{aligned} w' &= w + \Delta w \\ s' &= s - \Delta w \end{aligned} \tag{2.76}$$

where w is determined with the help of (2.28). The effective permittivity with the correction for the finite thickness of the metal coating t was derived by correcting the capacity C_1 and empirical fitting of results calculated by the integral equations method (Kitazawa *et al.*, 1976), so that

$$\varepsilon_{ef}(t) = \varepsilon_{ef_1} - \frac{0.7\left(\varepsilon_{ef_1} - 1\right)\frac{t}{s}}{\frac{K(k)}{K'(k)} + 0.7\frac{t}{s}} \tag{2.77}$$

where $\varepsilon_{ef1}(t)$ is given by (2.60). Then the characteristic impedance is

$$Z_{0_1}(t) = \frac{30\pi}{\sqrt{\varepsilon_{ef_1}(t)}} \frac{K'(k_e)}{K(k_e)} \tag{2.78}$$

where

$$k_e = \frac{w'}{w' + 2s'} \tag{2.79}$$

The finite thickness t of strip conductors reduces the values of ε_{ef} and Z_{01} which is in agreement with Kitazawa *et al.* (1976).

For two-conductor coplanar line the values w', s' are obtained from relationships (2.76) and (2.28), respectively. Then

$$\varepsilon^*_{ef_1}(t) = \varepsilon^*_{ef_1} - \frac{1.4\left(\varepsilon^*_{ef_1} - 1\right)\frac{t}{s}}{\frac{K'(k)}{K(k)} + 1.4\frac{t}{s}} \tag{2.80}$$

and

$$Z^*_{0_1}(t) = \frac{120\pi}{\sqrt{\varepsilon^*_{ef_1}(t)}} \frac{K(k_e)}{K'(k_e)} \tag{2.81}$$

where

$$k^*_e = \frac{s'}{s' + 2w'} \tag{2.82}$$

The specific attenuation α_C of a coplanar waveguide of the type with the top screening for $h_2 \Rightarrow \infty$ resulting from a finite conductivity of strip conductors can be determined similarly as for the microstrip as follows (Hoffman, 1983)

$$\alpha_C = \frac{2.18 R_S \sqrt{\varepsilon_{ef}}}{\eta_0 d K(k) K(k') \left[1 - \left(\frac{w}{d}\right)^2\right]} \left\{ \frac{2d}{w} \left[\pi + \ln\left(\frac{4\pi w\left(1 - \frac{w}{d}\right)}{t\left(1 + \frac{w}{d}\right)}\right)\right] + \right.$$

$$+2\left[\pi + \ln\left(\frac{4\pi d\left(1-\frac{w}{d}\right)}{t\left(1+\frac{w}{d}\right)}\right)\right]\Bigg\} \tag{2.83}$$

where $d = 2s + w$, $k = w/d$, R_S is determined from (2.84) and η_0 is the wave impedance of the material. The formula (2.83) holds when $t > 3\delta_n$ and $t \ll w, s$.

2.1.5 Slot line

This line is formed by a slot of width s in the metal coating of the dielectric substrate as shown in Fig. 2.21. In order that it can be implemented as a transmission line, it is necessary to minimize its losses by radiation. This can be achieved by using dielectric substrate with a sufficiently high permittivity ($\varepsilon_r > 10$). The slot mode field rapidly vanishes with increasing distance r. Electric field force lines are closed about the slot, those of the magnetic field are perpendicular to the slot (Fig. 2.21). In a longitudinal section the

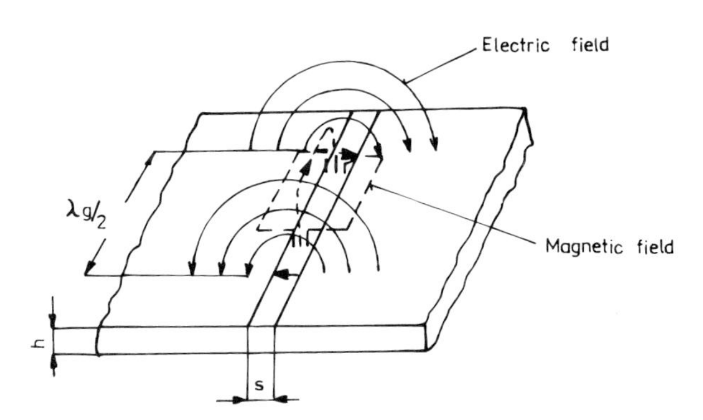

Fig. 2.21 Slotted-line and static field lines.

magnetic field force lines are always closed by steps $\lambda_g/2$ where λ_g is the wavelength in the line. The slot mode does not have the nature of TEM and thus, these basic parameters are dependent on the frequency already in the first approximation and thus, they are different from those of microstrip lines with the quasi-TEM mode. In contrast to waveguides, they have no cut-off frequency.

The formulation (2.6b) is suitable for the definition of the characteristic impedance. The slot line configuration is suitable for connecting parallel elements. A combination is frequently advantageous with the microstrip line on one side and slot line on the opposite side.

The analytical approximation of the second order investigates the waveguide model of the slot line by the transverse resonance method (Cohn, 1969). The line is closed in a rectangular waveguide (Fig. 2.22) in such a way that the walls perpendicular to the line

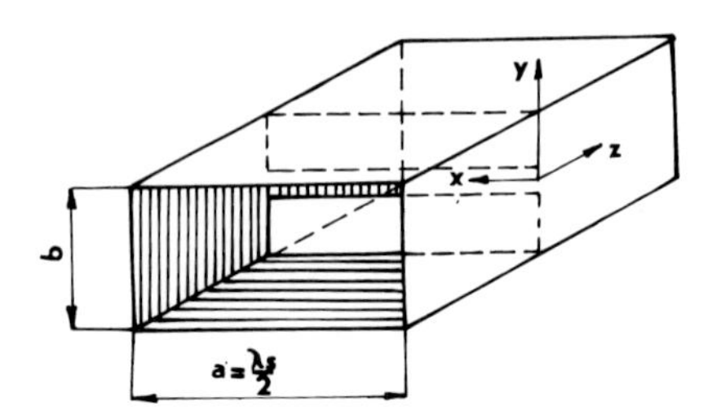

Fig. 2.22 Waveguide model of slot-line.

are separated from each other by $\lambda_g/2$ and those parallel with the slot and perpendicular to the substrate by $\pm b/2$. Since the fields are tightly coupled in the vicinity of the slot, the effect of walls is negligible for a sufficiently large height of the waveguide b. By the numerical approximation of result of this method the following relationships were obtained (Garg and Gupta, 1976):

$$\frac{\lambda_g}{\lambda_0} = 0.923 - 0.448 \log \varepsilon_r + 0.2\frac{s}{h} - \left(0.29\frac{s}{h} + 0.047\right) \log\left(\frac{h}{\lambda_0} \cdot 10^2\right) \tag{2.84}$$

$$\begin{aligned} Z_0 = {} & 72.62 - 35.19 \log \varepsilon_r + 50\frac{\left(\frac{s}{h} - 0.02\right)\left(\frac{s}{h} - 0.1\right)}{\frac{s}{h}} \\ & + \log\left(\frac{s}{h} \cdot 10^2\right)(44.28 - 19.58 \log \varepsilon_r) \\ & - \left[0.32 \log \varepsilon_r - 0.11 + \frac{s}{h}(1.07 \log \varepsilon_r + 1.44)\right] \\ & \cdot \left(11.4 - 6.07 \log \varepsilon_r - \frac{h}{\lambda_0} \cdot 10^2\right)^2 \end{aligned} \tag{2.85}$$

$$\begin{aligned} \frac{\lambda_g}{\lambda_0} = {} & 0.987 - 0.483 \log \varepsilon_r + \frac{s}{h}(0.111 - .0022\varepsilon_r) \\ & - \left(0.121 + 0.094\frac{s}{h} - 0.0032\varepsilon_r\right) \log\left(\frac{h}{\lambda_0} \cdot 10^2\right) \end{aligned} \tag{2.86}$$

$$
\begin{aligned}
Z_0 &= 113.19 - 53.55 \log \varepsilon_r + 1.25 \frac{s}{h} (114.59 - 51.88 \log \varepsilon_r) \\
&+ 20 \left(\frac{s}{h} - 0.2 \right) \left(1 - \frac{s}{h} \right) \\
&- \left[0.15 + 0.23 \log \varepsilon_r + \frac{s}{h} (-0.79 + 2.07 \log \varepsilon_r) \right] \\
&\cdot \left[10.25 - 5 \log \varepsilon_r + \frac{s}{h} (2.1 - 1.42 \log \varepsilon_r) - \frac{h}{\lambda_0} \cdot 10^2 \right]^2
\end{aligned} \quad (2.87)
$$

These relationships hold within a precision range of $\pm 2\%$ for $9.7 \leq \varepsilon_r \leq 20$, $0.02 \leq s/h \leq 1$, $0.01 \leq h/\lambda_o 0.25/\sqrt{\varepsilon_r - 1}$.

The dependence of parameters of the slot line on its dimensions for $\varepsilon_r = 10$, $0.05 \leq s/h \leq 1$ and $0.01 \leq h/\lambda_o \leq 0.08$ is shown in Fig. 2.23. For a non-zero thickness of metal coating t the phase constant is reduced (i.e. λ_g/λ_o increases) with enlarging the metal coating thickness. For $\varepsilon_r = 20$ the reduction of the phase constant makes of about 1% for $t/w = 0.02$.

The conductivity losses of the slot line α_C are

$$
\alpha_C = \frac{R_S A_C}{Z_0 h} \quad (2.88)
$$

where the attenuation coefficient

$$
A_C = f\left(\frac{s}{h}, \frac{t}{h}, \frac{h}{\lambda_0} \varepsilon_r \right) \quad \text{[dB]} \quad (2.89)
$$

and high-frequency resistance

$$
R_S = \sqrt{\pi f \mu \rho} \quad (2.90)
$$

The dependence of the factor A_C for ceramic material Al_2O_3 ($\varepsilon_r = 9.7$) of thickness $h = 0.635$ mm on the slot size and on the frequency is shown in Fig. 2.24. For dielectric losses α_D

$$
\alpha_D = \tan \delta \; F\left(\frac{s}{h}, \frac{t}{h}, \varepsilon_r, \frac{h}{\lambda_0} \right) \quad (2.91)
$$

The effect of the slot thickness on the magnitude of dielectric losses is very small. For $s/h \ll 1$ and high frequencies, α_D can be roughly estimated with the help of (2.12) with inserting $(\varepsilon_r + 1)/2$ for ε_{ef}.

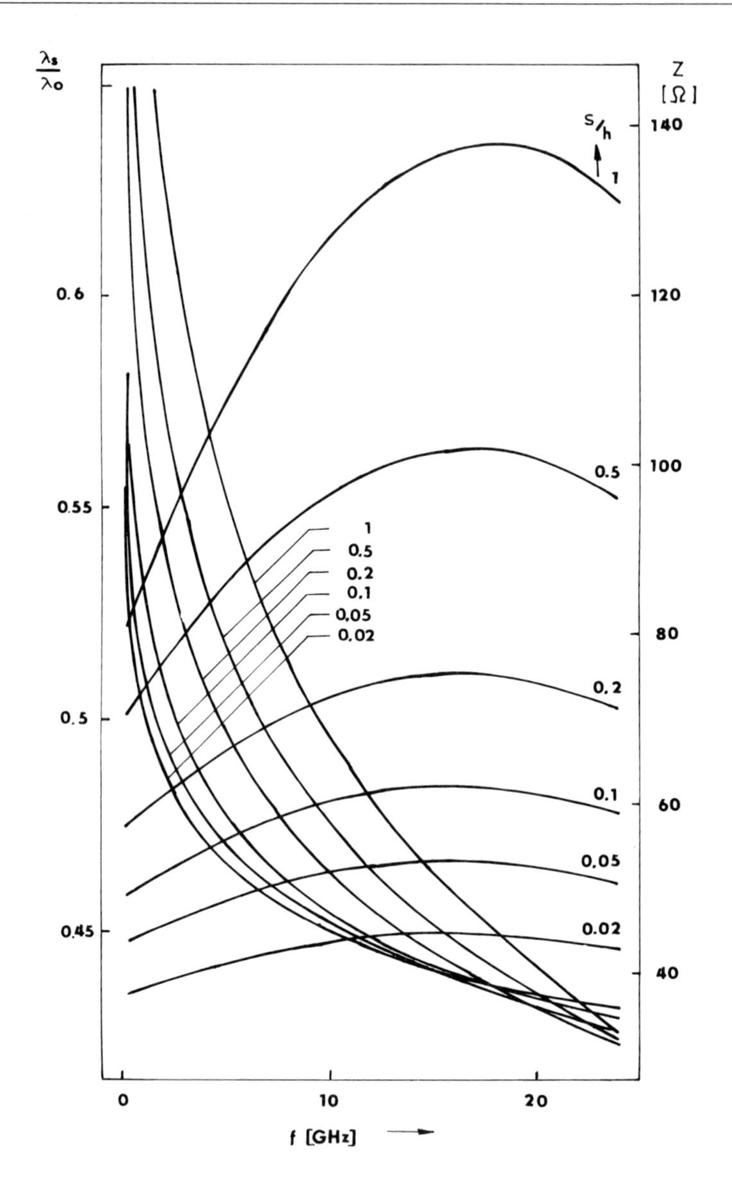

Fig. 2.23 Dispersion dependences of slotted-line parameters on corrund substrate ($\varepsilon_r = 10$, $h = 1\,\text{mm}$).

2.1.6 Fin-line

The fin-line is an ideal type of line for circuits in the band of millimetre waves. It exerts no special requirements for allowances, it is

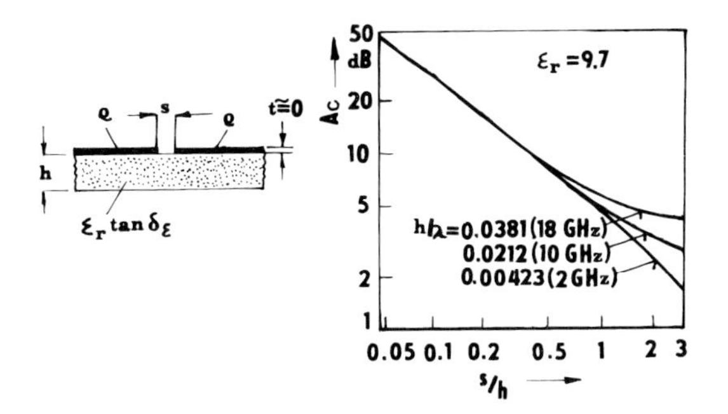

Fig. 2.24 Frequency and slot size dependences of the factor A_c.

compatible with further types of circuits and elements and it has a wide band for the single-mode activity, moderate attenuation, low dispersion in the frequency region of use in practice and it is cheap in the mass production.

The dispersion of line parameters is determined by the method of analysing eigen modes in the spatial or spectral region, by a method of finite elements or by a TLM two-dimensional method. These methods are very precise, however, they result in complicated programs for computers.

Besides the above mentioned accurate methods, the line parameters have also been approximated by different methods in computerized design of circuits. Closed relationships are sufficiently precise, derived by the static solution and approximation of numerical results of the spectral method (Pramanick and Bhartia, 1985).

The unilateral fin-line is shown in Fig. 2.25a. It is formed by a closed rectangular waveguide with transverse dimensions a, b which has in its E-plane a planar dielectric substrate of the thickness h which has on its one side deposited conductors separated by a longitudinal slot of the width s, attached to the top and bottom walls of the waveguide. The wavelength

$$\lambda_g = \frac{\lambda_0}{\sqrt{\varepsilon_{ef}(f)}} \tag{2.92}$$

where λ_0 is the wavelength in the free space and $\varepsilon_{ef}(f)$ is the frequency-dependent effective dielectric constant of the line given by

$$\varepsilon_{ef}(f) = k_e - \left(\frac{\lambda_0}{\lambda_{ca}}\right)^2 \tag{2.93}$$

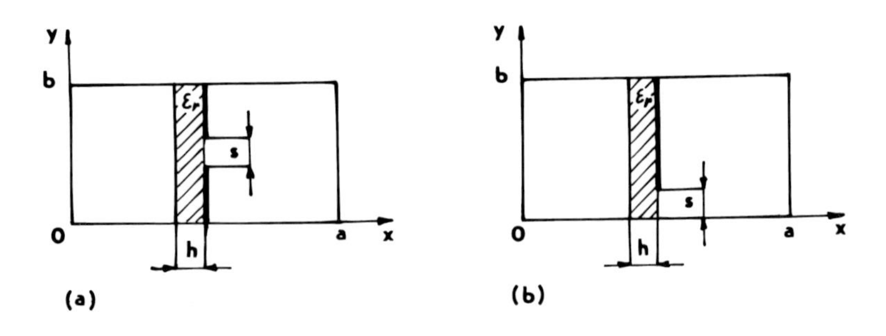

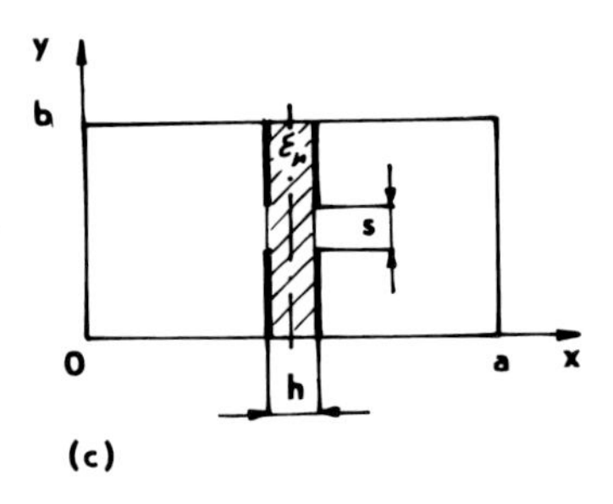

Fig. 2.25 Transverse cross-section (a) of one side (b) complementary one side (c) double side fin line.

where k_e is the equivalent dielectric constant at a frequency f and λ_{ca} is the cut-off wave length of the line of the same dimensions filled with air. It can be precisely expressed by (Pramanick and Bhartia, 1985)

$$\lambda_{ca} = 2a\sqrt{1 + \frac{4xb}{\pi a}\left(1 + 0.2\sqrt{\frac{b}{a}}\right)} \tag{2.94}$$

where

$$x = \ln\left[\csc\left(\frac{\pi s}{2b}\right)\right] \tag{2.95}$$

The equivalent dielectric constant k_e can be expressed as (Pramanick *et al.*, 1987)

$$k_e = \varepsilon_r \frac{\varepsilon_r - k_c}{1 + \left(\frac{s}{a}\sqrt{\frac{f-f_c}{f_c}}\right)^{1+0.6\frac{b}{s}}} \tag{2.96}$$

where f_c is the cut-off frequency in the line and

$$k_c = \left(\frac{\lambda_{cf}}{\lambda_{ca}}\right)^2 \tag{2.97}$$

The cut-off wave length λ_{cf} in the line is determined from relationship (2.97) with the use of (2.101) to (2.104).

The fin-line can be considered as a combination of a fin-waveguide (Fig. 2.26a) and waveguide loaded by a dielectric partition in the E-

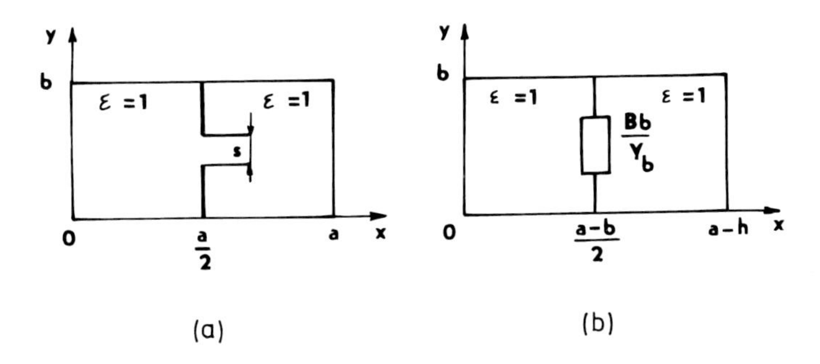

Fig. 2.26 (a) fin waveguide (b) equivalent circuit of doble side fin line.

plane. The cut-off wave length for a waveguide with a thin dielectric E-plane insert is determined with the help of a stationary formula obtained by the variation technique, so that

$$\lambda_{cd} = 2a\sqrt{1 + 0.5\left[\frac{2h}{a} + \frac{1}{\pi}\sin\left(\frac{2ph}{a}\right)\right](\varepsilon_r - 1)} \tag{2.98}$$

assuming the field distribution in a transverse section in the form

$$\vec{E}_y = \vec{y}\sin\left(\frac{\pi x}{a}\right) \tag{2.99}$$

where $\vec{y}$ is a unit vector in direction of y-axis. For a general distribution of the field in the waveguide cross-section the relationship (2.96) can be written in the form

$$\lambda_{cd} = 2a\sqrt{1 + F\left(\frac{h}{a}\right)(\varepsilon_r - 1)} \tag{2.100}$$

where the function F depends on the field distribution. Equation (2.100) has a stationary form and thus it does not vary for small changes of ε_r, h/a and frequency. The imprecision of (2.100) is smaller than $\pm 0.8\%$ for $h/a \leq 1/4$.

In a combination of both types of waveguides the field is concentrated in the vicinity of the E-plane insert, the field distribution in the cross-section becomes a function of s/b and the function $F(h/a)$ is modified to the function $F(h/a, s/b)$. The cut-off wavelength λ_{cf} for the unilateral fin-line is determined from relationship (2.97), where

$$k_c = 1 + F\left(\frac{h}{a}, \frac{s}{b}\right)(\varepsilon_r - 1) \tag{2.101}$$

By using precisely determined values of λ_{cf} obtained with the help of the spectral method, values of the function $F(h/a, s/b)$ were determined from (2.101) for several combinations of parameters h/a, s/b. The dependence on these parameters can be suitably approximated as

$$F\left(\frac{h}{a}, \frac{s}{b}\right) = \left[a_1\left(\frac{h}{a}\right)x + b_1\left(\frac{h}{a}\right)\right]\frac{h}{a} \tag{2.102}$$

where for $\varepsilon_r = 2.22$ and $b/a = 0.5$ the coefficients $a_1(h/a)$, $b_1(h/a)$ are determined as follows

$$a_1\left(\frac{h}{a}\right) = 0.4020974r^2 - 0.7684487r + 0.3932021 \tag{2.103}$$

$$b_1\left(\frac{h}{a}\right) = 2.4334\sin\left(0.05016r^2 + 0.2977r + 0.26\right) \tag{2.104}$$

where $r = \ln(a/h)$. When the slot is not centrally situated in the partition, then the following relationship is used for the calculation of the parameter x

$$x = \ln\left\{\csc\left(\frac{\pi s}{2b}\right)\csc\left[\frac{\pi}{2}\left(1 - \frac{2e}{b}\right)\right]\right\} \tag{2.105}$$

where e is the distance of the slot centre from the waveguide centre. The nature of the characteristic impedance in the fin-line has not been yet unambiguously determined. On the basis of experiments with chip resistors connected into the slot and theoretical arguments (Jansen, 1979), formulation (2.6b) seems to be the most suitable one for determining the characteristic impedance Z_0, the voltage U being defined as a line integral of the electric field between conductors of the line along the shortest track on the substrate surface and the power P is the total mean power propagating through the line. For known $\varepsilon_{ef}(f)$

$$Z_0 = \frac{Z_{0\infty}}{\sqrt{\varepsilon_{ef}(f)}} \tag{106}$$

where Z_0 is the characteristic impedance of the fin-line waveguide for the frequency $f \Rightarrow \infty \cdot Z_{0\infty}$ moderately increases with increasing frequency, since the increase of the electric field density leads to increasing the voltage on the slot. The following relationship was derived by approximating the spectral method results (Pramanick and Bhartia, 1985):

$$Z_0 = \frac{240\pi^2 (px + q) \frac{b}{a}}{(0.385x + 1.7621)^2 \sqrt{\varepsilon_{ef}(f)}} \tag{2.107}$$

where

$$\begin{aligned} p &= -0.763 \left(\frac{b}{\lambda}\right)^2 + 0.58\frac{b}{\lambda} + 0.0775r^2 - 0.668r + 1.262 \\ q &= 0.372\frac{b}{\lambda} + 0.914 \qquad for\ \frac{s}{b} > 0.3 \end{aligned} \tag{2.108}$$

where $r = \ln(a/h)$, and

$$\begin{aligned} p &= 0.17\frac{b}{\lambda} + 0.0098 \\ q &= 0.138\frac{b}{\lambda} + 0.873 \qquad for\ \frac{s}{b} \leq 0.3 \end{aligned} \tag{2.109}$$

These results were derived for $\varepsilon_r = 2.22$, however, the nature of relationship (2.101) is stationary and it can be used as a general equation for low values ε_r ($\varepsilon_r < 4$). The agreement with theoretical results is within the range of parameters according to Fig. 2.27 better than $\pm 2\%$.

The above described closed relationships can be simply used for complementary unilateral fin-line (Fig. 2.25b) where b will be the doubled height of the waveguide and s the doubled width of the slot.

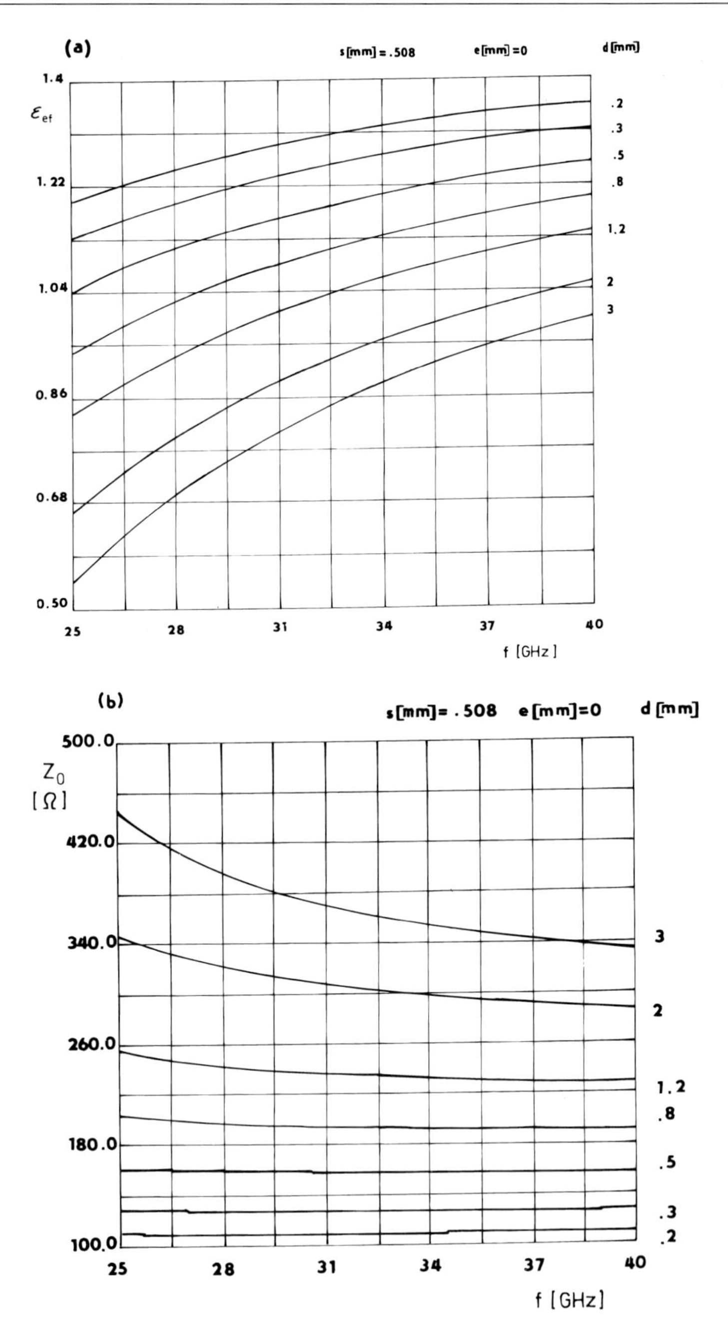

Fig. 2.27 Frequency dependences of parameters of one side fin-line on RT-duroid substrate ($\varepsilon_r = 2.2$, $s = 0.508$) in waveguide WR-28 ($7.112 \times 3.556\,\text{mm}$) (a) $\varepsilon_{ef}(f)$ (b) $Z_0(f)$.

Equation (2.101) will remain unaltered, however, the characteristic impedance will be one half of that in the normal fin-line.

Simulated curves of parameters of unilateral fin-line for substrate of a thickness of 0.127 mm, $\varepsilon_r = 2.22$ situated in the waveguide WR28 (7.112×3.556 mm) are shown in Fig. 2.28 (Pramanick and Bhartia, 1986) depending on the parameter s/b in the frequency band of 25 to 40 GHz.

The bilateral fin-line is shown in Fig. 2.25c. In contrast to the unilateral fin-line, conductors of this line are deposited on both sides of the substrate. The wavelength of the air-filled fin-waveguide is given by relationship (2.94) and the corresponding standardized susceptance of the slot is given as

$$\frac{B_a}{Y_a} = \frac{4bx}{\lambda_{ca}} \tag{2.110}$$

The cut-off wave length λ_{cf} is determined on the basis of an equivalent circuit of the line (Fig. 2.26) simulated as a waveguide with a wider dimension $a - h$ and narrower dimension b loaded by the susceptance B_b/Y_b. It holds that

$$\frac{B_b}{Y_b} = \frac{4b}{\lambda_{cf}} \left[x + \varepsilon_r G_s\right] \tag{2.111}$$

where the term G_s including the substrate effect is given by

$$G_s = \eta_s \arctan\left(\frac{1}{\eta_s}\right) + \ln\left(\sqrt{1 + \eta_s^2}\right) \tag{2.112}$$

$$\eta_s = \frac{h}{s} \tag{2.113}$$

Equation (2.94) for the waveguide filled with air has similarly the form

$$\lambda_{cf} = 2\,(a - h)\left[1 + \frac{4}{\pi}\frac{b}{a - s}\left(1 + 0.2\sqrt{\frac{b}{a - s}}\right) x_b\right] \tag{2.114}$$

where for x_b it follows from comparing (2.109) and (2.110)

$$x_b = x + \varepsilon_r G_s \tag{2.115}$$

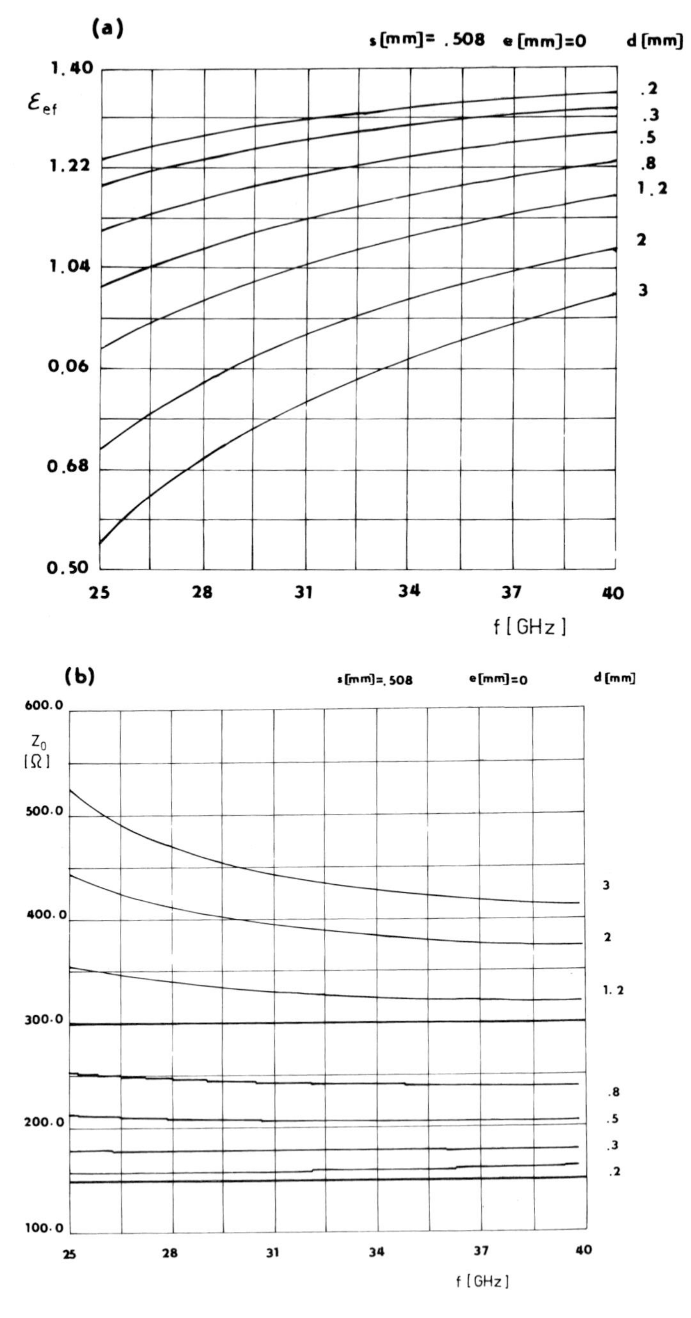

Fig. 2.28 Frequency dependences of parameters of double side fin line on RT-duroid substrate ($\varepsilon_r = 2.2$, $s = 0.508\,\text{mm}$) in waveguide WR-28 (a) $\varepsilon_{ef}(f)$, (b) $Z_0(f)$.

By the approximation of results of the spectral method (Pramanick and Bhartia, 1986) the following relationship was obtained for the characteristic impedance within a precision limit of $\pm 2\%$ for $s/b \leq 0.2$, $0.2 \leq b/\lambda \leq 0.6$ and $h/a \leq 0.05$

$$Z_0 = \frac{240\pi^2 (px' + q) \frac{b}{a-s}}{(0.385x' + 1.7621)^2 \sqrt{\varepsilon_{ef}(f)}} \tag{2.116}$$

where

$$x' = x_b - G_s (\varepsilon_r - 1) \tag{2.117}$$

$$p = 0.097 \left(\frac{b}{\lambda}\right)^2 + 0.01\frac{b}{\lambda} + 0.04095 \tag{2.118}$$

$$q = 0.0031\frac{b}{\lambda} + 0.89 \tag{2.119}$$

The simulated curves ε_{ef} for bilateral fin-line on a substrate of a thickness of $h = 0.508\,\text{mm}$, $\varepsilon_r = 2.2$ centrally situated in the waveguide WR28 are shown in Fig. 2.28 for different widths of slots d, depending on the frequency in a band of 25 to 40 GHz.

For an optimum proposal of the line it is of importance to know the losses. It is necessary to consider that with reducing the slot, conductivity losses increase which even exceed losses in a microstrip line designed for the same band. The calculation of dielectric losses is usually based on the perturbation method for low-loss materials. The resulting expression is in the form (Mirshekar-Syahkal and Davies, 1982)

$$\alpha_d = \frac{\omega \sum_{i=1}^{n} \varepsilon_i \tan\delta_i \int_{S_i} \left|\vec{E}_0\right|^2 dS}{2\Re \int_S \vec{E}_0 \times \vec{H}_0^* \, d\vec{S}} \tag{2.120}$$

where $\vec{E}_0$ and $\vec{H}_0$ are intensities of the electric and magnetic field, respectively, before introducing dielectric losses by means of the loss coefficient $\tan\delta_i$. The areas S and S_i represent the total area of the cross-section and the area of dielectric regions, respectively.

The conventional relationship for conductivity losses of the line with high-conductivity conductors is in the form (Mirshekar-Syahkal and Davies, 1982)

$$\alpha_c = \frac{R_S \int_c \left|\vec{H}_t\right|^2 dl}{2\Re \int_S \vec{E}_0 \times \vec{H}_0^* \, d\vec{S}} \tag{2.121}$$

where R_S is the surface resistance and $\vec{H}_t$ is the tangential component of the magnetic field along the circumference C of the conductor for the loss-free case.

For the dependence of conductivity and dielectric losses on the parameter s/b, for a symmetrically situated substrate of a thickness of $h = 0.127\,\text{mm}$ ($\varepsilon_r = 2.22$, $\tan\delta = 2 \times 10^{-4}$, $\varrho = 3 \times 10^{-8}\,\Omega\text{m}$) in the waveguide WR-28 see Fig. 2.29 (Mirshekar-Syahkal and Davies, 1982)

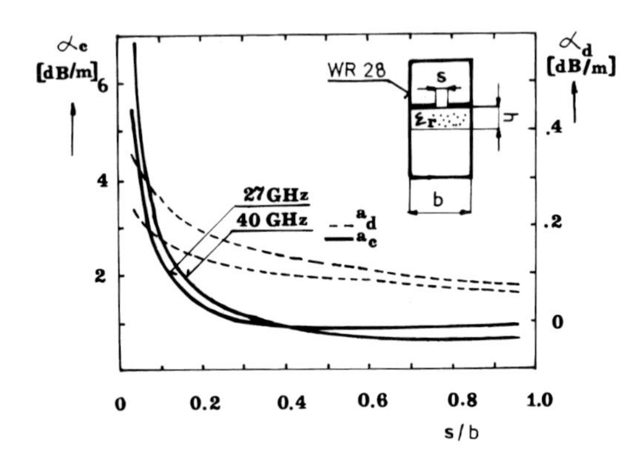

Fig. 2.29 Dependence of conductive and dielectric losses on parameter $\frac{s}{b}$.

2.1.7 Coupled lines

Coupled lines are used for designing directional coupling elements, filters, matching networks, and delay lines. The most important type is the coupled microstrip line (Fig. 2.30). It is formed by two strip

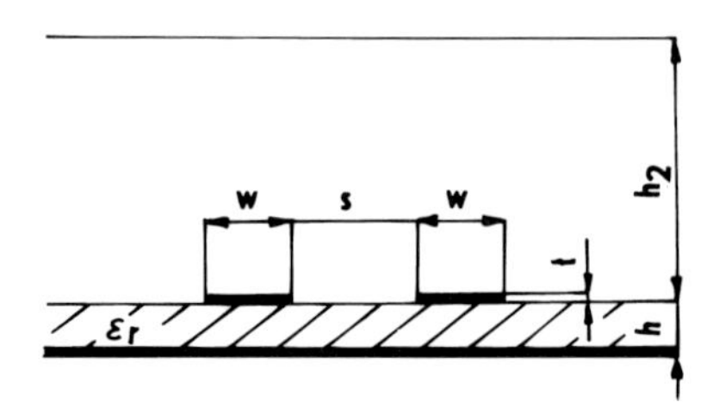

Fig. 2.30 Coupled microstrip line.

conductors of a width w and thickness t, separated from each other by a slot of width s, which are deposited on one side of a planar dielectric substrate of thickness h with a relative dielectric constant ε_r. The second side is metal-coated. The lines can be open, shielded or covered in a conductive package.

The coupled microstrip lines were investigated by a number of methods and authors. The work (Bryant and Weiss, 1968) served as a reference static numerical solution for this purpose. In further years an analysis based on the spectral method was developed, which can be used for the calculation of line parameters dependent on the frequency with an arbitrarily chosen precision (Jansen, 1978; 1979). The analytical expressions presented below (Kirchning and Jansen, 1984) were obtained by the numerical approximation of these results for a range of standardized parameters

$$0.1 \leq u \leq 10; \qquad 0.1 \leq g \leq 10; \qquad 1 \leq \varepsilon_r \leq 18 \tag{2.122}$$

where $u = w/h$ and $g = s/h$.

The expression for the static value of the effective relative permittivity of the even mode $\varepsilon_{efe}(0)$ was taken from Hammerstad and Jensen (1980)

$$\varepsilon_{ef_e}(0) = \frac{\varepsilon_r + 1}{2} + \frac{\varepsilon_r - 1}{2}\left(1 + \frac{10}{v}\right)^{-a_e(v)b_e(\varepsilon_r)} \tag{2.123}$$

where

$$v = \frac{u\left(20 + g^2\right)}{10 + g^2} + ge^{-g} \tag{2.124}$$

$a_e(v)$ and $b_e(\varepsilon_r)$ being determined by relationships (2.25) and (2.26), respectively.

The static value of the effective relative permittivity for the odd mode $\varepsilon_{efo}(0)$ is modelled by the relationship

$$\varepsilon_{ef_o}(0) = \left[\frac{\varepsilon_r + 1}{2} + a_o(u, \varepsilon_r) - \varepsilon_{ef}(0)\right] e^{-c_0 g^{d_0}} + \varepsilon_{ef}(0) \tag{2.125}$$

where

$$\begin{aligned} a_0(u, \varepsilon_r) &= 0.7287\left(\varepsilon_{ef}(0) - \frac{\varepsilon_r + 1}{2}\right)\left(1 - c^{-0.179u}\right) \\ b_0(\varepsilon_r) &= \frac{0.747\varepsilon_r}{0.15 + \varepsilon_r} \\ c_0 &= b_0(\varepsilon_r) - \left(b_0(\varepsilon_r) - 0.207\right) e^{-0.414u} \\ d_0 &= 0.593 + 0.694e^{-0.562u} \end{aligned} \tag{2.126}$$

and $\varepsilon_{ef}(0)$ is the coefficient of the effective relative permittivity for a simple microstrip of width w and thickness $t = 0$ given by (2.23). The dispersion for both modes is introduced with the help of generalized Getsinger's relationships such as equation

$$\varepsilon_{ef_{e,o}}(f_n) = \varepsilon_r - \frac{\varepsilon_r - \varepsilon_{ef_{e,o}}(0)}{1 + F_{e,o}(f_n)} \tag{2.127}$$

where f_n is the standardized frequency ($f_n = fh$ [GHz mm]). The frequency dependence of terms $F_e(f_n)$ and $F_o(f_n)$ is described by the relationships

$$F_e(f_n) = P_1 P_2 \left[(P_3 P_4 + 0.1844 P_7) f_n\right]^{1.5763} \tag{2.128}$$

where

$$\begin{aligned}
P_1 &= 0.27488 + \left[0.6315 + \frac{0.525}{(1 + 0.0157 f_n)^{20}}\right] u - 0.065683 e^{-8.7513u} \\
P_2 &= 0.33622\left(1 - e^{-0.03442\varepsilon_r}\right) \\
P_3 &= 0.0363 e^{-4.6u}\left[1 - e^{-\left(\frac{f_n}{38.7}\right)^{4.97}}\right] \\
P_4 &= 1 + 2.751\left[1 - e^{-\left(\frac{\varepsilon_r}{15.961}\right)^{8}}\right] \\
P_5 &= 0.334 e^{-3.3\left(\frac{\varepsilon_r}{15}\right)^3} + 0.746 \\
P_6 &= P_5 e^{-\left(\frac{f_n}{18}\right)^{0.368}} \\
P_7 &= 1 + 4.069 P_6 g^{0.479} e^{-1.347 g^{0.595} - 0.17 g^{2.5}}
\end{aligned} \tag{2.129}$$

and the term $F_o(f_n)$ is modelled as

$$F_O(f_n) = P_1 P_2 \left[(P_3 P_4 + 0.1844) f_n P_{15}\right]^{1.5763} \tag{2.130}$$

where

$$P_8 = 0.7168\left[1 + \frac{1.076}{1 + 0.0576\,(\varepsilon_r - 1)}\right]$$

$$P_9 = P_8 - 0.7913\left(1 - e^{-\left(\frac{f_n}{20}\right)^{1.424}}\right)\arctan\left[2.481\left(\frac{\varepsilon_r}{8}\right)^{0.946}\right]$$

$$P_{10} = 0.242\,(\varepsilon_r - 1)^{0.55}$$

$$P_{11} = 0.6366\left(e^{-0.3401 f_n} - 1\right)\arctan\left[1.263\left(\frac{u}{3}\right)^{1.629}\right]$$

$$P_{12} = P_9 + \frac{1 - P_9}{1 + 1.183u^{1.376}} \tag{2.131}$$

$$P_{13} = \frac{1.695 P_{10}}{0.414 + 1.605 P_{10}}$$

$$P_{14} = 0.8928 + 0.1072\left[1 - e^{-0.42\left(\frac{f_n}{20}\right)^{3.215}}\right]$$

$$P_{15} = \left|1 - \frac{0.8928\,(1 + P_{11})\,P_{12}e^{-P_{13}g^{1.092}}}{P_{14}}\right|$$

For the normalized frequency $f_n \leq 25$ GHz mm the maximum error of this approximation is smaller than 1.4%.

The concept of the characteristic impedance for the hybrid mode is not unambiguous. The experimental results of measuring the characteristic impedance indicate that the best agreement with the theory is achieved on the basis of formulation (2.6a). The following relationships were obtained by the numerical approximation for static values of the impedance of the even $Z_{0e}(0)$ and odd mode $Z_{0o}(0)$ (Kirschning and Jansen, 1984)

$$Z_{0_e} = \frac{Z_0\,(0)\sqrt{\frac{\varepsilon_{ef}(0)}{\varepsilon_{ef_e}(0)}}}{1 - \frac{Z_0(0)}{377}\sqrt{\varepsilon_{ef}\,(0)}Q_4} \tag{2.132}$$

where

$$Q_1 = 0.8695u^{0.194}$$

$$Q_2 = 1 + 0.7519g + 0.189g^{2.31} \tag{2.133}$$

$$Q_3 = 0.1975 + \left[16.6 + \left(\frac{8.4}{g}\right)^6\right]^{-0.387} + \frac{\ln\left[\frac{g^{10}}{1+\left(\frac{g}{3.4}\right)^{10}}\right]}{241}$$

$$Q_4 = \frac{\frac{2Q_1}{Q_2}}{e^{-g}u^{Q_3} + (2 - e^{-g})\, u^{-Q_3}}$$

Similarly for $Z_{0o}(0)$ it is possible to write

$$Z_{0_o} = \frac{Z_0\,(0)\sqrt{\frac{\varepsilon_{ef}(0)}{\varepsilon_{ef_o}(0)}}}{1 - \frac{Z_0(0)}{377}\sqrt{\varepsilon_{ef}\,(0)}Q_{10}} \tag{2.134}$$

where

$$\begin{aligned}
Q_5 &= 1.794 + 1.14 \ln\left(1 + \frac{0.638}{g + 0.517g^{2.43}}\right) \\
Q_6 &= 0.2305 + \frac{\ln\left(\frac{g^{10}}{1+\left(\frac{g}{5.8}\right)^{10}}\right)}{281.3} + \frac{\ln\left(1 + 0.598g^{1.154}\right)}{5.1} \\
Q_7 &= \frac{10 + 190g^2}{1 + 82.3g^3} \\
Q_8 &= e^{-6.5-0.95\ln(g)-\left(\frac{g}{0.15}\right)^5} \\
Q_9 &= \ln\left(Q_7\right)\left(Q_8 + \frac{1}{16.5}\right) \\
Q_{10} &= \frac{Q_2Q_4 - Q_5e^{\ln(u)Q_6u^{-Q_9}}}{Q_2}
\end{aligned} \tag{2.135}$$

Within the range of parameters of (2.22) the precision of these static expressions is better than 0.6%. The dispersion of the impedance for the even mode is included in the form

$$Z_{0_e}\,(f_n) = Z_{0_e}\,(0)\,\frac{\left[0.9408\left(\varepsilon_{ef}\,(f_n)\right)^{C_e} - 0.9603\right]^{R_{17}}}{\left[(0.9408 - d_e)\left(\varepsilon_{ef}\,(0)\right)^{C_e} - 0.9603\right]^{R_{17}}} \tag{2.136}$$

where

$$
\begin{aligned}
C_e &= 1 + 1.275\left[1 - e^{-0.004625 p_e \varepsilon_r^{1.674} (f_n/18.365)^{2.745}}\right] \\
&\quad - Q_{12} + Q_{16} - Q_{17} + Q_{18} + Q_{20} \\
d_e &= 5.086 q_e \frac{r_e}{0.3838 + 0.386 q_e} \frac{e^{-22.2u^{1.92}}}{1 + 1.2992 r_e} \frac{(\varepsilon_r - 1)^6}{1 + 10(\varepsilon_r - 1)^6} \\
p_e &= 4.766 e^{-3.228 u^{0.641}} \qquad (2.137) \\
q_e &= 0.016 + (0.0514 \varepsilon_r Q_{21})^{4.524} \\
r_e &= \left(\frac{f_n}{28.843}\right)^{12}
\end{aligned}
$$

$$
\begin{aligned}
Q_{11} &= 0.893\left(1 - \frac{0.3}{1 + 0.7(\varepsilon_r - 1)}\right) \\
Q_{12} &= 2.121 \frac{\left(\frac{f_n}{20}\right)^{4.91}}{1 + Q_{11}\left(\frac{f_n}{20}\right)^{4.91}} e^{-2.87g} g^{0.902} \\
Q_{13} &= 1 + 0.038\left(\frac{\varepsilon_r}{8}\right)^{5.1} \\
Q_{14} &= 1 + 1.203\left(\frac{\varepsilon_r}{15}\right)^4 \left[1 + (\varepsilon_r\ 15)^4\right] \\
Q_{15} &= \frac{1.887 e^{-1.5 g^{0.84}} g^{Q_{14}}}{\frac{1 + 0.41\left(\frac{f_n}{15}\right)^3 + u^{\frac{2}{Q_{13}}}}{0.125 + u^{\frac{1.626}{Q_{13}}}}} \qquad (2.138) \\
Q_{16} &= \left[1 + \frac{9}{1 + 0.403(\varepsilon_r - 1)^2}\right] Q_{15} \\
Q_{17} &= 0.394\left(1 - e^{-1.47\left(\frac{u}{7}\right)^{0.672}}\right)\left(1 - e^{-4.25\left(\frac{f_n}{20}\right)^{1.87}}\right) \\
Q_{18} &= 0.61 \frac{1 - e^{2.13\left(\frac{u}{8}\right)^{1.593}}}{1 + 6.544 g^{4.17}} \\
Q_{19} &= \frac{0.21 g^4}{(1 + 0.18 g^{4.9})(1 + 0.1u^2)\left[1 + \left(\frac{f_n}{24}\right)^3\right]}
\end{aligned}
$$

$$Q_{20} = \left[0.09 + \frac{1}{1 + 0.1\left(\varepsilon_r - 1\right)^{2.7}}\right] Q_{19}$$

$$Q_{21} = \left|1 - \frac{42.54 g^{0.133} e^{-0.812 g} u^{2.5}}{1 + 0.033 u^{2.5}}\right|$$

In the above mentioned relationships $\varepsilon_{ef}(f_n)$ stands for the value of the effective dielectric constant for a simple microstrip of a width w obtained from relationship (2.36) and $Z_0(0)$ is the static value of its characteristic impedance. The dispersion of the characteristic impedance of the odd mode is modelled as

$$Z_{0_o}(f_n) = Z_0(f_n) + \frac{Z_0(0)\left(\frac{\varepsilon_{efo}(f_n)}{\varepsilon_{efo}(0)}\right)^{Q_{22}} - Z_0(f_n)\, Q_{23}}{1 + Q_{24} + (0.46g)^{2.2}\, Q_{25}} \tag{2.139}$$

where

$$\begin{aligned}
Q_{22} &= \frac{0.925\left(\frac{f_n}{Q_{26}}\right)^{1.536}}{1 + 0.3\left(\frac{f_n}{30}\right)^{1.536}} \\
Q_{23} &= 1 + \frac{0.005 f_n Q_{27}}{\left[1 + 0.812\left(\frac{f_n}{15}\right)^{1.9}\right]\left(1 + 0.025u^2\right)} \\
Q_{24} &= 2.506 Q_{28} u^{0.894} \frac{\left[\frac{(1+1.3u) f_n}{99.25}\right]^{4.29}}{3.575 + u^{0.894}} \\
Q_{25} &= \frac{0.3 f_n^2}{10 + f_n^2} \frac{1 + 2.333\left(\varepsilon_r - 1\right)^2}{5 + \left(\varepsilon_r - 1\right)^2} \\
Q_{26} &= 30 - 22.2 \frac{\left(\frac{\varepsilon_r - 1}{13}\right)^{12}}{1 + 3\left(\frac{\varepsilon_r - 1}{13}\right)^{12}} - Q_{29} \\
Q_{27} &= 0.4 g^{0.84} \frac{1 + 2.5\left(\varepsilon_r - 1\right)^{1.5}}{5 + \left(\varepsilon_r - 1\right)^{1.5}} \\
Q_{28} &= \frac{0.149\left(\varepsilon_r - 1\right)^3}{94.5 + 0.038\left(\varepsilon_r - 1\right)^3} \\
Q_{29} &= \frac{15.16}{1 + 0.196\left(\varepsilon_r - 1\right)^2}
\end{aligned} \tag{2.140}$$

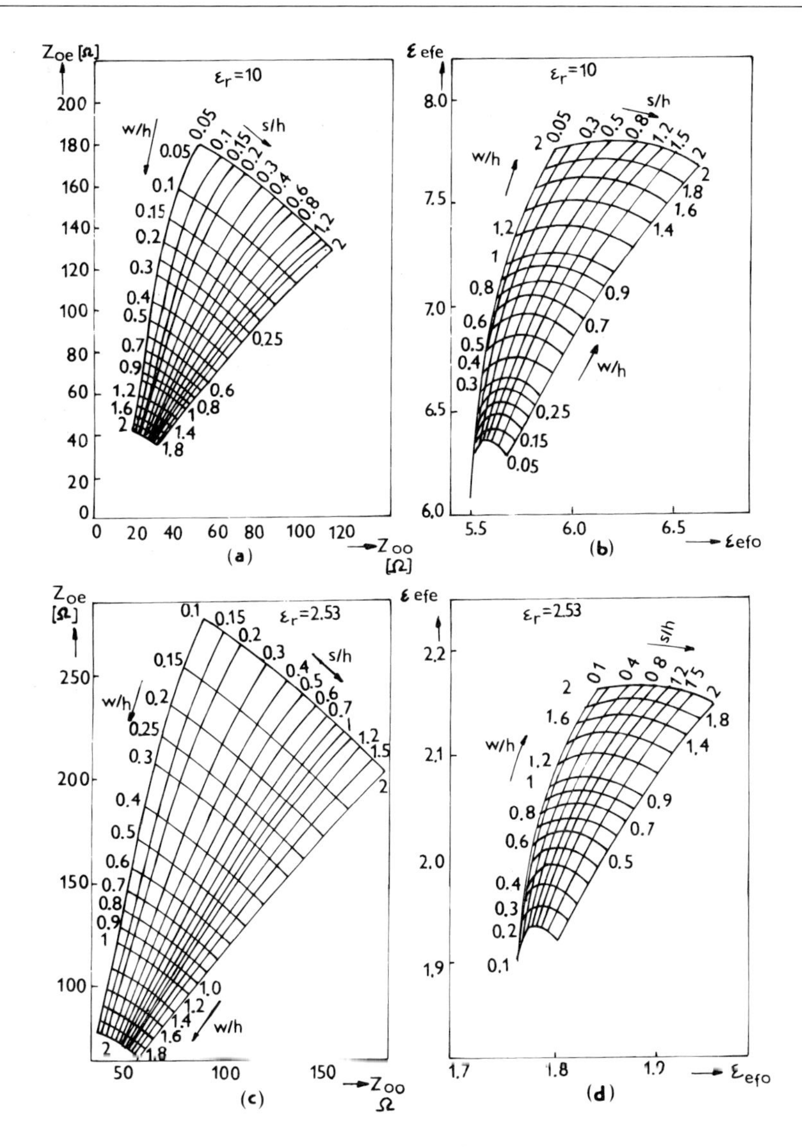

Fig. 2.31 Dependences of static parameters of coupled microstrip line Z_{0e}, Z_{0o}, ε_{efo}, ε_{efe} versus dimensions of structure.

The quantity $Z_0(f_n)$ is a frequency-dependent characteristic impedance of a simple microstrip of a width w and it is given by (2.40). The precision of modelled expressions within the range of parameters (2.122) for $f_n < 20\,\text{GHz mm}$ is better than 2.5%. For substrates with $\varepsilon_r \leq 12.9$ the dispersion dependences can be used with this precision up to $f_n = 15\,\mathrm{GHz\,mm}$. For substrates with $\varepsilon_r \leq 12.9$ and $f_n \leq 15\,\mathrm{GHz\,mm}$ the precision of the approximation is typically better than 1.5%.

The design curves for the static approximation for $\varepsilon_r = 10$ and $\varepsilon_r = 2.53$ are plotted in Fig. 2.31, the dispersion dependences for two values of the standardized microstrip width $u = 0.5$ and 1 are shown for both types of the substrate up to a value of $f_n = 30\,\mathrm{GHz\,mm}$ in Fig. 2.32.

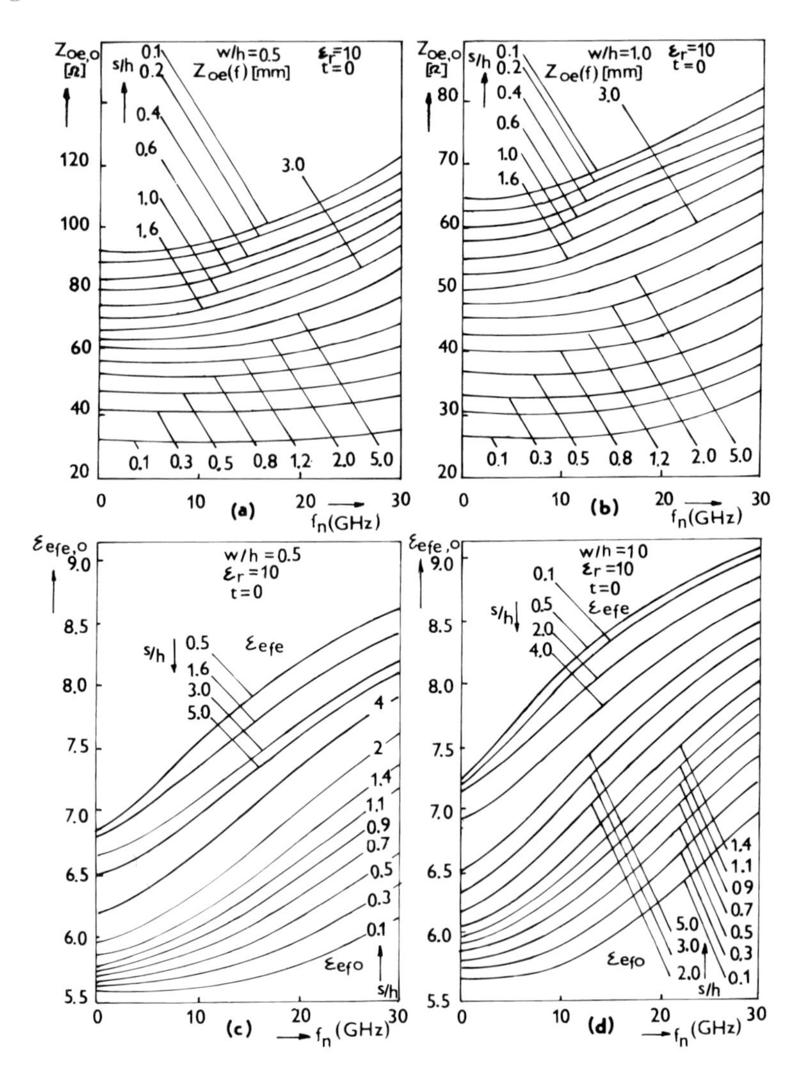

Fig. 2.32 Dispersion dependences of coupled microstrip line parameters.

The thickness of the metal coating can be *respected* by corrections according to Wheeler (1977) or March (1981). The effective width of strips is introduced for even and odd modes w_e, w_o assuming that $s \gg 2t$, i.e.

$$\begin{aligned} w_e &= w + \Delta w \left(1 - 0.5e^{\frac{-0.69\Delta w}{\Delta t}}\right) \\ w_o &= w_e + \Delta t \end{aligned} \tag{2.141}$$

where Δw is determined by relationship (2.28) and

$$\Delta t = \frac{th}{\varepsilon_r s} \tag{2.142}$$

The effect of the shielding plane at a distance of h_2 above the line can be included in a similar manner as for the microstrip (March, 1981). The value of the term q_T expressing the correction for the thickness of the line is given by relationship (2.34) for both modes. The correction for the finite distance of the shielding plane for the even mode is $h_2/h \leq 39$

$$\begin{aligned} q_{ce} &= \tanh\left(1.626 + 0.107\frac{h_2}{h} - 1.733\sqrt{\frac{h}{h_2}}\right) \qquad \text{for } \frac{h_2}{h} \leq 39 \\ q_{ce} &= 1 \qquad \text{for } \frac{h_2}{h} > 39 \end{aligned} \tag{2.143}$$

and for the odd mode

$$\begin{aligned} q_{co} &= \tanh\left[\frac{9.575}{7 - \frac{h_2}{h}} - 2.965 + 1.68\frac{h_2}{h} - 0.311\left(\frac{h_2}{h}\right)^2\right] \qquad \text{for } \frac{h_2}{h} \leq 7 \\ q_{co} &= 1 \qquad \text{for } \frac{h_2}{h} > 7 \end{aligned} \tag{2.144}$$

The filling factors q_e, q_o for particular modes are determined with the help of relationship (2.33). Correction terms for the characteristic impedance of the line in the air are in the form

$$\Delta Z_{0_{e,o}A} = f_{e,o}\left(u, \frac{h_2}{h}\right) g_{e,o}\left(g, \frac{h_2}{h}\right) \tag{2.145}$$

where the expressions for $f_{e,o}$ and $g_{e,o}$ are

$$f_{e,o}\left(u, \frac{h_2}{h}\right) = 1 - \arg\tanh\left[A + (B + Cu)\,u\right] \tag{2.146}$$

$$\begin{aligned} A &= -\frac{4.351}{\left(1 + \frac{h_2}{h}\right)^{1.842}} \\ B &= \frac{6.639}{\left(1 + \frac{h_2}{h}\right)^{1.861}} \\ C &= -\frac{2.291}{\left(1 + \frac{h_2}{h}\right)^{1.9}} \end{aligned} \tag{2.147}$$

$$g_{e,o}\left(g, \frac{h_2}{h}\right) = 270\left[1 - \tanh\left(D + E\sqrt{1 + \frac{h_2}{h}} - \frac{F}{1 + \frac{h_2}{h}}\right)\right] \tag{2.148}$$

$$\begin{aligned} &D = 0.747 \csc\frac{\pi x}{2}, \quad E = 0.725 \sin\frac{\pi y}{2}, \quad F = 10^{0.11 - 0,0947}, \\ &x = 10^{0.103 - 0.159}, \quad y = 10^{0.0492 - 0.073} \end{aligned} \tag{2.149}$$

$$f_0\left(u, \frac{h_2}{h}\right) = u^J \tag{2.150}$$

$$J = \tanh\frac{\left(1 + \frac{h_2}{h}\right)^{1.585}}{6} \tag{2.151}$$

$$\begin{aligned} g_0\left(g, \frac{h_2}{h}\right) &= 270\left[1 - \tanh\left(G + K\sqrt{1 + \frac{h_2}{h}} - \frac{L}{1 + \frac{h_2}{h}}\right)\right] \\ G &= 2.178 - 0.796g \end{aligned} \tag{2.152}$$

$$\begin{aligned} K &= 10^{20.491 g^{0.174}} && \text{for } g > 0.858 \\ K &= 1.3 && \text{for } g \leq 0.858 \\ L &= 2,51 g^{-0.462} && \text{for } g > 0.873 \\ L &= 2.674 && \text{for } g \leq 0.873 \end{aligned} \tag{2.153}$$

2.1.8 Methods of homogeneous line analysis

The following methods for determining parameters of homogeneous lines can be considered:

a) static methods, which are restricted to solving the Laplace's equation for the electric field potential with given boundary conditions and conditions in the interface, and

b) dynamic methods solving the electromagnetic field in the line at a chosen frequency and determining the propagation constant as a basic parameter.

A conformal mapping is frequently used for solving the static distribution of the field, particularly the Schwarz-Christoffel's transformation, which makes it possible to simplify the relatively complex geometry of the line and thus, to obtain more easily the solution of the two-dimensional Laplace's equation (Schneider, 1969a). The further method is a method of finite increments (Green, 1965), where the Laplace's equation is solved in its differential form or other method based on using the Green's function (Silvester, 1968). The results of static methods related to the characteristic impedance of the line and propagation velocity are further used either directly for the proposal of circuits, when the deviations for the given arrangement are not very large in the frequency region chosen, or after their correction for the dispersion with the help of a certain approximate method of the dispersion calculation. The static methods cannot be used for lines where the propagation of the quasi-TEM mode is not present.

The dynamic methods are of importance for determining characteristics of the line in a wide frequency band, for the study of higher modes and for the calculation of their cut-off frequencies. The choice of a suitable numerical method is of essential importance for the efficient analysis of the microwave line.

A very important method is the spectral method (Jansen, 1985), which is suitable for multiconductor microstrip structures with multilayer dielectrics. It is preferred due to its high efficiency, however, it is restricted to structures with infinitely thin conductors. The other method is a finite element method (Pantic, 1986) suitable for lines in general, with arbitrary numbers of conductors and with any shape of dielectric regions. It is frequently advantageous to solve the frequency characteristics of microstrip circuits in a time region, where it is unnecessary to solve the problem of the characteristic value. This is used for example in the TLM method (Transmission-Line Matrix) (Hoefer, 1985; Johns *et al.*, 1971). There are several modifications of these methods and of further methods suitable for determining the line parameters, as for example a method of sewing modes (Kowalski and Pregla, 1971) of lines (Worm and Pregla, 1984), momentum method (Harrington, 1968), which, however, will not be discussed here.

The method of conformal mapping. It is based on a transformation of the geometry of the problem to be solved to the geometry for which the solution is known. Particularly the transformation in the form of the Schwarz-Christoffel's integral is used

$$w = A \int_0^z \prod_{i=1}^{n} (z - a_i)^{\alpha_{i-1}} \, \mathrm{d}z + B \tag{2.154}$$

and the top half of the complex plane $z = x + jy$ is transformed to the interior of a polygon situated in the complex plane $w = u + jv$. The value a_i represents points on the real axis x, which correspond to vertices V_i of the polygon and exponents α_i determine the size of its vertex angles with respect to π. The complex constant A determine the sizes of the polygon and its rotation in plane w and the complex constant B determines a point of the polygon circumference, which is represented in the origin of plane z.

The problem in a half-plane is most frequently transformed to the problem of calculating the capacitance of a capacitor with parallel plates. First a representation is found, which transforms the given problem to the top half-plane and then, a suitable modification of the representation according to (2.154) is used. The vertex angles of the polygon represented by the Schwarz-Christoffel's integral are usually $\pi/2$ which means that the value of the coefficient α_i is of 0.5 and the transformation is determined by an elliptic integral. The capacitance per unit length of a plate capacitor, which is a final result of the geometry transformation, is obtained for the capacitor configuration in the plane according to Fig. 2.33 as follows:

$$C = \varepsilon \frac{2K(k)}{K'(k)} \tag{2.155}$$

where $K(k)$ is a complete first order elliptic integral and $K'(k)$ is a complementary first order elliptic integral. For the original line, it is

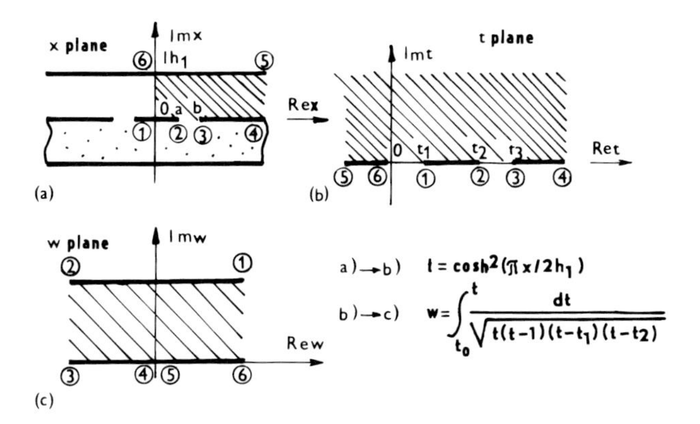

Fig. 2.33 Conformal mapping for coplanar waveguide on Fig. 2.17a and used transformation.

necessary to calculate contributions to the resulting capacitance for all the subregions transformed to the plate capacitor so that

$$C\left(\varepsilon_r\right)=\sum_{i=1}^{n}\varepsilon_i\frac{K\left(k_i\right)}{K'\left(k_i\right)} \tag{2.156}$$

where ε_i are values of the effective permittivity and k_i are moduli of the complete elliptic integral of the first order corresponding to particular transformations. The effective relative permittivity ε_{ef} is determined as

$$\varepsilon_{ef}=\frac{C\left(\varepsilon_r\right)}{C\left(1\right)}=1+q\left(\varepsilon_r-1\right) \tag{2.157}$$

where q is the filling factor. The characteristic impedance Z_o is determined by

$$Z_0=\frac{1}{v_f C\left(\varepsilon_r\right)} \tag{2.158}$$

where v_f is the phase velocity of the propagation of the electromagnetic wave along the line.

For multiconductor structures the capacity and subsequently impedance matrix of the structure are determined in the same way.

The method of finite increments (Gupta *et al.*, 1979). This method is based on a numerical solution of the Laplace's equation in its differential form. It is particularly suitable for closed MIC lines and the analysis can simply include the line thickness. The configuration

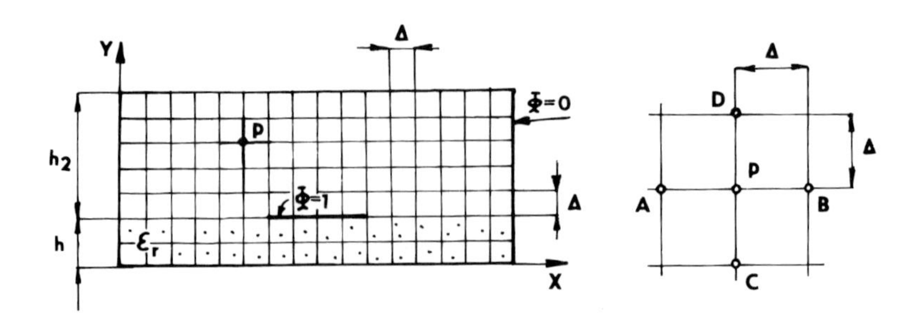

Fig. 2.34 Circuit configuration for method of finite increments analysis.

according to Fig. 2.34 is considered. The potential at points A, B, C, D can be expressed with respect to the potential ϕ_P at points P with the help of the Taylor's expansion as follows

$$\begin{aligned}
\phi_A &= \phi_P - \Delta\partial_x\phi + \frac{\Delta^2}{2!}\partial_{xx}\phi - \frac{\Delta^3}{3!}\partial_{xxx}\phi \\
\phi_B &= \phi_P + \Delta\partial_x\phi + \frac{\Delta^2}{2!}\partial_{xx}\phi + \frac{\Delta^3}{3!}\partial_{xxx}\phi \\
\phi_C &= \phi_P - \Delta\partial_y\phi + \frac{\Delta^2}{2!}\partial_{yy}\phi - \frac{\Delta^3}{3!}\partial_{yyy}\phi \\
\phi_D &= \phi_P + \Delta\partial_y\phi + \frac{\Delta^2}{2!}\partial_{yy}\phi + \frac{\Delta^3}{3!}\partial_{yyy}\phi
\end{aligned} \tag{2.159}$$

By neglecting fourth and higher orders of derivatives and by using

$$\partial_{xx}\phi + \partial_{yy}\phi = 0 \tag{2.160}$$

the equation

$$\phi_A + \phi_B + \phi_C + \phi_D \approx 4\phi_P \tag{2.161}$$

is obtained, which can be simply solved by the relaxation method. The process starts with initial values of potentials ϕ_P at points of the lattice. New ϕ_P values are obtained from the relationship

$$\phi'_P = \phi_P - \frac{\alpha\,(\phi_A + \phi_B + \phi_C + \phi_D)}{4} \tag{2.162}$$

The convergence rate is determined by the magnitude of the constant α. The relaxation condition should be modified for points in the vicinity of the interface of dielectrics and in corners. On the basis of this, it is possible to calculate the field distribution and the charge Q on the strip

$$Q = \varepsilon_0 \varepsilon_r \int E_n \, ds \tag{2.163}$$

where the integration path is a suitable curve surrounding the strip conductor. The capacitance is determined from the charge-to-voltage ratio.

The method of integral equations. The quasistatic analysis can be formulated in the form of integral equations. The Poisson's equation is considered with respect to the charge distribution in the line and the Green's function is defined

$$\nabla_\perp^2 G(x, y; x', y') = -\frac{1}{\varepsilon_0 \varepsilon_r} \delta(y - y') \, \delta(x - x') \tag{2.164}$$

where $[x', y']$ are coordinates of the charge and $[x, y]$ is the point of the field. $G(x, y; x', y')$ is the potential at point $[x, y]$ resulting from a unit charge at point $[x', y']$ of the line. With the help of the Green's function, the integral equation is expressed in the following form:

$$\phi(x, y) = \int G(x, y; x', y') \, \rho(x', y') \, dx' dy' \tag{2.165}$$

where ϕ is the potential and ρ is the charge distribution. It is integrated on the surface of the microstrip conductor. The analysis can be divided into two parts as follows (1) construction of a suitable Green's function and (2) solution of the integral equation (2.165).

The spectral method (Jansen, 1985) is a very efficient method for the analysis of planar microstrip structures. It is based on an integral transformation of the wave equation into the spectral region. By the analytical adjustment one integral equation in closed form is obtained, which has a dimension lower by one in comparison with the original problem. After their solutions the results are back transformed into the spatial region.

The problems of homogeneous transmission lines are formulated with respect to time harmonic fields. It is assumed that the line consists of M dielectric layers without losses and N conductive strips with an infinite conductivity σ and negligible thickness t. Losses are typically considered at the end of the procedure with the help of the

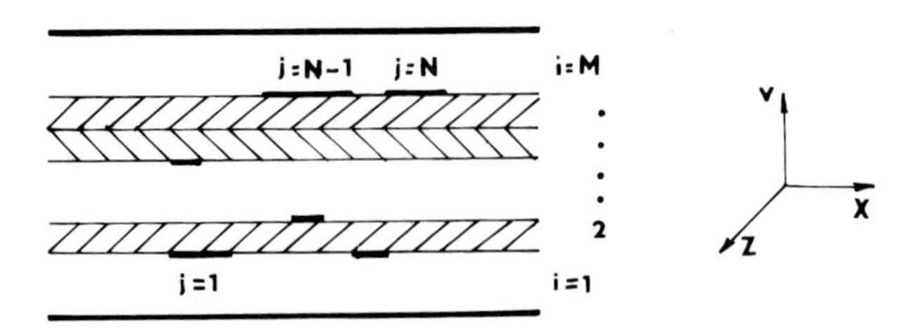

Fig. 2.35 Shielded transmission line structure for spectral-domain method of analysis (M-number of dielectric layer, N-number of conductive strip).

perturbation method. The line can be screened, shielded or open. The transverse cross-section through the general structure of the shielded line is shown in Fig. 2.35.

The electromagnetic field is expressed by a superposition of TE and TM modes described on the basis of Hertz's vectors

$$\Pi_i^j = z\psi_i^j(x, y)\, e^{-j\beta z} \tag{2.166}$$

for $i = 1, 2, \ldots, \mathrm{M}$ and $j = e,\ m$, where the scalar functions ψ^e, ψ^m have to adhere to the homogeneous Helmholtz's equation

$$\left[\Delta + \left(k_i^2 - \beta^2\right)\right]\psi^j = 0 \tag{2.167}$$

where k_i is the wave number of the i-th layer. The homogeneous form of the equation is also used when solving the problem with sources. They are introduced as forced current densities or electric fields at the interfaces between layers. By the Fourier's transformation of scalar potentials ψ_i^j (2.167) and of boundary conditions of the problem the relationships

$$\widetilde{\psi}_i^{e,m}(\beta, y) = \int_{-\infty}^{\infty} \psi_j^{e,m}(x, y)\, e^{j\beta x}\, dx \tag{2.168}$$

$$\left(\frac{d^2}{dy^2} - \gamma_i^2\right)\widetilde{\psi}_i^{e,m} = 0 \tag{2.169}$$

are obtained where $\gamma_i^2 = \beta^2 - k_i^2$. The potentials $\widetilde{\psi}_i^{e,m}$ can be analytically determined in a general form

$$\widetilde{\psi}_i^j(\beta, y) = a_i^j(\beta)\cosh\left[\gamma_i(y - y_i)\right] + b_i^j(\beta)\sinh\left[\gamma_i(y - y_i)\right] \tag{2.170}$$

where $i = 1, 2, \ldots, L_i$, and $j = e, m$. The conditions at the interface for scalar potentials correspond to conditions for transformed components of the field E_{yi} and H_{yi} so that

$$\begin{aligned} j\omega\varepsilon_i \widetilde{E}_{y_i} &= \beta^2 \widetilde{\psi}_i^e \\ j\omega\mu_i \widetilde{H}_{y_i} &= \beta^2 \widetilde{\psi}_i^m \end{aligned} \tag{2.171}$$

and in the spectral region they can be formulated as follows

$$\begin{aligned} \widetilde{E}_{x_i} &= \widetilde{E}_{x_{i+1}} \\ \widetilde{E}_{z_i} &= \widetilde{E}_{z_{i+1}} \\ \widetilde{H}_{x_i} - \widetilde{H}_{x_{i+1}} &= -\widetilde{J}_{z_i} \\ \widetilde{H}_{z_i} - \widetilde{H}_{z_{i+1}} &= \widetilde{J}_{x_i} \end{aligned} \tag{2.172}$$

where $\widetilde{J}_{xi}$, $\widetilde{J}_{zi}$ are Fourier's representations of current density components. By an analytical arrangement it is possible to eliminate all the unknown constants a_i^j, b_i^j. Last it is also necessary to fulfil conditions on electric walls, which represent strip conductors. The electric field component E_t, which is a tangent line on the strip surface and the current density J_t should vanish in complementary parts of the interface. In this way a system of equations is obtained for components of the current density in particular layers, which is numerically solved for the spectral region in a standard manner — by the Galerkin's method. The non-trivial solution of the system of homogeneous linear algebraic equations for the coefficients of a number of Fourier's transformations of suitably chosen basic functions $\widetilde{J}_{xin}(\beta)$ and $\widetilde{J}_{zin}(\beta)$ is obtained by attributing a zero value to their determinant, which occurs for the given frequency for a certain value of the propagation constant β.

The method of finite elements (MFE). The preceding methods are restricted in their applications to structures with a negligible thickness of conductive strips and with planar interfaces between dielectric materials. The effect of the finite thickness can be included into these methods on account of an increased tediousness of the numerical processing. By the MFE (Pantic and Mitra, 1986) it is, however, possible to investigate lines including any number of conductors and homogeneous dielectric regions of any shape. Their principle is explained in the following example.

A general line is considered, which is homogeneous in direction of the z-axis, where the TEM mode is a prevalent mode of the transmission. Under these conditions, the problem is simplified to searching for a skalar potential ϕ, which corresponds to the Poisson's equation in region S

$$\Delta\phi = -\frac{\rho}{\varepsilon}, \tag{2.173}$$

where ρ is the charge density and ε is the material permittivity, and which adheres to boundary conditions

$$\begin{aligned} \phi &= \phi_k && \text{on the surface } k\text{-th line} \\ \frac{\partial\phi}{\partial n} &= 0 && \text{on the boundary of dielectrics} \end{aligned} \tag{2.174}$$

A variational expression F is formed in such a way that (2.173) is the Euler's equation of the functional F, which adheres to conditions (2.174)

$$F = \frac{1}{2}\int_S \varepsilon \left|\nabla\phi\right|^2 dS - \int_S \rho\phi \, dS \tag{2.175}$$

The region S will be divided in an arbitrary manner to finite elements in such a way that all the dielectric interfaces coincide with sides of these elements. Triangular elements of the first order are usually chosen. The order of the element is determined by the degree of the polynomial, which is used for the approximation of the potential ϕ inside of the element. In the case of a higher order of elements, the precision and the speed of the calculation at a given number of nodes are increased. When simulating complicated interface shapes, it is more advantageous to use a higher number of simple elements.

Field singularities on margins of the line transverse cross-section call for the use of a very fine network or for introducing singular elements (Fig. 2.36). In the element, triangular polar coordinates ρ, σ are introduced. Cartesian coordinates of points inside of the triangle are as follows:

$$\begin{aligned} x &= x_1 + \rho\left[(x_2 - x_1) + \sigma\,(x_3 - x_2)\right] \\ y &= y_1 + \rho\left[(y_2 - y_1) + \sigma\,(y_3 - y_2)\right] \end{aligned} \tag{2.176}$$

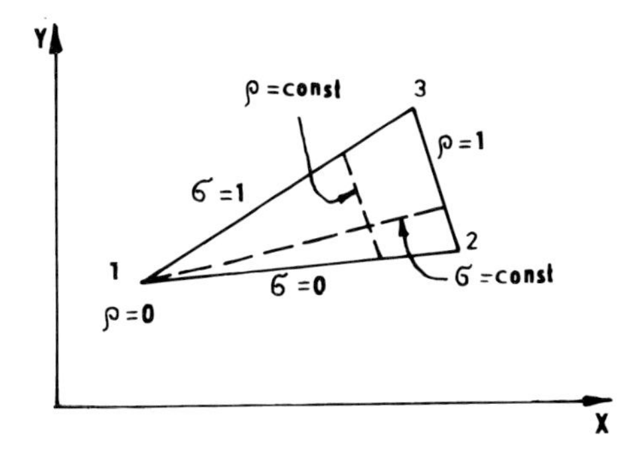

Fig. 2.36 Singular element with marked nodes and triangular axis system.

The distribution function of the scalar potential, which takes into account the field singularity at node 1 ($\rho = 0$) is in the form

$$\phi = \phi_1 \left(1 - \rho^\lambda\right) + \phi_2 \rho^\lambda \left(1 - \sigma\right) + \phi_3 \rho^\lambda \sigma \qquad \text{for } \frac{1}{2} < \lambda < 1 \quad (2.177)$$

where the parameter λ is chosen in accordance with the Meixner's boundary condition and ϕ_1, ϕ_2, ϕ_3 are corresponding node potentials. The singular element is adjusted with respect to the common element of the first order with the help of the linear variation of the scalar potential along the side 2–3.

The efficiency of the calculation for open structures, such as a system of conductors between two conductive walls or above ground plane, can be essentially improved by introducing two different types of infinite elements. The whole region is divided into the near and distant zones. The near zone is divided in a common manner to rectangular elements and the distant zone is divided to infinite elements of two types. The elements of the first type (Fig. 2.37a) shares two nodes (1,2) with a certain element of the close zone and two sides with a fictive point of intersection in infinity. Normalized coordinates are introduced sides with a fictive point of intersection in infinity. Normalized coordinates are introduced

$$\zeta = \frac{x}{x_1}, \quad \eta = \frac{(y - y_1)}{(y_2 - y_1)}, \quad x_1 < x < \infty, \quad y_1 \leq y \leq y_2 \qquad (2.178)$$

The field in the distant zone is equivalent to a dipole field due to the vicinity of ground plane. The scalar potential of the dipole ϕ_D,

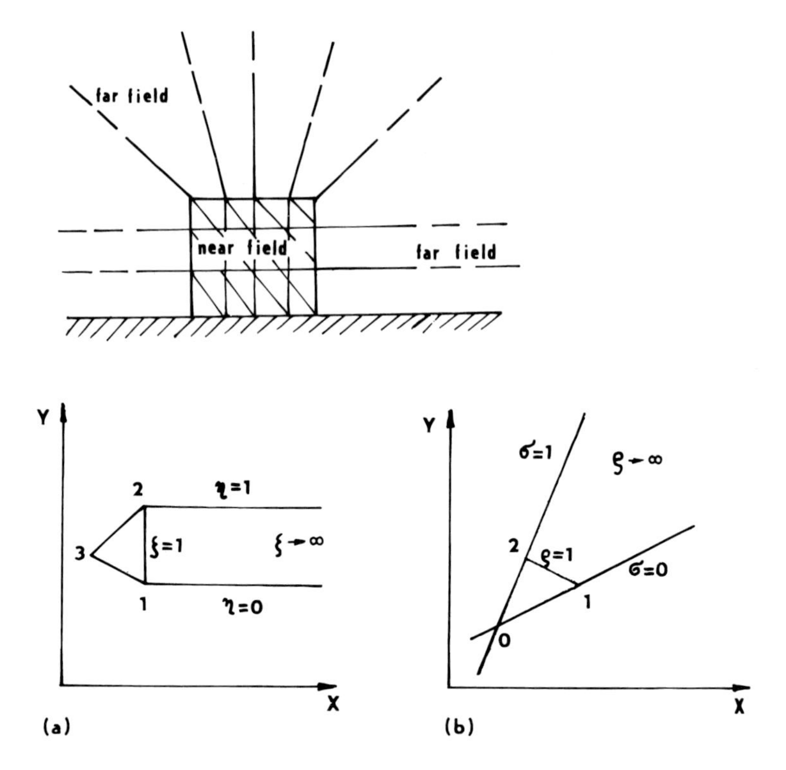

Fig. 2.37 Transmission line up the grounded plane and using infinite elements for far field (a) infinite element 1st order, (b) infinite element 2nd order.

whose moment p is oriented in direction of the y-axis can be written in cylindrical coordinates (r, Θ) as

$$\phi_D = \frac{\vec{p}}{2\pi\varepsilon} \frac{\sin\Theta}{\vec{r}} \tag{2.179}$$

In normalized coordinates the potential ϕ with respect to nodal potentials ϕ_1 and ϕ_2 can be expressed as

$$\phi = \frac{1}{\zeta^2} \left[\phi_1 (1 - \eta) + \phi_2 \eta\right] \tag{2.180}$$

The element of the second type (Fig. 2.37b) shares two nodes (1,2) with a certain element of the near zone and the radial arms, which are intersected at point $[x_0, y_0]$. Triangular polar coordinates ρ, σ are introduced. Then the Cartesian coordinates of points inside of the element are as follows

$$
\begin{aligned}
x &= x_0 + \rho\left[(x_1 - x_0) + \sigma\,(x_2 - x_1)\right] \\
y &= y_0 + \rho\left[(y_1 - y_0) + \sigma\,(y_2 - y_1)\right]
\end{aligned}
\tag{2.181}
$$

The scalar potential ϕ with respect to the nodal potentials ϕ_1, and ϕ_2 is approximated by:

$$
\phi = \frac{1}{\rho}\left[\phi_1\,(1-\sigma) + \phi_2\sigma\right] \tag{2.182}
$$

For both types of infinite elements a linear variation of the potential function is provided along the side 1–2. These elements are compatible with common elements of the first order and they are complete, finite and they adhere to the radiation condition.

For a known distribution of the potential, the capacity C per unit length is determined from the variation expression

$$
C = \frac{2W}{V^2} \tag{2.183}
$$

where W is the conduction energy per unit length and V is the line potential. The upper and lower limit for this capacity is obtained by solving the original and dual problems. The effective permittivity and the characteristic impedance are determined in the same way as in the static solution.

The TLM method (Transmission-Line Matrix) (Hoefer, 1985) represents the analysis in the time region. It is based on the Huygens's principle of the wave propagation. In a two-dimensional region, a Cartesian network of nodes is formed in such a way that the distance between neighbouring nodes is of Δd. In a given point at a given time point $t = 0$ this conduction is excited by the Dirac voltage pulse. This pulse achieves the next node in a time of $\Delta t = \Delta d/c$. It is furthermore assumed that the pulse enters the node in the direction of the positive x-axis (Fig. 2.38). In accordance with the Huygens's

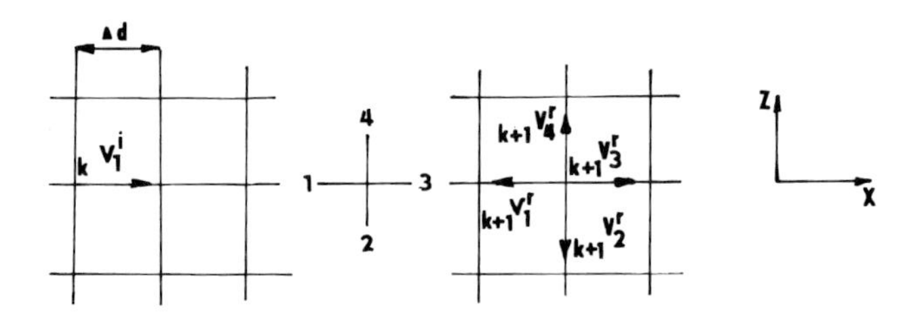

Fig. 2.38 Discretized Huygens wave model.

principle, the pulse energy is isotropically distributed into all the four directions and during this, each of these pulses will have 1/4 of the energy of the initial pulse. The corresponding quantities will have an one half amplitude and the reflection coefficient for the initial pulse will be negative, in order that the field continuity condition in the node could be adhered to. When denoting voltage pulses entering the node along lines m ($m = 1, 2, 3, 4$) in time $t = k\Delta t$ by $V_{k,m}^i$, when the total voltage $V_{k+1,n}^r$ reflected in direction of the conduction n ($n = 1, 2, 3, 4$) in time $t = (k+1)\Delta t$ will be of

$$V_{k+1,n}^r = \frac{1}{2}\left[\sum_m V_{k,m}^i\right] - V_{k,n}^i \tag{2.184}$$

In addition, each pulse leaving the node with coordinates $[x, z]$ simultaneously becomes a pulse entering the next node

$$\begin{aligned} V_{k+1,1}^i(x,z) &= V_{k+1,3}^r(x-\Delta d, z) \\ V_{k+1,2}^i(x,z) &= V_{k+1,4}^r(x, z-\Delta d) \\ V_{k+1,3}^i(x,z) &= V_{k+1,1}^r(x+\Delta d, z) \\ V_{k+1,4}^i(x,z) &= V_{k+1,2}^r(x, z+\Delta d) \end{aligned} \tag{2.185}$$

By the use of this scattering mechanism, a pulse response is obtained in network points chosen, which is transferred into the frequency region with the help of the Fourier's transformation.

The discrete nature of the network results in a dispersion of the wave front velocity. Precise results are obtained for the frequency values, for which it is possible to neglect this dispersion.

A concentrated equivalent of the basic structural element of the two-dimensional TLM network is shown in Fig. 2.39. By comparing

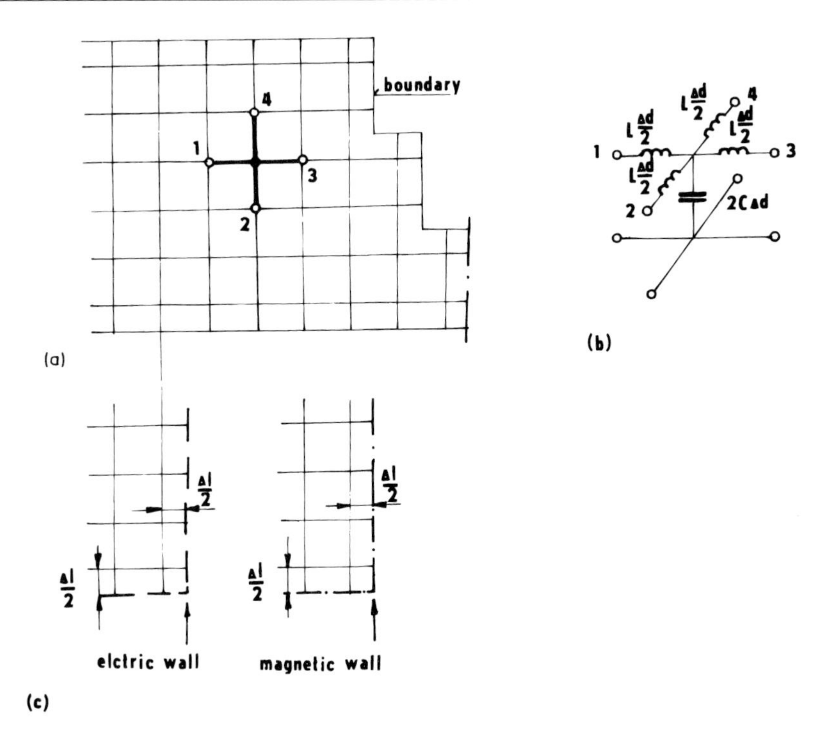

Fig. 2.39 (a) Boundary representation of TLM network, (b) Equivalent lumped model of basic element, TLM (c) Boundary of TLM network represented by electric and magnetic walls.

relationships between voltages and currents in the equivalent circuit, for relationships between the components H_z, H_x and E_y of the wave TE_{mo} in an rectangular waveguide

$$\begin{aligned} &E_y = V_y, \quad -H_x = (I_{z_2} - I_{z_4}), \quad -H_z = (I_{x_3} - I_{x_1}) \\ &\mu = L, \quad \varepsilon = 2 \times C \end{aligned} \tag{2.186}$$

where L, C are the inductance and capacity, respectively, per unit length. The whole circuit consisting of blocks according to Fig. 2.39 is a medium with a permittivity of $\varepsilon = 2\varepsilon_o$. This means that the propagation velocity in the TLM network

$$v = \frac{c}{\sqrt{2}}$$

When considering the network as a periodic structure, for the propagation in directions of main axes of the network the dispersion relationship is obtained (Johns and Beurle, 1971)

$$\sin\left(\beta_n \frac{\Delta d}{2}\right) = \sqrt{2}\sin\left(\frac{\omega \Delta d}{2c}\right) \tag{2.187}$$

where β_n is the propagation constant in the circuit.

Electric and magnetic walls can be simulated with the help of short-circuited or open equivalent elements (Fig. 2.39). This means that the eye size should be included as an integer in transverse dimensions of the structure. Loss walls can be formed with the help of the reflection coefficient

$$\rho = \frac{R-1}{R+1} \tag{2.188}$$

where R is the standardized surface resistance of the wall. For a good conductor it is possible to approximately write

$$\rho \simeq -1 + 2\sqrt{\frac{\varepsilon_0 \omega}{2\sigma}} \tag{2.189}$$

To make possible the solution of problems in a discontinuous dielectric, parallel open line of a length of $\Delta d/2$ with selectable characteristic admittance Y_o is connected to each node of the network. When the voltage in the network simulates the electric field, then the open stub represents an additional capacity at the node and reduces the velocity v_n of the wave in the network with simultaneous adhering to the condition on the air/dielectric interface. Then at low frequencies

$$v_n^2 = \frac{0.5c^2}{1 + \frac{Y_0}{4}} \tag{2.190}$$

The dispersion relationship for the propagation in the direction of coordinates is in the form

$$\sin^2\left(\frac{\beta_n \Delta d}{2}\right) = 2\left(1 + \frac{Y_0}{4}\right)\sin^2\left(\frac{\omega \Delta d}{2c}\right) \tag{2.191}$$

and for the cut-off frequency

$$f_c = \frac{c}{\pi \Delta d}\arcsin\left(\frac{v_n}{c}\right) \tag{2.192}$$

The dispersion of the standardized velocity v_n/c. is shown in Fig. 2.40 (Johns and Beurle, 1971). The impedance of the line is obtained with the help of known relationships according to a suitable representation (2.4), (2.6a), (2.6b).

APPENDIX

Characteristics of substrates for use in MIC

Material	ε	$\tan\delta$ $\times 10^4$	$\varkappa$ [$Wm^{-1}K^{-1}$]	α_1 [$ppm K^{-1}$]	α_{ε_r} [$ppm K^{-1}$]
AL_2O_3	$9.6 \div 10$	$1 \div 5$	37	6.3	136
Flint	3.78	1	1.7	0.55	13
Sapphire	9.4 ; 11.6	0.7	42	6	110 ; 140 *
BeO cer.	6.3	60	210	6.1	107
Teflon	2.08	3	0.2	106	350
Teflon filled by glass	2.33 2.2	9 9			** ***
Teflon filled by ceramics	13; 10.3				****
TiO_2	85	40	5	7.5	–575
$BaTi_4O_9$	37	8	2	9.4	–26
PTFE	2.34; 2.15				
Nerafen	2.61	20			
Cuprextit	$3.4 \div 4.8$	$40 \div 250$			
GaAs	12.9	20	46	5.7	
Si . ($\rho = 10\,\Omega m$)	11.9	150	145	4.2	

* — anisotropy; ** — RT-duriod 5870; RT-duroid 5880; *** — epsilam 10, anisotropy; **** — CuClad 217

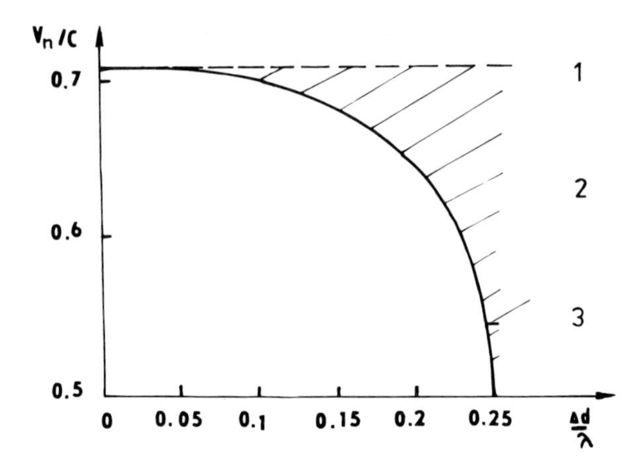

Fig. 2.40 Dispersion of the standardized velocity $\frac{v_n}{c}$. (1 — diagonal direct of propagation, 2 — propagation at any angle, 3 — axial direct of propagation.)

2.2 DISCONTINUITIES

M. Pavel

2.2.1 Characteristics of discontinuities and their use in MIC

The structure of the passive MIC network in general consists of a series of dielectric layers with different permittivities which have on their boundaries patterns of thin metallic foils. The whole system can be either open or closed in metallic shielding. The complex solution of characteristics of this general circuit or solution of the propagation of the electromagnetic field in this structure is a very complicated problem. With respect to a rapid progress in the field of the computing technique this problem has been principally managed, however, it would be very inefficient to continue in this way. In practice the circuit is considered as a system of interconnected segments of homogeneous lines, which considerably simplified the situation.

Sites of contacts of particular lines are named discontinuities of the line. In regions of these discontinuities, changes in the distribution of the electromagnetic field occur in comparison with the distribution of the field of homogeneous lines. The effects of these deviations on characteristics of circuits can be neglected only in the cases when the energy accumulated in these disturbed parts of the line is negligible in comparison with the energy transferred by the line. This is adhered to when the size of discontinuities is negligible in comparison with the wavelength of the electromagnetic radiation propagating through the circuit.

In the most simple approximation the discontinuities can be characterized by simple LC circuits. When taking into account ohmic losses, series resistances are introduced into the equivalent diagram. Losses in the dielectric can be simulated by parallel conductivities. Characteristics of particular RLC circuits, however, correspond to characteristics of discontinuities only in regions of lower frequencies. With increasing the frequency, it would be necessary to consider particular elements as frequency-dependent. For the characterization of discontinuities, frequency-dependent dissipative, impedance or admittance matrices are usually used. When selecting any form of the representation of discontinuities, it is necessary to remember that it is obtained assuming the isolated nature of the discontinuity. In other words, when solving the discontinuities, it is always assumed that they are formed by semi-infinite segments of homogeneous lines. This simplification is justified only in cases when the regions of disturbed fields of particular discontinuities do not overlap. This condition can be considered as satisfied to when the sizes of discontinuities are small in comparison with distances between them. When this is not the case then for precise calculations it is necessary to consider the whole system of simpler close discontinuities as one discontinuity with a complicated geometry.

2.2.2 Methods of the discontinuity analysis

For the determination of characteristics of discontinuities many calculation methods have been elaborated. Their mutual target is to find of the difference in the distribution of the charge, current or electromagnetic field in the given structure of the circuit in comparison with the distribution of the same quantity in the case of infinite homogeneous circuits and determination of elements characterizing the discontinuity on the basis of this difference.

The most simple method of solving the discontinuities in the MIC is a conversion of the given type of line to another type, which can be more easily solved. This method is mainly used for symmetric and non-symmetric microstrip lines. Most authors consider either directly or indirectly the work by M. Altschuler and A. Oliner (1960) which deals with the solution of basis discontinuities of the symmetric line. The basic concept is the replacement of joining segments of the microstrip line by connecting segments of fictitious waveguides with magnetic and electric walls. The problem is then processed by methods common in waveguide theory (Collin, 1960; Kvasil, 1957). The method of the replacement of microstrip lines by waveguides with the help of the Schwarz–Christoffel's transformation is described in detail in Section 2.1.

The serious drawback of this method is the fact that it adopts unallowable mathematical operations. As a matter of fact the Schwarz–Christoffel's transformation is discontinuous in the case of the line non-homogeneity. The equations for the Hertz vector are also not invariant with respect to the transformation used in nonhomogeneous lines. Thus, conditions are not adhered to for the conformal mapping, so that the solution of the waveguide discontinuity actually does not correspond to that of the original problem. The reason for the approximate agreement between the theoretical calculations and experiment is a similarity of the electromagnetic field distribution in the real circuit and fictitious waveguide, which holds particularly for the case of wide microstrip symmetric lines.

From the mathematical standpoint, more accurate results are achieved by different methods using the quasistatic approximation to the distribution of electromagnetic fields. These are methods assuming only quasi-TEM modes to be propagated through the lines, which makes it possible to solve only the static distribution of charges and currents.

An example is the determination of the effect of changes of the charge distribution in the discontinuity region for the most frequently used type of the line — non-symmetric microstrip lines — in comparison with the homogeneous line. Let us consider $q_0(x, y, z)$ as a charge density, which is present in the discontinuity region in the case of no changes on inlet lines in comparison with homogeneous lines and $q_c(x, y, z)$ as the actual value of the charge. The excess density of the charge is expressed by the quantity $q_e(x, y, z)$

$$q_c\,(x, y, z) = q_0\,(x, y, z) + q_e\,(x, y, z) \tag{2.193}$$

The characteristics of a discontinuity resulting from the presence of the excess charge $q_e(x, y, z)$ can be characterized by one or several concentrated capacitances, whose value is given by

$$C_{d_j} = \frac{\int_{V_j} q_e\left(x, y, z\right) dV}{\phi_p} \tag{2.194}$$

where V_j is a part of the discontinuity volume where the excess charge is concentrated, whose effect on characteristics of the circuit is described by the capacitance C_{d_j} and ϕ_p is the potential difference between the microstrip and earthing plane. The relationship between the potential $\phi(x, y, z)$ and charge density $q(x, y, z)$ is given by the Poisson equation

$$\nabla^2 \phi\left(x, y, z\right) = \frac{q\left(x, y, z\right)}{\varepsilon\left(x, y, z\right)} \tag{2.195}$$

determined by the discontinuity geometry for the boundary conditions. This equation can be solved by different methods. One of the most frequently used approaches is the solution using the Green's function. The Green's function of the Poisson equation is from the physical standpoint a contribution to the potential at point (x, y, z) resulting from the unit charge at point (x', y', z'). From this definition it can be seen that for the charge $q(x, y, z)$ and for the corresponding potential it is possible to write

$$\phi\left(x, y, z\right) = \int_V G\left(x, y, z, x', y', z'\right) q\left(x', y', z'\right) dx'dy'dz' \tag{2.196}$$

The three-dimensional integral can be reduced to a two-dimensional one under the assumption of an infinitely conductive microstrip of a zero thickness. Then the charge density depending on co-ordinates can be written in the form

$$q\left(x, y, z\right) = q\left(x, y\right) \delta\left(z - z_0\right) \tag{2.197}$$

There is such an orientation of the co-ordinate system that z_o is the co-ordinate of the microstrip plane and the earthing plane has co-ordinate $z = 0$. Unless otherwise determined, this co-ordinate system will always be considered below.

Since in further relationships it will be always necessary to express the potential only for the co-ordinate $z = z_0$, the potential functions will be written in their two-dimensional form and co-ordinate z will be considered to equal z_0.

From the potential superposition principle it is obvious that potentials ϕ_c, ϕ_0 and ϕ_e can be attributed to charge densities q_c, q_0, q_e, respectively, for which

$$\phi_c(x, y) = \phi_0(x, y) + \phi_e(x, y) \tag{2.198}$$

During this, the potential ϕ_c on the microstrip equals the constant ϕ_p. The potential ϕ_0 is calculated in the following way under an assumption that the discontinuity was produced by combining N segments of homogeneous lines. The charge density in the i-th infinite line and the corresponding potential will be denoted by ${}^i q_\infty(x, y)$ and ${}^i\phi_\infty(x, y)$, respectively. The method of determining these functions is described in Section 2.1. A fictitious case of the charge identical with ${}^i q_\infty(x, y)$ is further considered, however, with a polarity change in the plane of the contact between the line and discontinuity. The potential ${}^i\phi_-(x, y)$ corresponds to this distribution. When the Green's functions are corresponding to the potential ${}^i\phi_\infty$ and ${}^i\phi_-$ denoted by symbols ${}^iG_\infty(x, y, x', y')$ and ${}^iG_-(x, y, x', y')$ then

$${}^i\phi_-(x, y) = \int_S {}^iG_-(x, y, x', y')\, {}^i q_\infty(x', y')\, dx'dy' \tag{2.199}$$

$${}^i\phi_\infty(x, y) = \int_S {}^iG_\infty(x, y, x', y')\, {}^i q_\infty(x', y')\, dx'dy' \tag{2.200}$$

where S is the area of the charge ${}^i q(x, y)$ distribution.

The Green's functions can be determined for example from their physical definitions by a method of mirror images. As shown for example in Silvester and Benedek (1973), for a line parallel with axis x, the form of Green's function is given by equations

$${}^iG_\infty(x, y, x', y') = \frac{1}{2\pi\varepsilon_0(1+\varepsilon_r)} \sum_{n=1}^{\infty} K^{n-1} \log \frac{4n^2 + \left(\frac{y-y'}{z_0}\right)^2}{4(n-1)^2 + \left(\frac{y-y'}{z_0}\right)^2} \tag{2.201}$$

$${}^iG_-(x, y, x', y') = \frac{1-K}{4\pi\varepsilon_0}\left[f(0) - (1-K)\sum_{n=1}^{\infty} K^{n-1} f(n)\right] \tag{2.202}$$

where

$$f(n) = \log \frac{\sqrt{(x - x_i)^2 + y - y' + 4n^2 z_0^2} + (x - x_i)}{\sqrt{(x - x_i)^2 + y - y' + 4n^2 z_0^2} - (x - x_i)} \tag{2.203}$$

$$K = \frac{1 - \varepsilon_r}{1 + \varepsilon_r} \tag{2.204}$$

Here the symbol x_i stands for the x-co-ordinate of the site of contact if the i-th segment of the homogeneous line with the discontinuity. For lines of directions different from directions parallel with x-axis it is possible to obtain analogous forms of co-ordinate transformations. The above mentioned forms of Green's functions are correct under an assumption that the lines are open, i.e. without shielding. This assumption remains in force for further forms of Green's functions mentioned in this chapter. In the case of the presence of shielding, the forms of the Green's functions would be complicated due to adding further mirror charges.

For the potential ${}^i\phi_0(x, y)$ of a semi-infinite i-th segment of the line with a charge distribution as in the case of the infinite line, based on the superposition principle it holds that

$${}^i\phi_0(x, y) = \frac{1}{2}\left[{}^i\phi_\infty(x, y) +^i \phi_-(x, y)\right] \tag{2.205}$$

and for $\phi_0(x, y)$

$$\phi_0(x, y) = \sum_{i=1}^{N} {}^i\phi_0(x, y) \tag{2.206}$$

Since the potential $\phi_c(x, y)$ on the microstrip equals the constant ϕ_p, it follows from (2.198) and (2.206) that

$$\phi_e(x, y) = \phi_p - \sum_{i=1}^{N} {}^i\phi_0(x, y) \tag{2.207}$$

The distribution of the charge $q_e(x, y)$ can be obtained by solving the integral equation

$$\phi_e(x, y) = \int_S G_e(x, y, x'y')\, q_e(x', y')\, dx'dy' \tag{2.208}$$

where S is the discontinuity area. For this it is first necessary to know the Green's function $G_e(x, y, x', y')$. It can be obtained by the method of mirror images, again, as shown for example in Silvester and Benedek (1973). For basic types of discontinuities, such as rectangular bending of the line, impedance step T-connection and crossed lines it is in the form (Silvester and Benedek, 1973; Benedek and Silvester, 1973)

$$G_e(x, y, x'y') = \frac{1}{2\pi\varepsilon_0(1+\varepsilon_r)}\left[f(0) - (1-K)\sum_{n=1}^{\infty} K^{n-1} f(n)\right] \tag{2.209}$$

where K is given by relationship (2.204) and for the functions $f(n)$ it holds that

$$f(n) = \frac{1}{\sqrt{4n^2 + \left(\frac{x-x'}{z_0}\right)^2 + \left(\frac{y-y'}{z_0}\right)^2}} \tag{2.210}$$

for rectangular bending

$$f(n) = \frac{1}{\sqrt{4n^2 + \left(\frac{x-x'}{z_0}\right)^2 + \left(\frac{y-y'}{z_0}\right)^2}} + \frac{1}{\sqrt{4n^2 + \left(\frac{x+x'}{z_0}\right)^2 + \left(\frac{y-y'}{z_0}\right)^2}} \tag{2.211}$$

for the impedance jump and T connection and

$$\begin{aligned} f(n) = {} & \frac{1}{\sqrt{4n^2 + \left(\frac{x-x'}{z_0}\right)^2 + \left(\frac{y-y'}{z_0}\right)^2}} + \frac{1}{\sqrt{4n^2 + \left(\frac{x-x'}{z_0}\right)^2 + \left(\frac{y+y'}{z_0}\right)^2}} \\ & + \frac{1}{\sqrt{4n^2 + \left(\frac{x+x'}{z_0}\right)^2 + \left(\frac{y-y'}{z_0}\right)^2}} + \frac{1}{\sqrt{4n^2 + \left(\frac{x+x'}{z_0}\right)^2 + \left(\frac{y+y'}{z_0}\right)^2}} \end{aligned} \tag{2.212}$$

for crossed lines. Relationships (2.210 – 2.212) hold under an assumption that the discontinuities are oriented in the co-ordinate system as shown in Fig. 2.41.

Equation (2.208) can be solved for example by the Galerkin's method (Miklin and Smolitsky, 1967). Based on the knowledge of

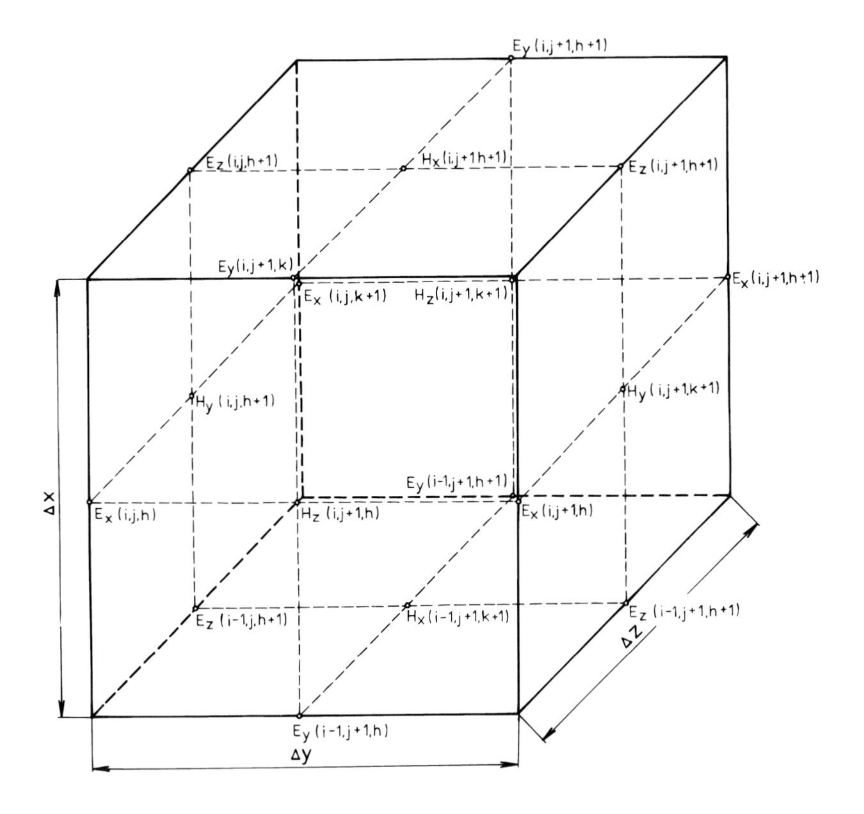

Fig. 2.41 Spatial network.

the distribution of the charge $q_e(x, y)$ the values C_{d_j} are obtained with the help of (2.194).

Besides capacity elements of the equivalent circuit of a discontinuity, inductive elements can also be obtained by quasistatic methods, whose existence results from changes in the distribution of currents and from their interactions. The general discontinuity of the microstrip line can serve as an example, again. For the magnetic potential A at point $[x, y, z]$ the general relationship

$$A(x, y, z) = \mu_0 \int_V G(x, y, x', y', z')\, J(x', y', z')\, dx'dy'dz' \qquad (2.213)$$

is valid where J is the current density vector and G is the Green's function. From Maxwell's equations and Ohm's law it follows that

$$J(x, y, z) = -\sigma \frac{\partial A(x, y, z)}{\partial t} - \sigma \nabla \phi(x, y, z) \qquad (2.214)$$

where σ is the conductivity and ϕ is the internal scalar potential. By the substitution from (2.213) the equation

$$J(x,y,z) + \mu_0 \sigma \frac{\partial}{\partial t} \int_V G(x,y,z,x',y',z') J(x',y',z') \, dx'dy'dz' = = -\nabla \phi (x,y,z) \tag{2.215}$$

is obtained. After considering the very high conductivity of the microstrip, this equation is changed to the form

$$\mu_0 \frac{\partial}{\partial t} \int_V G(x,y,z,x'y'z') J(x',y',z') \, dx'dy'dz' = -\nabla \phi (x,y,z) \tag{2.216}$$

For a very small thickness of the strip and for a constant current density within the whole thickness

$$J(x,y,z) = J(x,y) \, \delta (z - z_0) \tag{2.217}$$

in the arrangement of the co-ordinate system which was considered in the analysis of capacitance elements. Thus, the Green's function is in the form (Thompson and Gopinath, 1975)

$$G(x,y,x',y') = \frac{1}{4\pi \sqrt{(x-x')^2 + (y-y')^2}} - \frac{1}{4\pi \sqrt{(x-x')^2 - (y-y')^2 + 2z_0^2}} \tag{2.218}$$

In this case when considering the propagation of a signal with its time dependence of the type $e^{j\omega t}$ (2.216) is changed to

$$j\omega\mu_0 \int_S G(x,y,x',y') J(x',y') = -\nabla \phi (x,y) \tag{2.219}$$

The forced potential ϕ on the strip adheres to the Laplace's equation

$$\nabla^2 \phi (x,y) = 0 \tag{2.220}$$

The solution of (2.219) and (2.220) for ϕ and J makes it possible to obtain the inductance L from Thompson and Gopinath (1975)

$$L = \frac{\int_S A(x,y) J(x,y) \, dxdy}{\left[\int_S J(x,y) \, dxdy\right]^2} \tag{2.221}$$

When calculating with the help of (2.221) the total inductance of the discontinuity L_T and when submitting from it inductance values $L_{\infty i}$, which would have segments of inlet lines to the discontinuity if there were a part of an isolated homogeneous infinite line, the inductance L_D resulting from the presence of the discontinuity is obtained

$$L_D = L_T - \sum_{i=1}^{N} L_{\infty i} \tag{2.222}$$

The inductance L_D is typically divided into the sum of the inductances corresponding to particular regions of the discontinuity and these partial inductances form elements of the substitution diagram of the discontinuity.

In this simple form the calculation would encounter a considerable error due to subtracting large and very similar values in (2.222). Thus, a modified method is usually used, introducing failure currents. Thus, an analogy is used of the above mentioned method for the calculation of capacitance elements of the discontinuity equivalent circuit. This method is more detailed for the case of the rectangular line bend (Thompson and Gopinath, 1975) and for the microstrip step discontinuity (Gopinath *et al.*, 1976).

The methods described offer results which are in agreement with the experiment in lower bands of microwave frequencies. At higher frequencies effects, which are neglected here, start to occur more considerably. These are particularly dispersion characteristics of discontinuities connected with the excitation of higher modes, losses resulting from radiation and losses induced by the excitation of surface waves. Several methods have been elaborated for the description of these effects. One of them is for example the moment method (Katehi and Alexopoulos, 1985) based on solving the Pocklington's integral equation

$$E(r) = \int_S G(r, r') \, J(r') \, ds' \tag{2.223}$$

which expresses the dependence of the electric field vector E at point r on the current density J with the help of the dyadic Green's function (Katehi and Alexopoulos, 1985; Collin, 1960) of the type

$$G(r, r') = \int_0^\infty \left(\omega^2 T + \Delta\right) J_0 \left[\omega \left(|r - r'|\right)\right] F(\omega) \, d\omega \tag{2.224}$$

The symbol T stands for the unit dyadic function, J_0 is the Bessel's function of the wavelength and $F(\omega)$ is the dyadic function in the form (Alexopoulos *et al.*, 1985)

$$F(\omega) = \frac{A(\omega, \varepsilon, h)}{f_1(\omega, \varepsilon, h) f_2(\omega, \varepsilon, h)} \tag{2.225}$$

The functions f_1, f_2 are analytical functions of the wavelength, material permittivity and substrate thickness. Their zero values correspond to poles of the function $F(\omega)$ and they are related to the excitation of TE and TM surface waves.

Equation (2.223) can be solved for example in such a way that the discontinuity region is divided into a series of segments to which a system of base functions of $f_n(r)$ is related. The distribution of currents $J(r)$ is expressed with the help of these functions in the form

$$J(r) = \sum_{n-1}^{N} J_n f_n(r) \tag{2.226}$$

In a suitable choice of these base functions with a possible use of their relevant weights, the Galerkin's method can be employed to arrange relationship (2.223) to a matrix equation in the form

$$V_m = Z_{m,n} J_n \tag{2.227}$$

where J_n is a vector of coefficients J_n, $Z_{m,n}$ is the impedance matrix and V_m is the excitation voltage vector depending on the method of exciting the lines considered. The excitation is usually considered in the form of the quasi-TEM mode. As shown in Katehi and Alexopoulos (1985), it can be simulated by a fictitious voltage generator situated in the inlet homogeneous line. Then for a suitable choice of base functions for elements V_m

$$V_m = \delta_m \tag{2.228}$$

where $\delta_m = 1$ for the case of a generator situated in the m-th segment of the discontinuity and $\delta_m = 0$ for all the other cases.

Based on a knowledge of the electric field distribution in the case of suitably chosen excitations, it is necessary to determine S-coefficients of the discontinuity depending on the frequency and on the basis of these coefficients it is possible to obtain the frequency-dependent equivalent circuit.

A further very efficient method is the method of finite elements in the time domain (Zangh and Kenneth, 1988). It is based on a direct numerical solution of Maxwell's equations

$$\begin{aligned} \frac{\partial E}{\partial t} &= \frac{1}{\varepsilon_i} \nabla \times H \\ \frac{\partial H}{\partial t} &= -\frac{1}{\mu_0} \nabla \times E \end{aligned} \tag{2.229}$$

where the subscript $i = 1, 2$ stands for the substrate and free space respectively. For unambiguous solution, the value of the magnetic and electric field is given at time $t = 0$ and on boundaries of the region of interest at $t > 0$.

The numerical solution procedure itself simulates the propagation of an electromagnetic pulse in the space. For this purpose, a spatial network of points is selected, where particular components of fields (Fig. 2.39) are followed. Each component of the field H can be obtained from the line integral of the field E with using values of this field in neighbouring points of the spatial network. The same procedure can also be implemented for components of the field E calculated from values of the field H at neighouring points. During this, the relevant fields E calculated at time $n\,\Delta t$ are expressed with the help of fields H given at time $(n - 0.5)\,\Delta t$ and field E at time $(n - 1)\,\Delta t$ where n is an integer and Δt is an interval of the time variable discretization.

This procedure leads to a change from the system of differential equations to algebraic equations of the type

$$\begin{aligned} E_x^{n+1}(i,j,k) &= E_x^n(i,j,k) + \\ &\frac{\Delta t}{\varepsilon} \left[\frac{H_z^{n+0.5}(i,j+1,k) - H_z^{n+0.5}(i,j,k)}{\Delta y} - \right. \\ &\left. - \frac{H_y^{n+0.5}(i,j,k+1) - H_y^{n+0.5}(i,j,k)}{\Delta z} \right] \end{aligned}$$

$$E_y^{n+1}(i,j,k) = E_y^n(i,j,k) + \frac{\Delta t}{\varepsilon}\left[\frac{H_x^{n+0.5}(i,j,k+1) - H_x^{n+0.5}(i,j,k)}{\Delta z} - \frac{H_z^{n+0.5}(i+1,j,k) - H_z^{n+0.5}(i,j,k)}{\Delta x}\right]$$

$$E_z^{n+1}(i,j,k) = E_z^n(i,j,k) + \frac{\Delta t}{\varepsilon}\left[\frac{H_y^{n+0.5}(i+1,j,k) - H_y^{n+0.5}(i,j,k)}{\Delta x} - \frac{H_x^{n+0.5}(i,j+1,k) - H_x^{n+0.5}(i,j,k)}{\Delta y}\right]$$

$$H_x^{n+0.5}(i,j,k) = H_x^{n-0.5}(i,j,k)\frac{\Delta t}{\mu}\left[\frac{E_z^n(i,j,k) - E_z^n(i,j-1,k)}{\Delta y} - \frac{E_y^n(i,j,k) - E_y^n(i,j,k-1)}{\Delta z}\right]$$

$$H_y^{n+0.5}(i,j,k) = H_y^{n-0.5}(i,j,k)\frac{\Delta t}{\mu}\left[\frac{E_x^n(i,j,k) - E_x^n(i,j,k-1)}{\Delta z} - \frac{E_z^n(i,j,k) - E_z^n(i-1,j,k)}{\Delta x}\right]$$

$$H_z^{n+0.5}(i,j,k) = H_z^{n-0.5}(i,j,k)\frac{\Delta t}{\mu}\left[\frac{E_y^n(i,j,k) - E_y^n(i-1,j,k)}{\Delta x} - \frac{E_x^n(i,j,k) - E_x^n(i,j-1,k)}{\Delta y}\right]$$

(2.230)

where x, y, z are intervals of the discretization of spatial co-ordinates and i, j, k are integers determining the situation of the spatial point $r \equiv (ix, jy, kz)$. The above mentioned relationships hold for a homogeneous region of the space. At sites of conductive areas it is necessary to nullify tangent components of the electric fields and a perpendicular component of the magnetic field. The effect of the interface of dielectric media can be considered, as shown in Zangh and Kenneth (1988), by replacing the permittivity ε_i in relationships for the calculation of tangent components of the field E at points on this interface by the permittivity $(\varepsilon_1 + \varepsilon_2)/2$.

With respect to the fact that for numerical calculations it is necessary to restrict the space for performing the calculations, on the boundaries of this space it is necessary to define conditions for fields E and H. The method of finite elements is very sensitive to the choice of these conditions. In principle it is necessary to define these conditions in such a way, which can prevent the reflection back to the field of the calculation (absorption conditions). Particular approaches to the choice of these conditions can be different and they are detailed in Zangh and Kenneth (1988). The effect of their imperfect determination can be partially eliminated by completing the calculation for a suitably chosen value of the time co-ordinate.

The pulse of the component E_x of the electric field is usually selected as an excitation pulse. It is defined as constant at a given time on a rectangular area delimited by the strip of the inlet line, earthing area and perpendicular lines directed from the strip margin to the earthing plane. The zero field is defined in addition to this rectangle. The Gaussian course of the pulse is assumed in the form

$$E_x = e^{-\frac{[t - t_0 - (z - z_0)]^2}{T^2}} \tag{2.231}$$

where v is the phase velocity of the pulse propagation in the given medium. The constants z_o, t_o and T_1 similarly as the discretization intervals Δt, Δx, Δy and Δz are determined in a specific way for each case solved with respect to a sufficiently precise description of the configuration of the discontinuity, to saving the stability of the solution and to achieving a sufficiently wide spectrum of the pulse making possible the characterization of the discontinuity in the frequency band required.

It should be noted that the boundary conditions given at the time of the pulse generation for the inlet plane delimiting the space of the calculation (plane where the excitation pulse is generated) must be changed to absorption conditions similar to those for the other limit planes at the time when the reflected wave arrives at this plane.

For obtaining the time course of the response of the circuit investigated to the excitation pulse, the characteristic of this circuit in the frequency region is obtained by the Fourier's transformation and it is then used for obtaining relevant S-parameters (Zangh and Kenneth, 1988).

The methods mentioned describe the behaviour of the discontinuity in an unbounded space. For microstrip discontinuities this is the most general case from the standpoint of their nature. Sometimes, it is, however, also necessary to know the effect of shielding, which is actually present in almost all the real circuits. In many types of lines the shielding is even their essential functional part (e.g. in the fin-line type). In the case of a space bounded by conductive walls representing the shielding, it is necessary to take the shielding into account by including further mirror images of the current into the form of the Green's function in the case of the moment method, and by a suitable choice of boundary conditions in the case of the method of finite elements. Many other methods are also used for the solution of discontinuities in a space bounded by screening conductive areas, as e.g. a method by (Jansen 1985) in the spectral domain, variation methods (Webb and Mittra, 1985), etc.

2.2.3 Microstrip discontinuities

The microstrip open end discontinuity is the most simple and obviously most frequently occurring type of the discontinuities in the MIC. For its geometry arrangement and equivalent circuit diagram see Fig. 2.42. The dominant phenomenon determining its characteristics is the scattering electric field, which may be characterized by a capacitance element C. From the practical standpoint, this capacitance is frequently replaced by a fictitious extending line of a length Δl, which may be expressed by

$$\Delta l = C Z_0 \frac{c}{\sqrt{\varepsilon_{ef}}} \tag{2.232}$$

where c is the velocity of light, Z_0 the characteristic impedance of the line and ε_{ef} the effective permittivity of the material. This expression directly determines the compensation of the scattering capacitance. On the base of calculations in a quasistatic approximation, (Hammerstadt, 1975) obtained by an approximation the relationship for Δl is given by

$$\frac{\Delta l}{h} = 0.412 \frac{\varepsilon_{ef} + 0.3\frac{w}{h} + 0.264}{\varepsilon_{ef} - 0.258\frac{w}{h} + 0.8} \tag{2.233}$$

which gives a higher accuracy than 4% in comparison with values calculated for $2 \leq \varepsilon_r \leq 50$ at $w/h \in \langle 0.2; 10 \rangle$.

However, a more precise, also more complicated form was presented in Kirschning, Jansen and Koster (1981)

$$\frac{\Delta l}{h} = \frac{\xi_1 \xi_3 \xi_5}{\xi_4}$$

where

$$\xi_1 = 0.434907 \frac{\varepsilon_e^{0.81} + 0.26 \left(\frac{w}{h}\right)^{0.8544} + 0.236}{\varepsilon_{ef}^{0.81} - 0.189 \left(\frac{w}{h}\right)^{0.8544} + 0.87}$$

$$\xi_2 = 1 + \frac{\left(\frac{w}{h}\right)^{0.371}}{2.358\varepsilon_r + 1}$$

$$\xi_3 = 1 + \frac{0.5274 \arctan \left[0.084 \left(\frac{w}{h}\right)^{\frac{1.9413}{\xi_2}}\right]}{\varepsilon_{ef}^{0.9236}}$$

$$\xi_4 = 1 + 0.0377 \arctan \left[0.067 \left(\frac{w}{h}\right)^{1.456}\right] \left[6 - 5e^{0.036(1-\varepsilon_r)}\right]$$

$$\xi_5 = 1 - 0.218 e^{-7.5\frac{w}{h}} \tag{2.234}$$

The accuracy of this relationship is better than 2.5% for $\varepsilon_r \leq 50$ and $w/h \in \langle 0.01; 100 \rangle$. Relationships (2.233, 2.234) can be properly adopted up to about 18 GHz. Above this limit more considerable radiation into the space and generation of surface waves occur. These effects are manifested by a decrease of the coefficient of the discontinuity reflection. This behaviour can be described by adding the conductivity G to the capacitance C.

The capacitance C also varies due to the dispersion. Typical frequency dependences of these quantities (Jansen, 1984; Gupta, Garg and Chadha, 1981) are shown in Fig. 2.43. In Gupta, Garg and Chadha (1981) it is shown that the increase of losses with increasing ratio h/λ and with increasing ε_r mainly results from growing losses induced by the excitation of surface waves.

The microstrip gap is a discontinuity which is similar as to its character to the microstrip open circuit. Similar physical effects occur, complemented by coupling between inlet strips facilitated by scattering fields. This phenomenon is used in the design of filters and for the

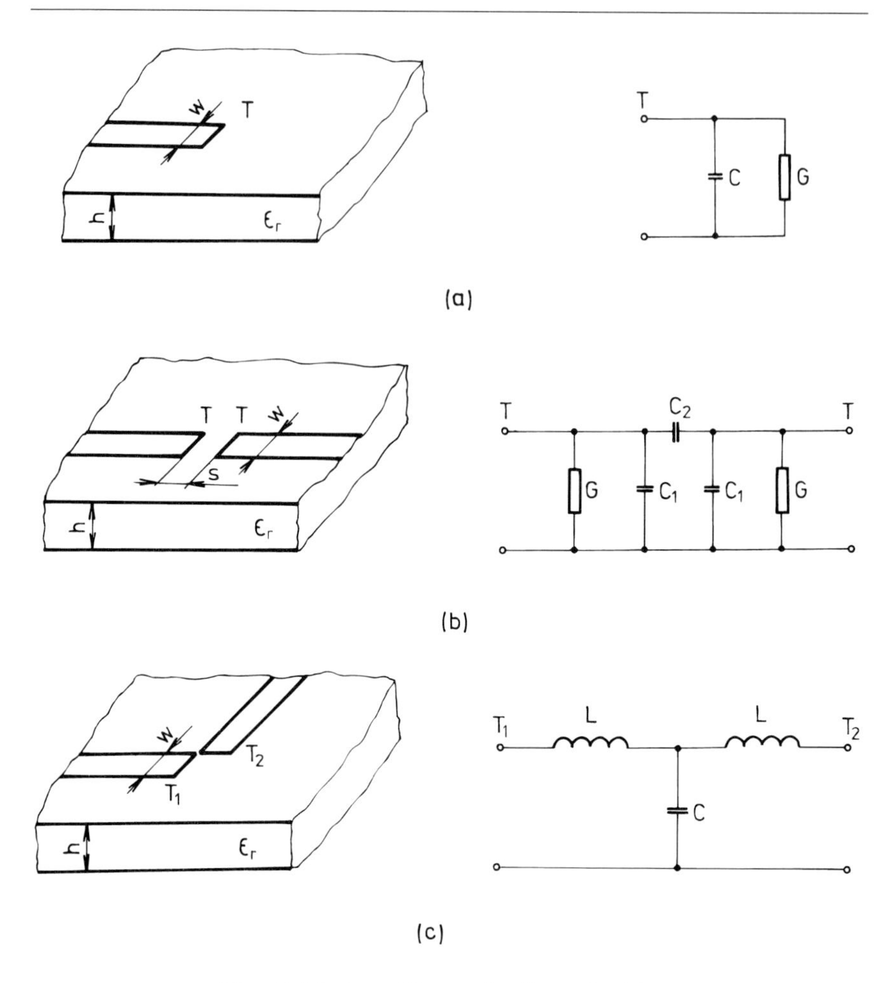

Fig. 2.42a,b,c Microstrip discontinuities.

separation of direct currents in microwave circuits. The microstrip gap, together with equivalent diagrams, is shown in Fig. 2.42. The basic elements of the equivalent diagram are two identical capacitors representing the scattering and capacitances of inlet lines. They have somewhat smaller capacitances than the corresponding ends of the line, which results from shielding by the second inlet line. In a static approximation, these scattering capacitances together with the coupling capacitance are calculated from two auxiliary arrangements, with symmetric and antisymmetric excitations, which correspond to applying an infinite magnetic or infinite electric wall perpendicularly to the line in the centre of the slot. When denoting C_e the

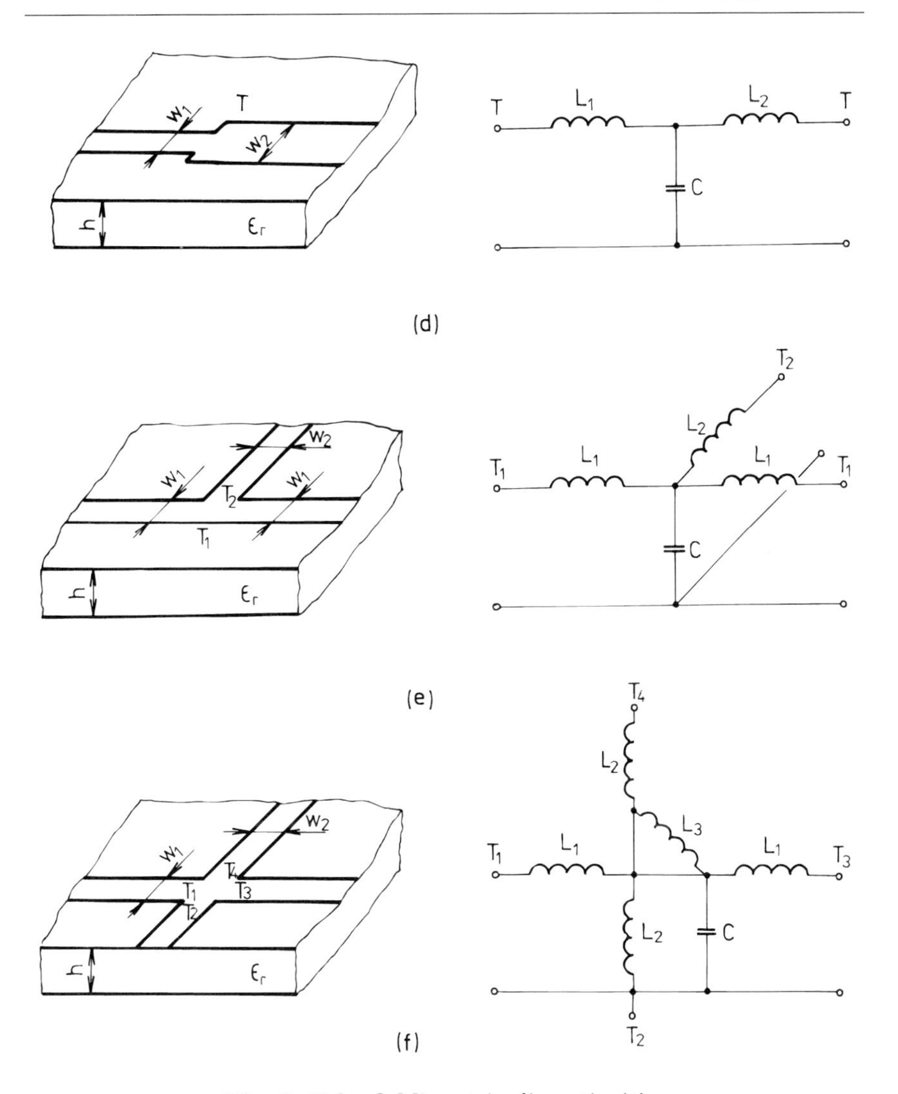

Fig. 2.42d,e,f Microstrip discontinuities.

capacitance of a discontinuity in the symmetric excitation and C_0 the capacity at the antisymmetric excitation, then

$$
\begin{aligned}
C_1 &= \frac{1}{2} C_e \\
C_{12} &= \frac{1}{2}\left(C_0 - \frac{1}{2} C_e\right)
\end{aligned}
\tag{2.235}
$$

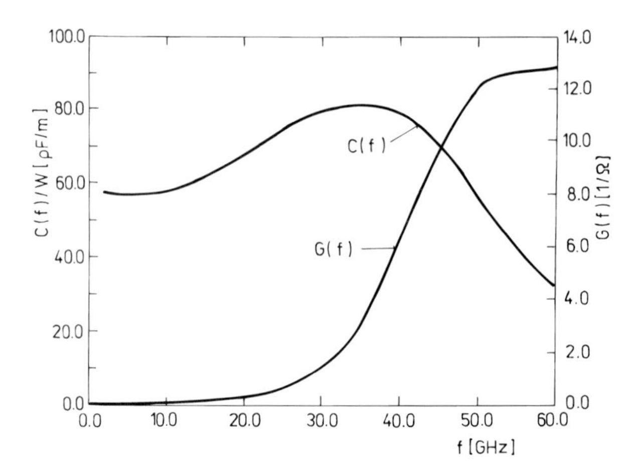

Fig. 2.43 Typical frequency dependences of C and G.

For the values of C_0 and C_e the approximate relationships were derived (Gupta *et al.*, 1981)

$$\begin{aligned} C_e &= C_{e0}\left(\frac{\varepsilon_r}{9.6}\right)^{0.9} \\ C_0 &= C_{00}\left(\frac{\varepsilon_r}{9.6}\right)^{0.8} \end{aligned} \tag{2.236}$$

which hold for ε_r in a range of 2.5 to 15. The values of capacitances C_{e0} and C_{00} for width of strips satisfying to the condition $0.5 \leq w/h \leq 2$ are given by

$$\begin{aligned} C_0 &= w\left(\frac{s}{w}\right)^{m_0} e^{K_0} \\ C_e &= w\left(\frac{s}{w}\right)^{m_0} e^{K_e} \end{aligned} \tag{2.237}$$

where

$$\begin{aligned} m_0 &= \frac{w}{h}\left(0.267\log\frac{w}{h} - 0.3853\right) \\ K_0 &= 4.26 - 0.631\log\frac{w}{h} \qquad \text{for } 0.1 \leq \frac{s}{w} \leq 1 \end{aligned}$$

$$
\begin{aligned}
m_e &= 0.8675 \\
K_e &= 2.043\left(\frac{w}{h}\right)^{0.12} \qquad \text{for } 0.1 \leq \frac{s}{w} \leq 0.3 \\
m_e &= \frac{1.565}{\left(\frac{w}{h}\right)^{0}.16} - 1 \\
K_e &= 1.97 - \frac{0.03h}{w} \qquad \text{for } 0.3 \leq \frac{s}{w} \leq 1
\end{aligned}
\tag{2.238}
$$

In these relationships w is the strip width, s the slot width and h is the thickness of the substrate with a relative permittivity of ε_r. The dimensions are given in metres and capacitances in pF. The accuracy of the approximation is higher than 7%. The capacitors C_1 can be replaced by segments of the line, similarly as in the case of the microstrip open circuit, whose lengths can be calculated from relationship (2.232) where the capacitance C is replaced by the capacitance C_1.

In a higher frequency band it is necessary to take into account the increasing radiation of energy and generation of surface waves. Due to the effects, the capacitance C_1 decreases and the coupling capacitance C_{12} moderately increases. The losses can be included into the equivalent circuit by adding parallel conductivity elements to the capacitance elements. The nature of the frequency dependence of the value of individual elements is shown in Fig. 2.44 (Katehi and Alexopoulos, 1985).

Right-angle bend of the line. This type of the discontinuity relatively frequently occurs in the microwave integrated circuits due to attempts to minimize the dimensions of these circuits or possibly to optimize their shape with respect to the desired situations of inlet lines. The discontinuity and its equivalent circuit is shown in Fig. 2.42. The following approximate relationships were used on the basis of methods in the quasistatic approximation for elements of the equivalent diagram (Gupta, Garg and Chadha, 1981):

$$
\begin{aligned}
\frac{C}{w} &= \frac{(14\varepsilon_r + 12.5)\frac{w}{h} - (1.83\varepsilon_r - 2.25)}{\sqrt{\frac{w}{h}}} + \frac{0.02\varepsilon_r}{\frac{w}{h}} \qquad \text{for } \frac{w}{h} < 1 \\
\frac{C}{w} &= (9.5\varepsilon_r + 1.25)\frac{w}{h} + 5.2\varepsilon_r + 7 \qquad \text{for } \frac{w}{h} \geq 1
\end{aligned}
\tag{2.239}
$$

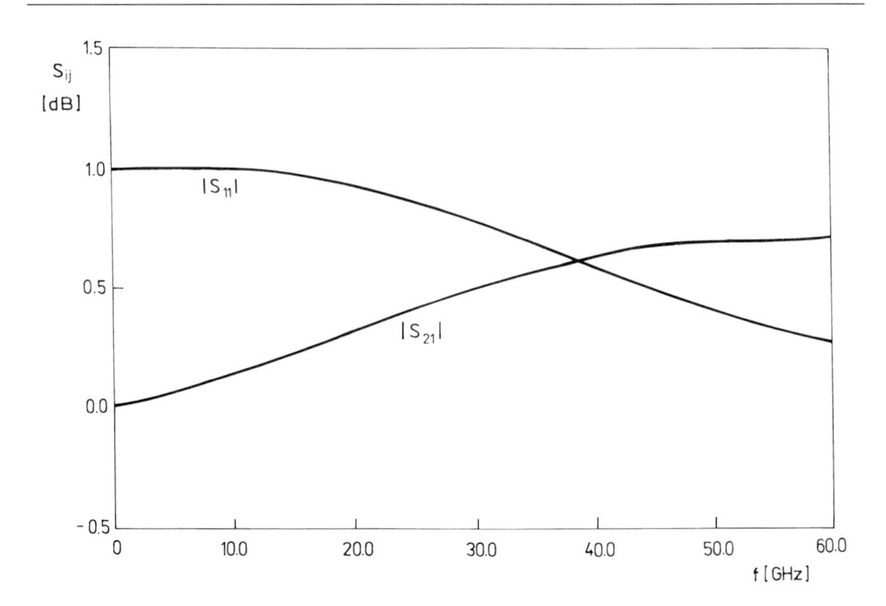

Fig. 2.44 Frequency dependences of S_{ij}.

and

$$\frac{L}{h} = 50\left(4\sqrt{\frac{w}{h}} - 4.21\right)$$

where C/w is given in [pF.m^{-1}] and L/h in [nH.m^{-1}]. The accuracy of the relationships is about 5% within a range of $2.5 \leq \varepsilon_r \leq 15$ and $0.1 \leq w/h \leq 5$. The capacitance C can be divided into three parallelly connected capacitance elements for which

$$C = C_1 + 2C_2 \tag{2.240}$$

The value C_2 is chosen in such a way that it corresponds to the capacitance of the segment Δl of a homogeneous line of a width w, with an inductance L. Then, the elements L and C_2 can be replaced in the equivalent circuit by line segments Δl and C remains only as the parallel element C_1. Its excess capacitance can be compensated by a mitre. In this way, the inductance L is reduced to a certain extent, which means necessary changes in the distribution of the capacitance C to components C_1 and C_2. The capacitance component of the right-angle line bend is, however, changed by the mitre with respect to the nature of the charge and current distribution, more

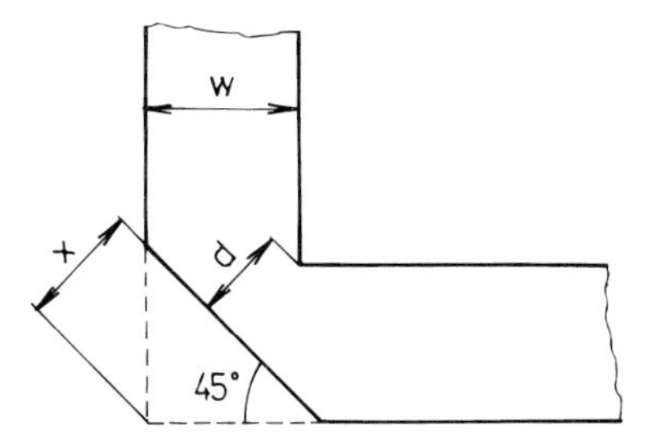

Fig. 2.45 Right-angle line band.

rapidly than the inductive component. This makes it possible to provide the reflectionless behaviour of the discontinuity. This method of the compensation is shown in Fig. 2.45. The measure of cutting for which the right-angle line bend is reflectionless is given by the experimentally established relationship (Douville and James, 1978)

$$\frac{x}{d} = 0.52 + 0.65e^{-1.35\frac{w}{h}} \tag{2.241}$$

This relationship is valid for $\varepsilon_r \leq 25$ and $w/h \leq 0.25$. As already mentioned, the line segments Δl are changed by the mitre. For optimum mitre they can be approximately characterized by the relationship

$$2\frac{\Delta l}{h} = 0.5968\frac{w}{h} - 0.4163 \tag{2.242}$$

All the above mentioned equations were derived on the basis of the quasistatic approximation or they were obtained from experiments in a lower microwave frequency band. With increasing frequency dispersion characteristic of the discontinuity starts to appear (Chadha and Gupta, 1982), which results in a growth of the length Δl and in a moderate increase of the capacitance C (see Fig. 2.41). In a region of mm-waves it is to be expected, similarly as in the other types of discontinuities, that there will be an increase of losses resulting from radiation and generation of surface waves. These losses can be simulated in an equivalent diagram by adding conductivity elements to the capacitance C.

The microstrip step discontinuity. This discontinuity most frequently occurs in impedance transformers, matching circuits and filters. The step geometry and equivalent diagram are shown in Fig. 2.40. Inductive elements of the equivalent diagram correspond to changes in the current distribution and the capacity represents the increase of the scattering electric field in the discontinuity region. On the basis of calculations carried out in the quasistatic approximation, the approximation relationships for C and L were obtained (Gupta *et al.*, 1981)

$$\frac{C}{\sqrt{w_1 w_2}} = (4.386 \ln \varepsilon_r + 2.33) \frac{w_2}{w_1} - 5.472 \ln \varepsilon_r - 3.17 \qquad (2.243)$$

$$\frac{L_1 + L - 2}{h} = 40.5 \left(\frac{w_2}{w_1} \right) - 32.57 \ln \frac{w2}{w1} + 0.2 \left(\frac{w_2}{w_1} - 1 \right)^2 \qquad (2.244)$$

for $\varepsilon_r \leq 10$ and $w_2/w_1 \in \langle 1.5; 3.5)$. The quantity $c/\sqrt{w_1 w_2}$ is in [pF.m^{-1}] and $(L_1 + L_2)/h$ in [nH.m^{-1}]. The accuracy of relationship (2.243) is about 10% and that of (2.244) at $w_1/h = 1$ and $w_2/w_1 \leq 5$ of about 5%. From the standpoint of including the effect of the discontinuity in the behaviour of the circuit and particularly for the compensation of this effect it is suitable to divide, similarly as in the case of the right-angle bend, the capacitance C into three parallel capacitances C_1, C_2 and C' in such a way that the capacity C_1 together with inductance L_1 corresponds to the segment of the line of width w_1 and length Δl_1 and the capacitance C_2 with inductance L_2 to the segment of the line of width w_2 and length Δl_2.

A problem remains of dividing the inductance $L = L_1 + L_2$ into partial inductances L_1, L_2. In Gupta (1981) it is recommended for the first approximation to divide this inductance in the ratio of inductances of homogeneous lines related to unit length. This corresponds to the condition $\Delta l_1 = \Delta l_2$. It is, however, shown, that this dividing corresponds to the real situation only for lower frequency bands. With increasing frequency, the length of the narrower strip moderately increases and that of the wider strip decreases. As it follows for example from Zangh and Kenneth (1985); Koster and Jansen (1986); Chadha and Gupta (1982), for commonly used dimensions of the line and materials in the MIC the amplitudes of scattering parameters of the step are almost constant in a wide frequency band. This corresponds to a very low capacitance C'. The small frequency dependence of the capacitance C was also checked experimentally (Jansen and Koster, 1982). Thus, parasitic elements of this discontinuity can be compensated by shortening the conduction by lengths Δl_1 and Δl_2. For frequency values of above 16 GHz it is also necessary to provide a mitre of the wider strip (Chadha and Gupta, 1982).

T-junction of lines. This is a discontinuity occurring in many types of microwave circuits. These are particularly hybrid couplers, matching circuits and filters. For the shape of the discontinuity and its equivalent circuit see Fig. 2.40.

The capacitive element C results from decreasing the charge density in the discontinuity region in comparison with the density in homogeneous lines. For this reason, it is negative for all the line dimensions. The magnitude of this capacitance can be expressed by an approximative relationship obtained from a quasistatic approximation. This relationship is in the form

$$\frac{C}{w_2} = \frac{100}{\tanh(0.0072 Z_{02})} + 0.64 Z_{02} - 261 - \frac{\sqrt{\varepsilon_{ef}}}{c Z_0} 10^{12} \qquad (2.245)$$

for $Z_{01} = 50\,\Omega$, $\varepsilon_r = 9.9$, $Z_{02} \in \langle 25; 100 \rangle$, where Z_{02} and Z_{01} are characteristic impedances of lines of widths w_2 and w_1, respectively. The relationship (2.245) is different by the last term from those given in Gupta, Garg and Chadha (1981) and Gupta, Garg and Bahl (1979). The error in relationships in the above mentioned literature resulted from omitting differences in the choice of reference phase planes in the work by Silvester and Benedek (1973), which served as a basis for the approximation.

The inductive elements L_1 and L_2 also have negative values for most configurations, which also results from reducing the current density in the discontinuity region in comparison with the density in homogeneous lines. The approximation relationships (Gupta, Garg and Chadha, 1981) obtained on the basis of calculations in the quasistatic approximation (Thompson and Gopinath, 1975) are in the form

$$\frac{L_1}{h} = -\frac{w_2}{h}\left[\frac{w_2}{h}\left(-0.016\frac{w_1}{h} + 0.064\right) + \frac{0.016 h}{w_1}\right]\frac{Z_{01}\sqrt{\varepsilon_{ef}}}{c} 10^9$$

for $0.5 \le w_1/h \le 2$; $0.5 \le w_2/h \le 2$ and (2.246)

$$\frac{L_2}{h} = \left[\left(0.12\frac{w_1}{h} - 0.47\right)\frac{w_2}{h} + 0.195\frac{w_1}{h} - 0.357 + A\right]\frac{Z_{02}\sqrt{\varepsilon_{ef}}}{c} 10^9$$

$$A = 0.0283 \sin\left(\pi\frac{w_1}{h} - 0.75\pi\right) \qquad (2.247)$$

for $1 \le w_1/h \le 2$; $0.5 \le w_2/h \le 2$. The quantities L_1/h, L_2/h and C/w_2 are given in [nH.m^{-1}] and [pF.m^{-1}], respectively. Similarly as

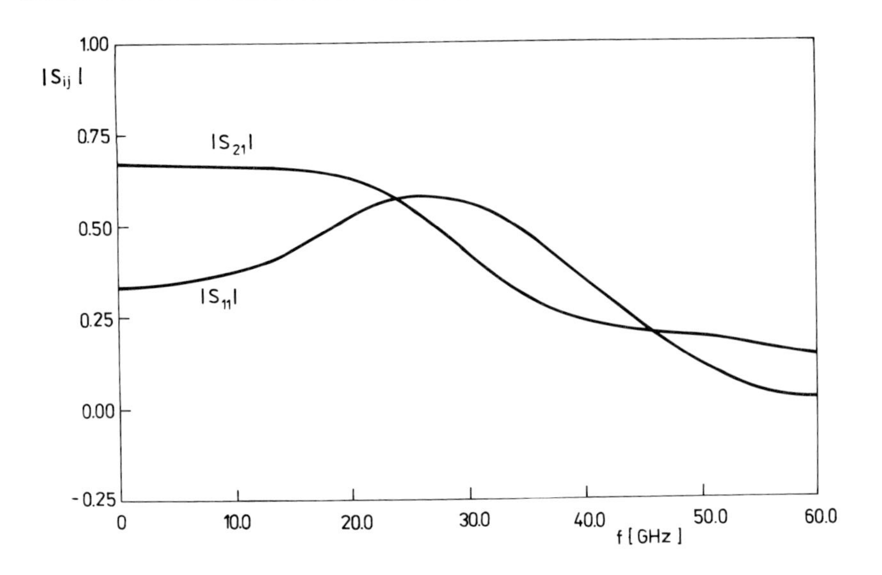

Fig. 2.46 Frequency dependence of S-parameters.

in the preceding discontinuities, here it is also possible to divide the capacitance C to parallel combined capacitances C_1, C_1, C_2 and C' in such a way that the capacitance C_1 corresponds to an identical segment of length Δl_1 of a homogeneous line of the width w_1 as the inductance L_1. The capacitance C_2 and inductance L_2 also correspond to an identical length Δl_2 of the line of a width of w_2. The equivalent diagram from Fig. 2.40 can then be replaced by the other one where the capacitance C' stands instead of the capacitance C and instead of the inductive L_1 and L_2 line segments Δl_1 and Δl_2 are used. The parasitic elements of the discontinuity can be partially compensated by reducing inlet lines of lengths Δl_1 and Δl_2. (The line will be actually mostly lengthened, since the resulting lengths are typically negative.) In a frequency band above 4 GHz, it is also necessary to consider the capacitance C', which is a reason for the frequency dependence of absolute values of S-parameters, Fig. 2.46. This capacitance can be eliminated by shaping the strip in the discontinuity region (Chadha and Gupta, 1982; Gupta, Garg and Bahl, 1979) as shown in Fig. 2.47. This compensation is efficient in a frequency region of about 0 – 12 to 20 GHz (depending on the line dimensions). At higher frequencies radiation from the discontinuity and generation of surface waves occur, which result in increases of losses, which could be characterized, again as in the preceding

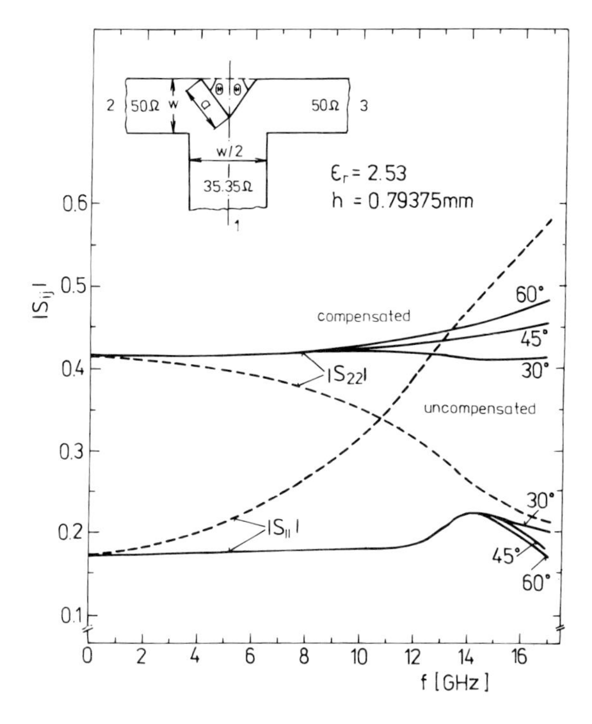

Fig. 2.47 Elimination of capacitance.

discontinuities by the conductivity G parallel with the capacity C'. No compensation for this effect is possible.

Cross junction. This discontinuity occurs in circuits of a similar types a9s in the case of the T-junction. For its equivalent diagram and geometric arrangement see Fig. 2.42. The capacitive element of the equivalent diagram corresponds to the capacitance of the middle part of the discontinuity delimited by phase planes T_1, T_2, T_3 and T_4, which is reduced by a value corresponding to decreasing the charge density in the inlet lines in the vicinity of the discontinuity. In this region, similar changes in the current distribution occur as in the T-junction. Thus, it would be possible to expect that the inductances L_1 and L_2 will be negative and the inductance L_3, which should represent currents in the central part of the discontinuity, will be positive. According to relationships (2.249) and (2.250) the situation is, however, quite the opposite. This paradox results from the fact that for the inductances L_1 and L_2 currents in the central part of the discontinuity are also included and the inductance L_3 is a compensation for this fact. This arrangement is suitable from the standpoint of determining the phase of the signal passing through the discontinuity in a direction in the case of coupling disconnected quarter-wave segments instead of transverse lines. For the capacitance C the approximation formula was derived (Gupta, Garg and Bahl, 1979)

$$\frac{C}{w_1} = \left\{\log\frac{w_1}{h}\left[86.6\frac{w_2}{h} - 30.9\sqrt{\frac{w_2}{h}} + 367\right] + \left(\frac{w_2}{h}\right)^3 + 74\frac{w_2}{h} + 130\right\}$$

$$\cdot\sqrt[3]{\frac{h}{w_1}} - 1.5\frac{w_1}{h}\left(1 - \frac{w_1}{h}\right) + \frac{2h}{w_2} - 240 \qquad (2.248)$$

which holds for $0.3 \le w_1/h \le 3.0$; $0.1 \le w_2/h \le 3.0$; $\varepsilon_r = 9.9$ and which was obtained by a calculation in the quasistatic approximation (Silvester and Benedek, 1973). The following relationships were similarly derived on the basic of quasistatic calculations (Gopinath *et al.*, 1976) for inductive elements

$$\frac{L_1}{h} = \left\{\frac{w_1}{h}\left[163.6\frac{w_2}{h} + 31.2\sqrt{\frac{w_2}{h}} - 11.8\left(\frac{w_2}{h}\right)^2\right] - 32\frac{w_2}{h} + 3\right\}$$

$$\left(\frac{w_1}{h}\right)^{-\frac{3}{2}} \qquad (2.249)$$

$$\frac{L_3}{h} = 5\frac{w_2}{h}cos\left[\frac{\pi}{2}\left(1.5 - \frac{w_1}{h}\right)\right] - \left(1 + \frac{7h}{w_1}\right)\frac{h}{w_2} - 337.5 \qquad (2.250)$$

valid for $0.5 \le w_1/h \le 2.0$. For the inductance L_2 a relationship is valid, which has formally the same form as equation (2.249), however, with changing subscripts 1 and 2 of quantities w_1 and w_2. The quantities C/h and L_i/h are given in [pF.m^{-1}] and in [nH.m^{-1}], respectively. The total capacitance C can be, similarly as in the preceding cases, divided into five components so that capacitance elements C_1 and C_2 are added to L_1 and L_2 and C' remains in the central part of the diagram. The values C_1 and C_2 are chosen in such a way that they form together with L_1, L_2 segments of the line of widths w_1, w_2 and lengths l_1, l_2. This new equivalent circuit can be used for a partial compensation of parasitic elements of the discontinuity by a suitable choice of reference phase planes.

The passage to the new equivalent diagram can seem to be incorrect with respect to the inductance L_3. It is, however, necessary to realize that the original equivalent diagram does not completely correspond to the physical structure of the discontinuity. The concentrated central capacity C also represents an imprecise approximation similarly as the classification according to the above mentioned key.

Fig. 2.48 shows a frequency dependence of the absolute values of S–parameters for a crossed 50–Ω line on a ceramic substrate of a thickness of 0.635 mm. It prevalently results from the radiation and generation of surface waves.

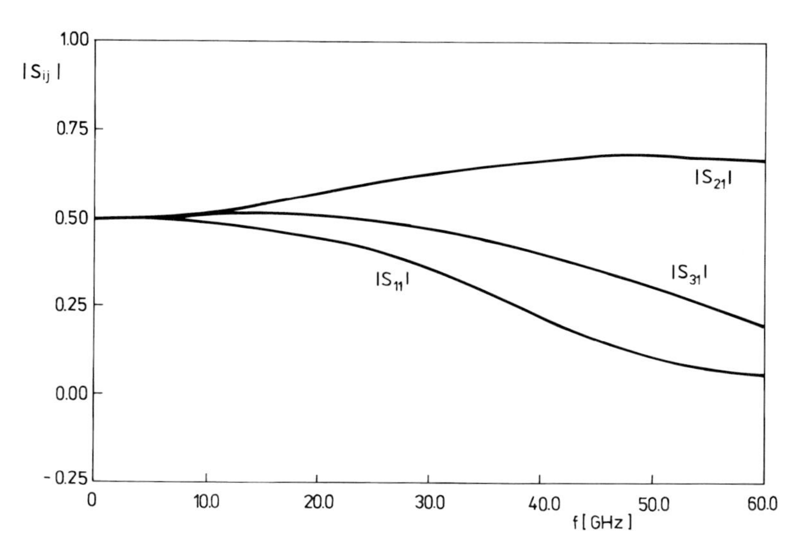

Fig. 2.48 Frequency dependences of $|S_{ij}|$.

The open end a coupled microstrip line. The coupled microstrip lines and their discontinuities are particularly used in filter structures and in different types of coupling circuits. The simplest type of discontinuity of the coupled microstrip line is its open end. For its geometric arrangement and equivalent circuit see Fig. 2.49a.

Similarly as with the open end of a simple line, the scattered electric field is a dominant phenomenon in this case, too. Its effect on the behaviour of the discontinuity can be described by three capacitance elements. The elements C_1 represent the scattering capacitance of ends of particular strips with respect to the earth and the capacitance C_{12} the scattering coupling capacitance between strips. It is advantageous to express the resulting effect by an equivalent extension of the line of the even (Δl_e) and odd (Δl_o) mode (Kirschning and Jansen, 1984)

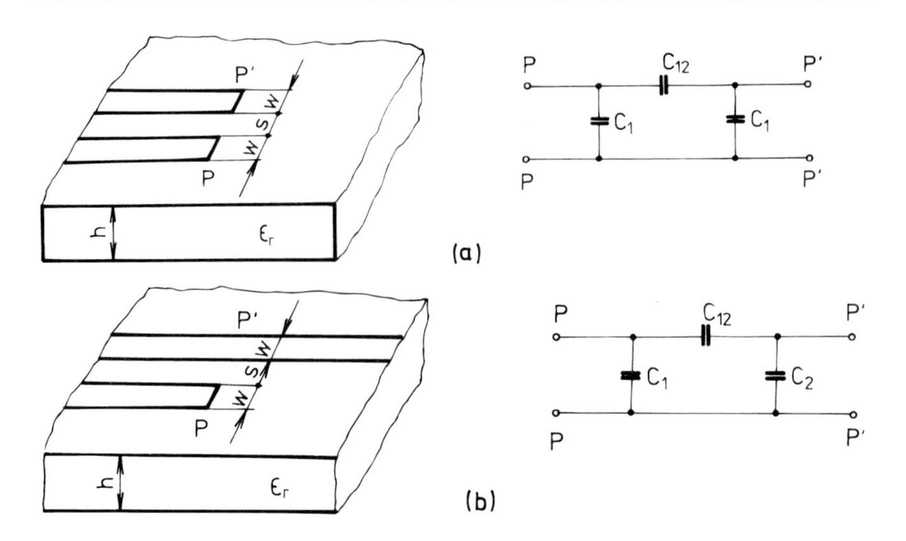

Fig. 2.49 The open end of a coupled microstrip line.

$$\Delta l_e = \left[\Delta l\left(\frac{2w}{h}, \varepsilon_r\right) - \Delta l\left(\frac{w}{h}, \varepsilon_r\right) + 0.0198h\left(\frac{s}{h}\right)^{R_1}\right] e^{-0.328\left(\frac{s}{h}\right)^{2.444}} + \Delta l\left(\frac{w}{h}, \varepsilon_r\right) \tag{2.251}$$

where

$$R_1 = 1.187\left[1 - e^{-0069\left(\frac{w}{h}\right)^2}\right]$$

$$\Delta l_o = \left[\Delta l\left(\frac{w}{h}, \varepsilon_r\right) - hR_2\right]\left(1 - e^{-R_4}\right) + hR_2 \tag{2.252}$$

where

$$R_2 = 0.2974\left(1 - e^{-R_3}\right)$$

$$R_3 = 0.343\left(\frac{w}{h}\right)^{0.6187} + \frac{0.45\varepsilon_r}{1+\varepsilon_r}\left(\frac{w}{h}\right)^{1.357+1.65(1+0.7\varepsilon_r)}$$

$$R_4 = (0.271 + 0.028\varepsilon_r)\left(\frac{s}{h}\right)^{\frac{1.167\varepsilon_r}{0.66-\varepsilon_r}} + \frac{1.025\varepsilon_r}{0.687+\varepsilon_r}\left(\frac{s}{h}\right)^{\frac{0.958\varepsilon_r}{0.706+\varepsilon_r}}$$

The quantity Δl stands for an equivalent extension corresponding to the open end of a simple microstrip line. The accuracy of the relationships is of about 5% for $w/h \in \langle 0.1; 10\rangle$, $s/h \in \langle 0.1; 10\rangle$,

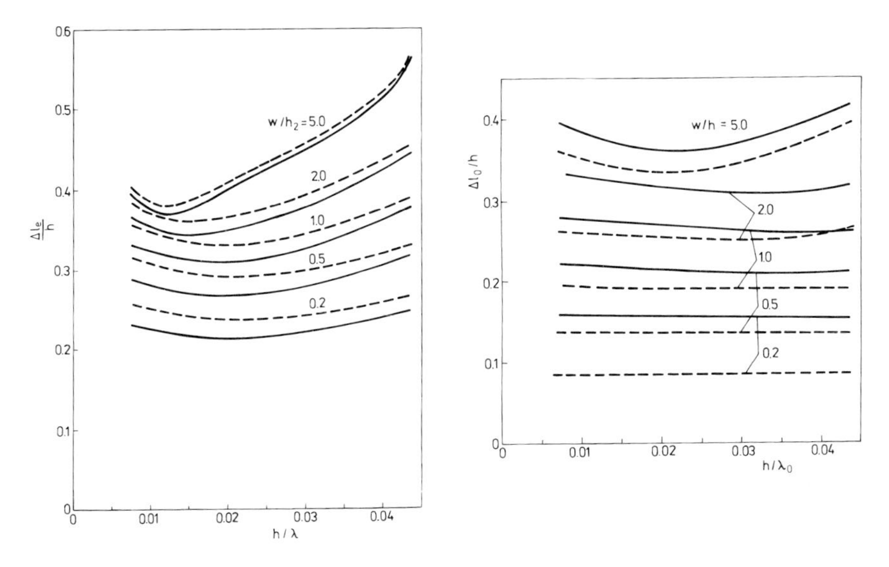

Fig. 2.50 Dependence of parameters.

$\varepsilon_r \in \langle 1; 18 \rangle$. The frequency dependence of elements of the equivalent circuit of this continuity was considered particularly by Jansen (1978; 1975). As can be seen from Fig. 2.50, this dependence in not very considerable in lower frequency bands. It is, however, necessary to consider that the results were obtained for the shielded line so that the effects of the radiation from the discontinuity and of losses resulting from the surface wave generation are not manifested.

The open end of a coupled microstrip line. This discontinuity and its equivalent scheme are shown in Fig. 2.49b. It very frequently occurs in filter structures. The capacitance elements C_1, C_{12} of the equivalent circuit are, similarly as for the open end of a coupled microstrip line, consequences of scattering electric fields. The negative capacitance C_2 results from a decrease of the charge density on the non-ended strip in the discontinuity region. The value of capacity elements of the equivalent diagram are plotted in the graphs of Fig. 2.51 for several widths of strips at $\varepsilon_r = 2.3$ and $\varepsilon_r = 9.6$ and slot widths in a range $0.1 \leq s/h \leq 1$.

The frequency dependences for this type of the discontinuity have not been previously published. It is, however, possible to expect their courses similar to that in the open end of a coupled line.

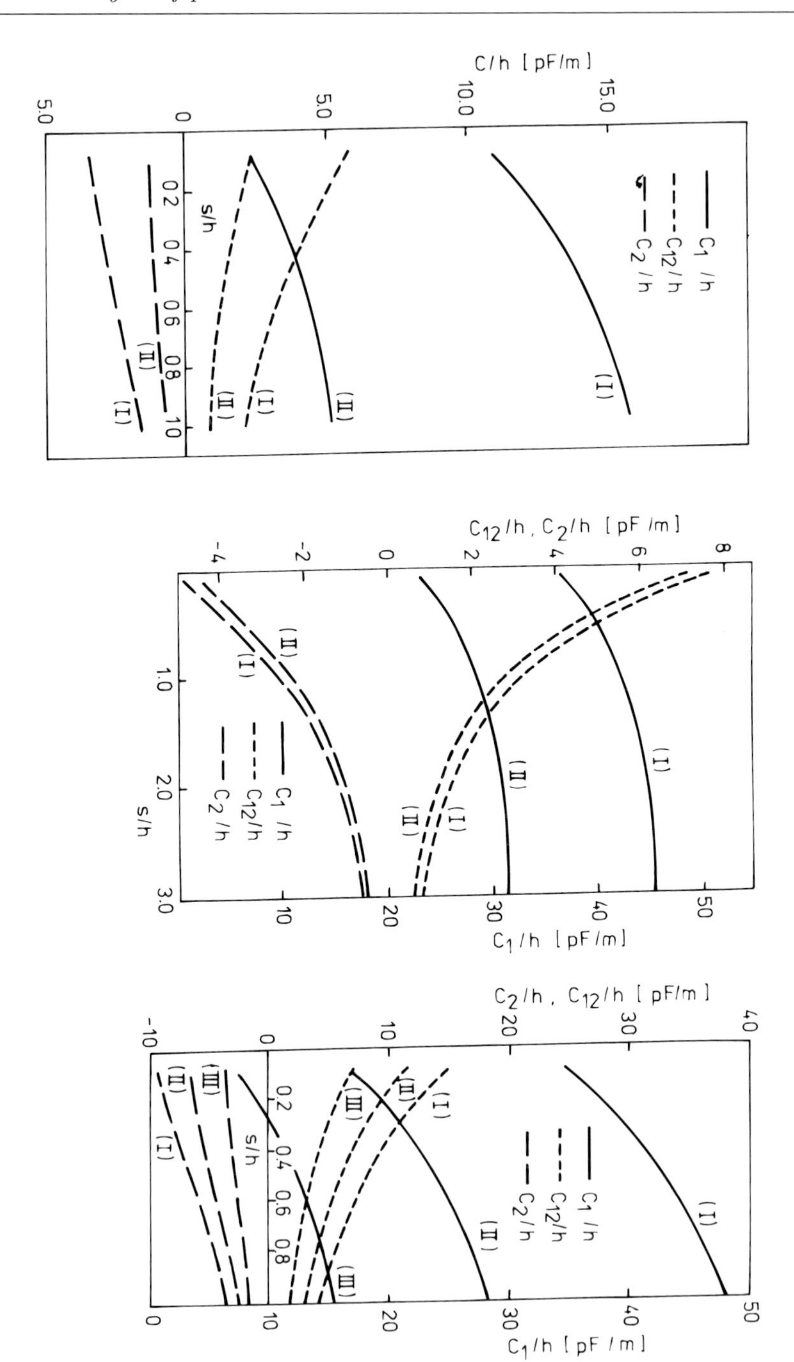

Fig. 2.51 Dependence of capacitive elements.

2.2.4 Discontinuities in other transmission lines for MIC

The following approximative relationship holds for the lengthening of the slot line depending on the frequency and slot width (Jansen, 1984)

$$\frac{\Delta l}{h} = -\left(0.855\frac{s}{h} + 1.283\right)\frac{f}{100} + 0.3545\ln\left(\frac{s}{h} + 0.89\right) + 0.905 \tag{2.253}$$

when $\varepsilon_r = 9.7$, $h = 0.635$, $f \in (3;18)\,\mathrm{GHz}$, $s/h \in \langle 0.2;1\rangle$, f being expressed in GHz. It is assumed that the slot line is covered in a waveguide whose width is of $d = s + 20h$ and its top and bottom walls are separated from the substrate by its tenfold thickness. The accuracy of the approximation is higher than 5% for a slot with dimensions $0.2 \leq s/h \leq 1.0$. In comparison with experiment (Knorr and Saenz, 1973) carried out on open lines the agreement is very good for lines with $s/h \leq 0.5$. For a wider line the measured value of Δl can be as much as by 20% higher than the calculated value.

The short circuit of a coupled slot line. The coupled slot line short circuit is obtained similarly as in the simple slot circuit (Fig. 2.52). From this it follows that the effects occurring in this

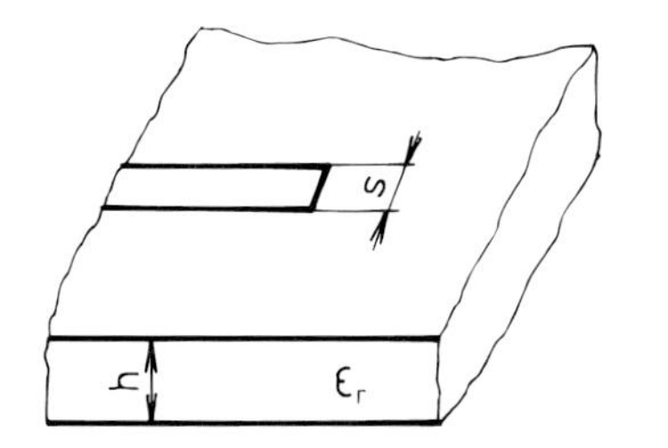

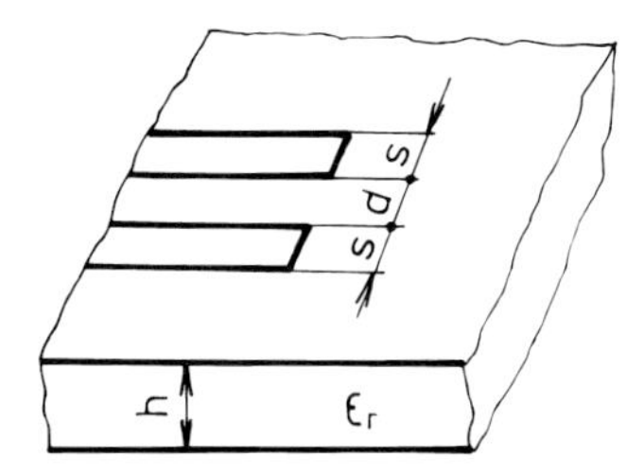

Fig. 2.52 The short circuit of a coupled slot line.

discontinuity are similar to those in the case of the simple slot and thus, the imperfectness of the short circuit is manifested as a line lengthening. The magnitude of this lengthening is different for symmetric and antisymmetric excitations of the coupled line. In antisymmetric excitation the odd mode is propagated through the line. The lengthening Δl_0 of the electric length of the line, as shown in Fig. 2.53a is in this case in a wide range of widths of slots s and

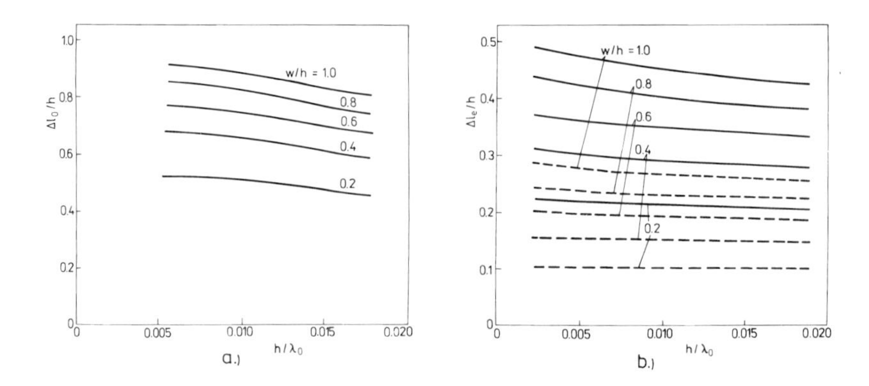

Fig. 2.53 The lengthening $\Delta l_{o,e}$ of the electric length of the line.

of their distances d almost identical with that induced by the short circuit of a simple slot line. Thus, it can be expressed by (2.253).

During the symmetric excitation, the coupled slot line is identical with the coplanar line. The equivalent lengthening of the line corresponding to the short circuit is several times smaller than that during the antisymmetric excitation, however, it is strongly dependent on the distance d between the slots. The frequency dependence of the line lengthening Δl_e is demonstrated in Fig. 2.53b.

The open end of the line on a suspended substrate. For the geometric arrangement of this discontinuity see a schematic diagram in Fig. 2.54a. It is actually an open end of the line on a two-layer substrate covered by shielding. In contrast to the open microstrip line losses by radiation and generation of surface waves cannot occur. In a higher frequency band, the effect of screening is, however, implemented in the generation of higher modes. For a typical frequency

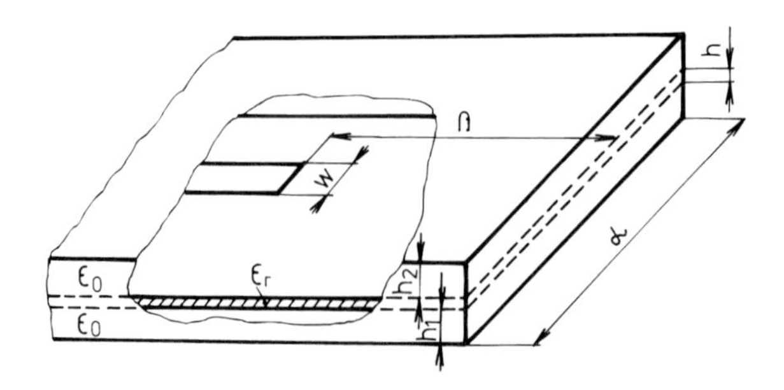

Fig. 2.54 (a) The open end of the line on a suspended substrate.

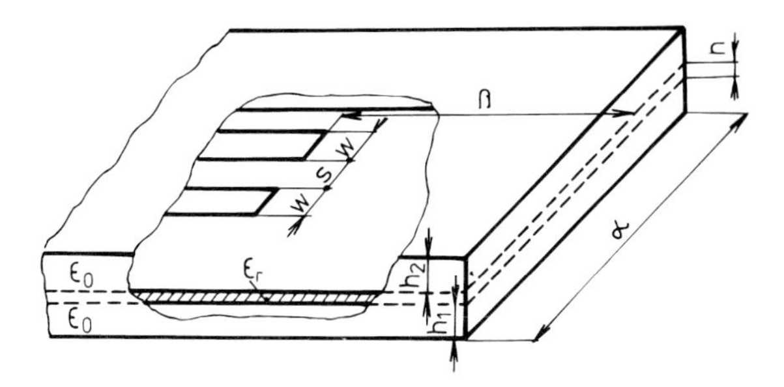

Fig. 2.54 (b) The open end of coupled lines on a suspended substrate.

dependence of the equivalent lengthening of the line corresponding to the scattering capacity of the open line see Fig. 2.55 (Jansen, 1978).

The open end of coupled lines on a suspended substrate. This discontinuity is shown in Fig. 2.54b. Similarly as in microstrip coupled lines, scattering fields are also formed inducing effects, which can be characterized by the line prolongation. In the symmetric excitation, the even mode will propagate through the line. In this case, the equivalent lengthening of the open end is almost independent of the slot s between the strips (Fig. 2.56a). The equivalent lengthening during the non-symmetric excitation, i.e. in the case of the odd mode propagation, is considerably smaller than that in the case of the even mode, however, with the presence of a strong dependence on the slot size s. The magnitude of equivalent lengthening of the line and of the frequency dependences can be seen from graphs (Jansen, 1978) shown in Fig. 2.56b. In contrast to coupled microstrip lines, due to the shielding, neither radiation losses nor losses resulting from the surface wave generation are implemented. At a higher frequency, the effect of higher modes in the shielded waveguide is, however, manifested.

The finline step discontinuity. This type of discontinuity belongs to the most commonly occurring discontinuities in circuits operating in a region of milimetre waves. It is usually a part of matching circuits, impedance transformers and filter structures. Its geometric arrangement is in Fig. 2.57a. The equivalent schematic diagram in part (b) of the Figure consists of two segments of the line Δl_1 and Δl_2 with slot widths s_1 and s_2 and the parallel capacitance C. The line segments are used to partially compensate for the discontinuity effect.

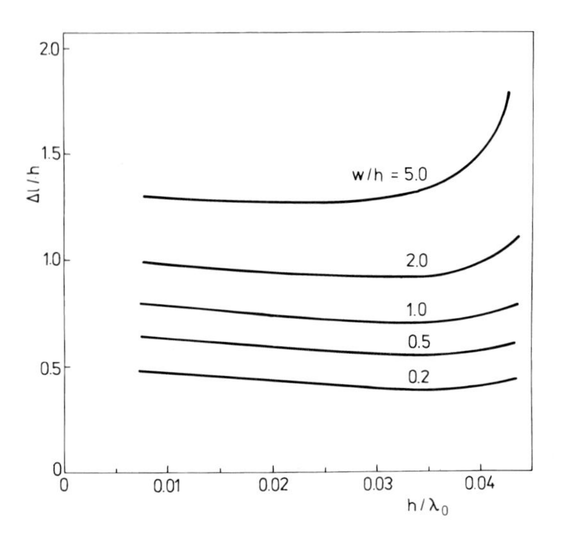

Fig. 2.55 Typical frequency dependence of the equivalent lengthening of the line.

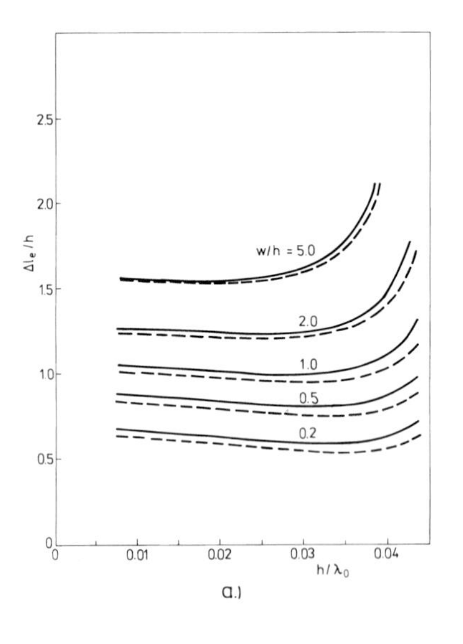

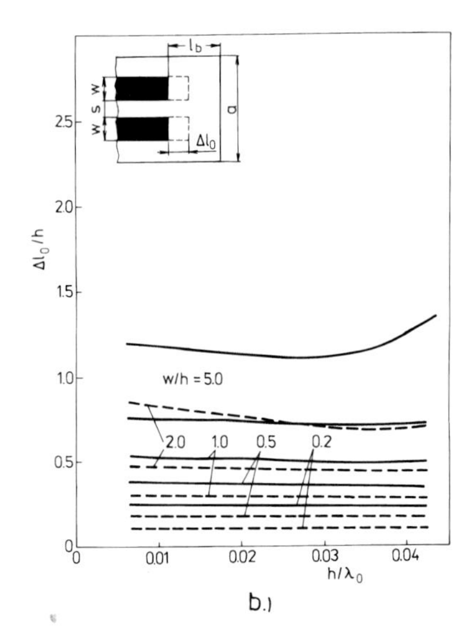

Fig. 2.56a Typical frequency dependence of the equivalent lengthening of the coupled lines.

Fig. 2.56b The magnitude of equivalent lengthening of the coupled lines.

The frequency dependences of elements of the equivalent diagram (Helard *et al.*, 1985) in Fig. 2.57b for the substrate $\varepsilon_r = 2.22$ and thickness $h = 0.254$ mm situated in the R28 waveguide see Fig. 2.58 for three different dimension variations of slots.

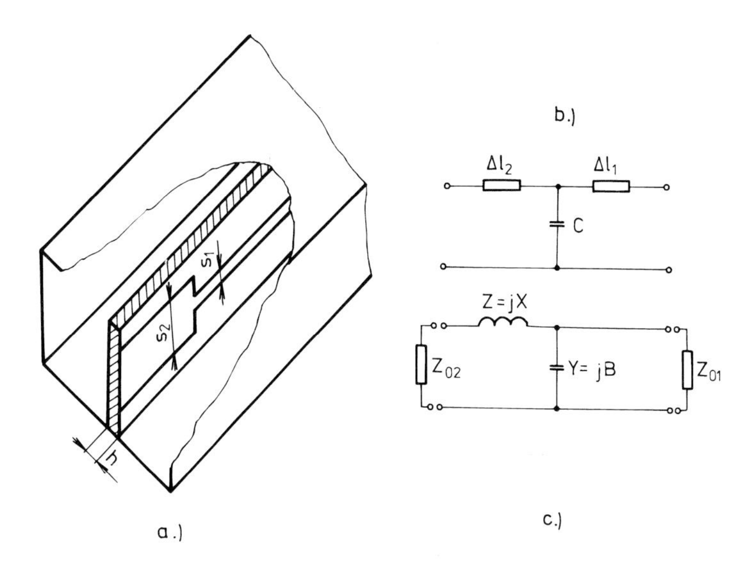

Fig. 2.57 The finline step discontinuity.

For elements of the equivalent diagram shown in Fig. 2.57c with elements corresponding to segments Δl_1 and Δl_2 included in the impedance Z with the capacity C corresponding to the admitance Y the following approximation relationships were derived (Pic and Hoefer, 1981)

$$
\begin{aligned}
Z_{01} &= 2\frac{\sqrt{s_1 s_2}}{\lambda_{g2}} \log \frac{2s_2}{\pi s_1} \\
\frac{X}{Z_{01}} &= 0.4\pi \frac{s_2 - s_1}{\lambda_{g1}}
\end{aligned}
\tag{2.254}
$$

The meaning of variables s_1, s_2 is obvious from Fig. 2.56. Z_{01} is the characteristic impedance of a line with the slot width s_1, λ_{gi} is the wave length on a line with the slot width s_i, B is the susceptance of the element Y and X is the reactance of the element Z in the equivalent scheme.

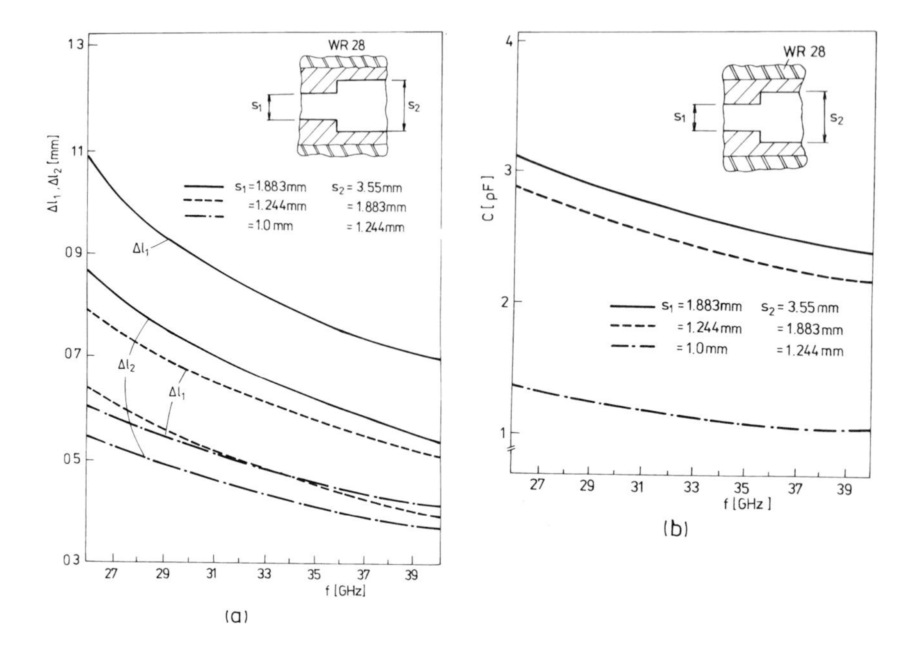

Fig. 2.58 The frequency dependences of elements of the equivalent diagram in Fig.2.57b for different dimension variations of slots.

The interrupted slot of a finline. For the geometric arrangement of the discontinuity together with the equivalent diagram see Fig. 2.59.

The dominant effects are currents originating in a conductive band interrupting the slot of the line. These currents induce a magnetic field. From the standpoint of the circuit, this discontinuity can be replaced by a T-circuit consisting of inductive elements. As can be seen from Fig. 2.60a, which demonstrates a typical course of normalized reactances through the relevant element of the equivalent diagram, depending on the strip width (Schiavon *et al.*, 1988), the coupling decreases with increasing strip width down to zero. Then the circuit is manifested as a perfectly short-circuited one, however, in a plane shifted with respect to the plane of the line slot interruption. This shift corresponds to the normalized reactance X_s/Z_0. As an illustration, its dependence on the frequency is shown in Fig. 2.60b for several widths of the strip. As shown in (Knorr, 1981), this reactance exerts on only weak dependence on the thickness h of the dielectric substrate.

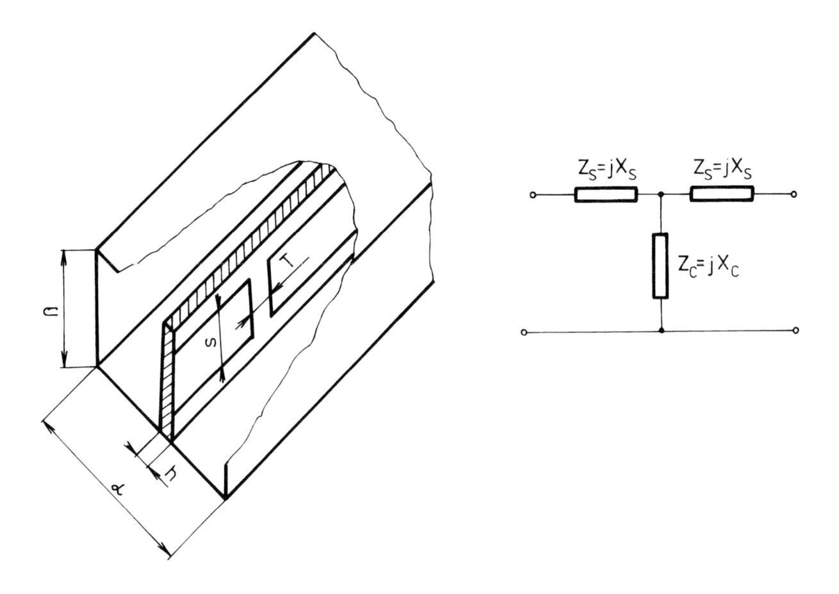

Fig. 2.59 The interrupted slot of a finline.

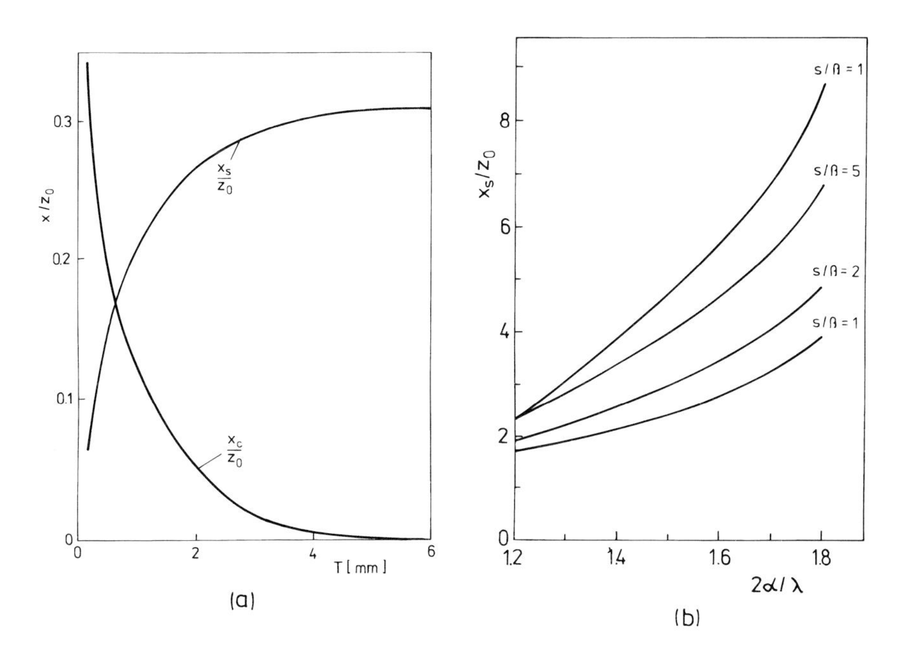

Fig. 2.60 Normalized reactances versus (a) strip width (b) frequency.

2.3 LUMPED ELEMENTS

2.3.1 Use of lumped elements

In microwave technique circuits are prevalently constructed on the basis of elements of lines with distributed parameters, however, in these elements, the dimensions are so large that they cannot be considered as negligible with respect to the wavelength of the signal processed. Thus, the lumped elements, where the dimensions can be neglected with respect to the wavelength have traditionally been used in frequency bands below 1 GHz. The reason for this is technology requirements of the production of miniature parts for higher frequency bands. With improving technology possibilities and with the development of monolithic microwave circuits a possibility, however, occurs of the miniaturization of electronic elements and thus also the extension of the use of lumped elements to the region of millimetre waves.

The lumped elements have several advantages in comparison with elements with distributed parameters. These are first of all an essentially lower frequency dependence, which makes possible an easier design of the wide-band circuits. The further advantages are miniature dimensions making possible the miniaturization of circuits. This is also connected with the third advantage that a lower energy is concentrated in the element region. This results in decreasing losses in dielectric media in certain types of lumped elements. Thanks to this in these elements it is principally possible to use even dielectric substrates of a lower quality. This possibility is, however, not frequently used in practice, since the lumped elements are prevalently produced on substrates together with line elements with distributed parameters. In these cases, it is necessary to use material suitable for all the components used.

2.3.2 Capacitive elements

The simplest capacitor in the MIC is the interrupted line as described in Section 2.2.3. In its simple form, it can, however, yield only very low values of the series capacitance. These values could be increased on account of an extreme reduction of the slot width s, which is, however limited by technology possibilities or by enlarging the slot

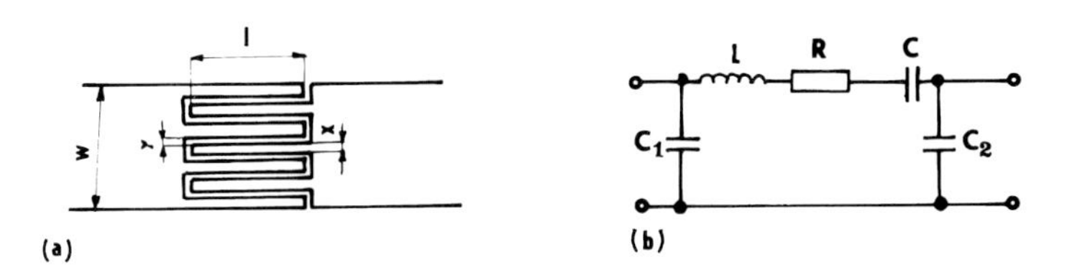

Fig. 2.61 Capacitive element (a) interdigital structure (b) equivalent diagram.

length, which would mean in this form a larger width of the strip w and thus also a change of the line impedance.

The enlargement of the slot length can, however, be achieved by changing the configuration of the slot interruption of the line — by forming an interdigital structure as shown in Fig. 2.61a. In this structure, it is, however, to expect several parasitic effects depreciating characteristics of the capacitor. For an equivalent diagram of this element see Fig. 2.61b. The dominant element is the capacitance C, which can be expressed by the approximate relationship (Gupta *et al.*, 1981)

$$C = [(N-3)\,0.0886 + 0.0993]\,\frac{\varepsilon_r + 1}{w}\,l \qquad [\text{pF; cm}] \qquad (2.255)$$

where N stands for the number of the structure digits. The importance of the quantity 1 is obvious from Fig. 2.61a.

Besides this dominant capacitance, parasitic elements occur as mentioned above. The capacitances C_1 and C_2 are given by the scattering capacitance with respect to the earthing metallization and their magnitude decreases with increasing substrate thickness, the resistance R resulting from a finite conductivity of metallic components of the element and the inductance L are due to the fact that the interdigital structures exert an inductive nature with respect to their non-zero length. At a higher frequency the parasitic elements exert effects simulating the frequency dependence of the capacitance C. A further increase of the frequency leads to the occurrence of the distributed nature of the element so that the frequency dependence is even potential. The frequency at which the effects other than the dominant concentrated capacitance C occur depends on the w/l ratio (Pengelly *et al.*, 1977).

The above mentioned configuration makes it possible to obtain capacitors with a capacity of several pF. The other method of obtaining microwave capacitors, which also makes it possible to achieve a higher capacity, is the construction of layer capacitors. They are used essentially in three basic types. These are capacitors of the type MIM (metal-insulator-metal), MIS (metal-insulator-semiconductor) and ceramic capacitors, which are actually a certain modification of the MIN capacitors.

The MIM capacitors are most frequently used in monolithic microwave integrated circuits. They are formed by two metallic layers separated from each other by a thin insulting oxide layer. In certain cases, multilayer structures can also be used for achieving a higher capacitance. Chip capacitors for the hybrid microwave integrated technique are, however, not produced in this form for technology reasons.

Capacitors of the MIS type are prevalently used for these applications. One of their electrodes is formed by a high-doped semiconductor with a non-rectifying metallic contact. The other electrode, separated by an insulation layer usually of silicon dioxide, is formed by a metallic layer. Similarly as in the simple MIM structure a capacitance as high as 500 pF is achieved with an electrode area of about $1\,\text{mm}^2$. They can be used in the whole range of centimetre waves. The frequency characteristics are, however, due to necessarily complicated contacting on the circumference, deteriorated in comparison with the MIM capacitors. The losses are also higher by losses in the high-doped semiconductor. The frequency and voltage dependence of the capacitance is, in addition, partially deteriorated by producing a spatial charge in the semiconductor region.

Ceramic capacitors for the hybrid microwave integrated technique have capacities of about 1000 pF. They are similar to the MIM capacitors the insulating layer being formed by a thicker layer of a dielectric with a high permittivity, which simultaneously form the supporting part of the chip. The dielectric is usually based on titanium dioxide. The higher capacitance is achieved on account of increased losses, which increase with increasing frequency. The frequency region, where these capacitors are applicable, depends on the dielectric used. For low requirements it, however, typically does not exceed the X band and for circuits with higher requirements the S band.

The capacitance of a simple rectangular layer capacitor can be quantitatively determined by considering it as a segment of a microwave homogeneous line with end scattering capacitances. The capacitance

$$C_{w,l} = \frac{l}{v_w Z_w} \tag{2.256}$$

corresponds to the homogeneous segment of the line of a width w and length l, where Z_w and v_w are the characteristic impedance and rate of propagation, respectively in the line of a width w.

Similarly for the line of a width 1 and length w it holds

$$C_{l,w} = \frac{w}{v_l Z_l} \tag{2.257}$$

where Z_1 and v_1 are the characteristic impedance and velocity of propagation in the line of a width 1, respectively.

From the relationships (2.256) and (2.257) it is possible to determine the capacitance C_{rw} and C_{lw} corresponding to sides of a length of w and l, respectively, so that

$$\begin{aligned} C_{rl} &= \frac{l}{2}\left(\frac{1}{v_w Z_w} - \frac{\varepsilon w}{h}\right) \\ C_{wl} &= \frac{w}{2}\left(\frac{1}{v_l Z_l} - \frac{\varepsilon l}{h}\right) \end{aligned} \tag{2.258}$$

where h is the dielectric thickness. With increasing thickness h and decreasing dimensions w and l the accuracy of relationship (2.258) decreases due to increasing mutual affecting of the charge distribution on capacitor edges.

At sufficiently high w/h and l/h ratios the capacitor capacitance can be expressed in the form

$$C = \frac{l}{v_w Z_w} + \frac{w}{v_l Z_l} - \varepsilon \frac{wl}{h} \tag{2.259}$$

The relationships for v_l, v_w, Z_l and Z_w are presented in Section 2.1. For a circular capacitor of a diameter r, for which $r/h > 1$, the following relationship was derived based on an assumption that the scattered fields are similar to those for a rectangular capacitor.

$$C = \frac{\varepsilon_0 \pi r^2}{h}\left\{1 + \frac{2h}{\pi r}\left[ln\left(\frac{\pi r}{2h}\right) + 1.7726\right]\right\} \tag{2.260}$$

When connecting the capacitor into a circuit, parasitic elements occur in addition to the capacitance C. For their arrangement in an

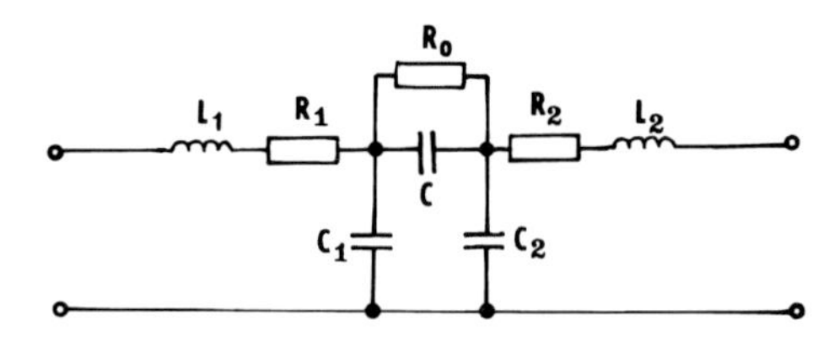

Fig. 2.62 Parasitic elements of capacitor connecting into circuit.

equivalent circuit see Fig. 2.62. The resistances R_1, R_2 correspond to losses in inlet lines and in capacitor electrodes. The resistance R_0 corresponds to losses in the dielectric. The inductances L_1, L_2 result from the contacting line. The capacitances C_1 and C_2 are scattering parasitic capacitances with respect to the earthing metallization of the circuit.

These parasitic elements, together with the fact that at higher frequencies non-zero dimensions of capacitors start to be manifested, lead to a frequency dependence of the capacitance of the capacitor with increasing frequency.

2.3.3 Inductive elements

The simplest inductive element used in many microwave integrated circuits is a thin wire used for example to connect chip elements in the circuit. The precise consideration of parameters of this element in the equivalent circuit is very difficult and problematic with respect to the fact that its shape is usually not precisely defined. For providing an estimate of the effect of the contacting wire on circuit characteristics it is possible to use its approximation by a segment of the homogeneous line with a certain cross-section (Fig. 2.63). When solving this structure by a conformal representation, then (Caverly,

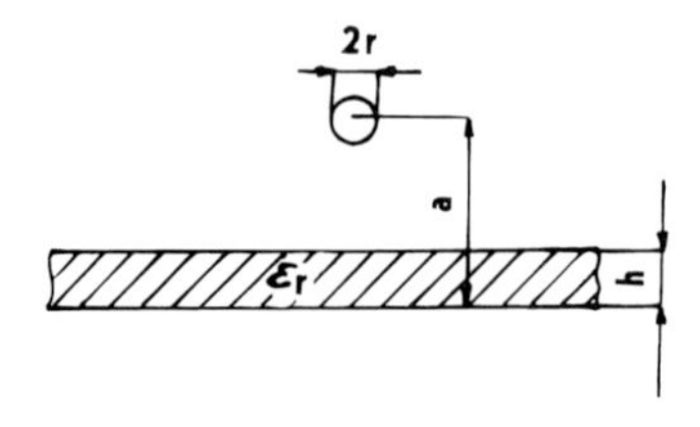

Fig. 2.63 Inductive element.

1986)

$$Z_0 = \frac{60}{\sqrt{\varepsilon_{ef}}}\left\{\cosh\left[\left(1-x^2\right)\frac{1}{2y-\frac{y}{2}}\right]\right\}^{-1}$$
$$x = \frac{1}{\left(\frac{2a}{r}\right)^2 - 1} \tag{2.261}$$
$$y = \frac{2}{\frac{4a}{r} - \frac{r}{a}}$$

where ε_{ef} is given by the expression

$$\varepsilon_{ef} = \frac{\ln\left(\frac{2a}{r}\right)}{\ln\left[\frac{2(a-h)}{r} + \frac{2h}{r\varepsilon_r}\right]} \tag{2.262}$$

The explanation of symbols a, h, r can be seen in Fig. 2.63.

With respect to given dimensions of the element, in addition to the distributed representation, it is possible to use the lumped representation. The equivalent circuit of the contacting wire is formed by a series connection of the inductance L and resistance R. For the inductance of the wire of a length L and diameter r in the free space it is possible to write

$$L = 0.2l\left[\ln\left(\frac{2l}{r} - 1\right)\right] \qquad \text{[nH;mm]} \tag{2.263}$$

The resistance R, which results from conductivity losses can be expressed by

$$R = \frac{R_s l}{\pi r} \tag{2.264}$$

where R_s is the surface resistance. The quality factor Q is determined by the equation (Gupta *et al.*, 1981)

$$Q = \frac{2\pi f L}{R} \tag{2.265}$$

When considering the effect of an earthing plane at a distance a from the wire, then for the inductance L (Gupta *et al.*, 1981)

$$L = 0.2l\left\{\ln\frac{2a}{r} + ln\left[\frac{1+\sqrt{l^2+r^2}}{1+\sqrt{l^2+4a^2}}\right] + \sqrt{1+\left(\frac{2a}{l}\right)^2}\right.$$

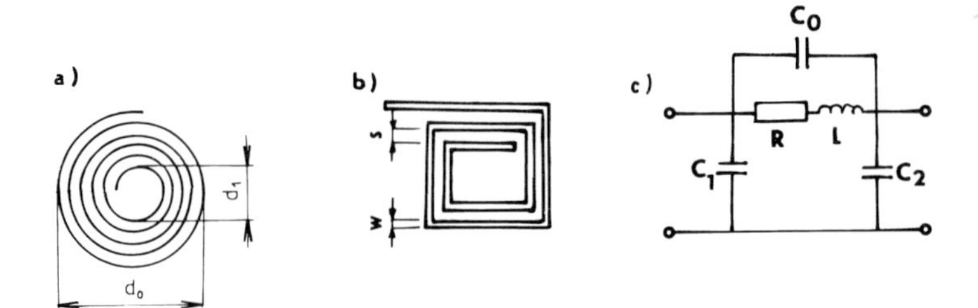

Fig. 2.64 Spiral configuration.

$$+\frac{r}{l}-\frac{2a}{l}-\sqrt{1+\left(\frac{r}{l}\right)^2}\Bigg\} \tag{2.266}$$

With thin wire, it is possible to obtain only very low inductance values — of about 1 nH. For obtaining higher values or for forming the inductance with a fixed magnitude corresponding to the design, a spiral configuration is used as shown in Fig. 2.64a. When considering this conductor in free space and when solving the problem in the quasistatic approximation, its inductance can be expressed in the form (Gupta *et al.*, 1981)

$$L = 0.3939\pi n^2\alpha\left\{\ln\frac{8\alpha}{\beta}+\frac{1}{24}\left(\frac{\beta}{\alpha}\right)^2\left(\ln\frac{8\alpha}{\beta}+3.583\right)-0.5\right\} \tag{2.267}$$

where

$$\alpha=\frac{d_0+d_1}{4}, \qquad \beta=\frac{d_0-d_1}{2}$$

The symbol n stands for the number of turns and the symbols d_0 and d_1 (mm) are explained in Fig. 2.64a.

Equation (2.267) can be within a accuracy range of several per cent replaced by a simple relationship (Gupta *et al.*, 1981)

$$L=\frac{39.39n^2\alpha^2}{8\alpha+11\beta} \tag{2.268}$$

The loss resistance is

$$R=K\pi n\alpha\frac{R_s}{w} \tag{2.269}$$

where w (mm) is the line width and K is a correction factor whose magnitude ranges between 1 and 2.5 and depends on the line thickness. The quality coefficient is determined from relationships (2.268) and (2.269)

$$Q = \frac{78.77 f w n \alpha}{K R_s (8\alpha + 11\beta)} \tag{2.270}$$

where f [GHz] is the frequency.

In an equivalent circuit, the parasitic capacitance elements shown result from the capacitive coupling between conductors (capacitor C_o) and capacitive coupling to the earthing metallization (capacitors C_1 and C_2). In practice square-shaped spirals are used as the spiral configuration shown in Fig. 2.64a (see Fig. 2.64b). The reason for this is on the one hand in a more easy implementation and, on the other hand in the fact that with the same number of turns rather higher inductance is achieved which means space saving in the integrated circuit. For more precise proposals, particularly in the high-frequency region, it is necessary to analyse spiral inductive elements as those with distributed parameters and then, on the basis of numerical calculations for a particular configuration to obtain by an optimization the values of lumped elements in the equivalent diagram. These methods are shown for example in Cabana (1983) and Parisot (1984). With the help of these methods, it is possible to design spiral inductances which agree with the experiment up to the frequency Ku band (see Fig. 2.65).

2.3.4 Resistive elements

In the microwave frequency region, this type of element is not used as much as in the low-frequency region. The reasons for this are specific features of the use of microwave signals. In most circuits of the centimetre band and of shorter waves, there is a considerable attempt to obtain a maximum use of signals with minimum possible losses. This is also connected with restricting the use of loss elements in these types of circuits. They are, however, irreplaceable in many cases. Most frequently, these are reflectionless ends, serving for suppressing undesirable signals or damping elements providing the separation of circuits from each other or reducing the power level. They are, however also used in the other applications.

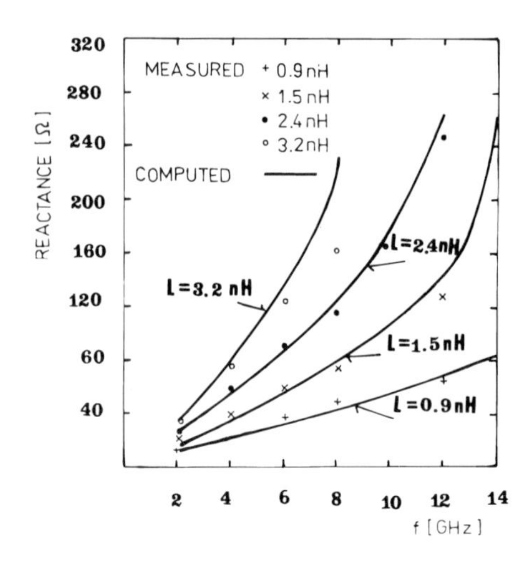

Fig. 2.65 Computed and measured results of dependence reactance versus frequency for spiral inductance.

The most frequently used type of resistive element in the MIC is a miniature line segment, where the metallization is formed by a loss material. The dimensions must be negligible in comparison with the wavelength to save the assumption of lumped elements.

For example of these elements see Fig. 2.66. The magnitude of the resistance for a rectangular configuration is given by the relationship

$$R = R_s \frac{l}{w} \tag{2.271}$$

where R_s is the resistance of the square of a given resistance layer, l is the length and w is the width of the resistive line.

In attempts to reduce parasitic phenomena accompanying discontinuities, which can be in certain cases connected with resistive elements, the other shapes of resistors are also used (Volman, 10982). For the trapezoidal type it is easy to derive the value of the resistance in the form

$$R = \frac{R_s l}{w_2 - w_1} \ln \frac{w_2}{w_1} \tag{2.272}$$

where R_s and l have the same meaning as in relationship 2.271 w_1 is the smallest width and w_2 is the largest width of the resistive line segment.

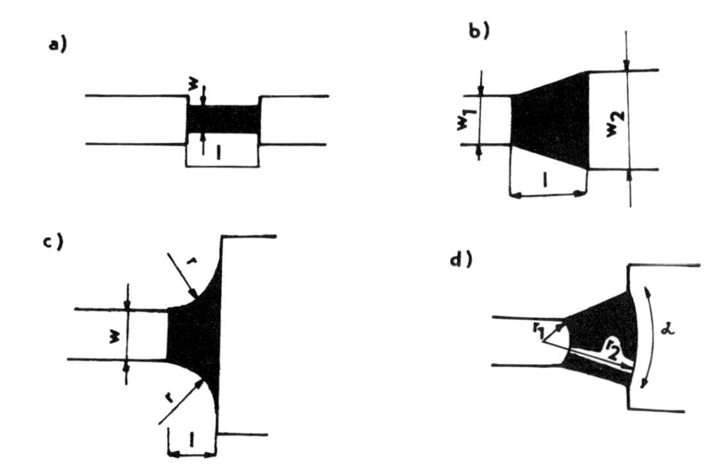

Fig. 2.66 The example of resistive elements in the MIC.

In the case of the resistance in the form shown in Fig. 2.66c the resistance value is of

$$R = R_s \left\{ \frac{1 + \frac{w}{2r}}{\sqrt{\left(1 + \frac{w}{2r}\right)^2 - 1}} \cdot \right. \tag{2.273}$$

$$\left. \cdot \arctan \left[\frac{2 + \frac{w}{2r}}{\sqrt{\left(1 + \frac{w}{2r}\right)^2 - 1}} \tan \left(\arcsin \frac{l}{r} \right) - \frac{1}{2} \arcsin \frac{l}{r} \right] \right\}$$

where symbols R_s, w and l stand for the same quantities as in preceding expressions and r is the radius of circles forming the resistance margins. The resistance in the form according to Fig. 2.66 is of

$$R = R_s \frac{\pi \alpha}{180} \ln \frac{r_2}{r_1} \tag{2.274}$$

where α, r_1, r_2 are dimensions obvious from the figure. All these relationships can be obtained by a simple integration.

Resistive elements of the same shapes as in Fig. 2.66 can be formed not only directly by the thin-layer or thick-layer technology in the microwave circuit but also in the form of chips which are subsequently inserted into the circuits.

Similarly as the other passive lumped elements, the resistors also exert features with increasing frequency suggesting the presence of

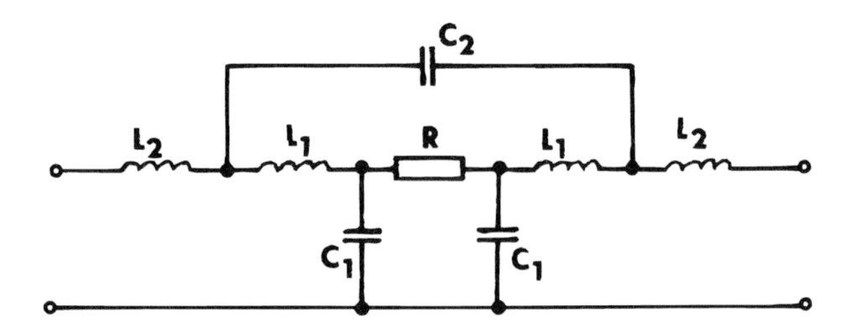

Fig. 2.67 Equivalent circuit of resistive element in the MIC.

parasitic elements. For a general equivalent scheme of the resistance with parasitic elements see Fig. 2.67. The capacity C_1 is particularly manifested, whose magnitude corresponds to a resistive line segment. The inductance L_1 corresponds to the inductance of this line. The capacity C_2 results from increasing the charge density in the region of discontinuities accompanying the connection of the resistor into the external circuit. The inductances L_2 manifested particularly when inserting chips are of a similar origin, i.e. due to the connection into the circuit.

For miniature resistive elements in the MIC, whose length does not exceed 0.2 mm, it is however, possible to consider with a very good approximation that in a band up to 18 GHz they represent a resistance without considerable parasitic effects. A rather larger parasitic effect is encountered in more complex resistive elements as reflectionless terminations and attenuating elements. The most common types of configuration of reflectionless terminations are shown in Fig. 2.68. For the reflection coefficient of a reflectionless termination it is possible to write

$$\Gamma = \frac{R + Z_1 - Z_0}{R + Z_1 - Z_0} \tag{2.275}$$

Where R is the resistive element resistance, Z_1 is the impedance of the load connected and Z_o is a characteristic impedance of the terminator line. The load Z_1 equals a perfect short-circuiting in an ideal case. It is implemented either by a quarter-wave open segment of the line in the case of a narrow-band element or by a short circuit formed by a metallized opening in the dielectric substrate or by its metallized edge in the case of a wide-band terminating resistance.

The wide-band short-circuit always exerts a moderately inductive nature, since it is formed by a miniature short-circuited segment of

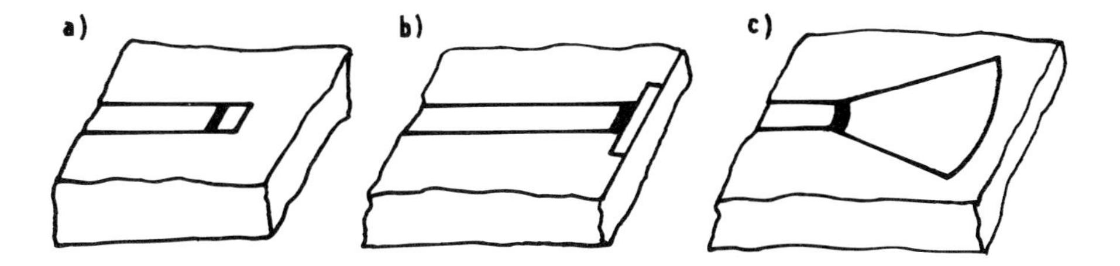

Fig. 2.68 The most common types of configuration of reflectionless terminations.

the line. It can, however, also be implemented in such a way that this parasitic element will not be considerably manifested in a band up to 18 GHz.

In some cases, the use of the direct short-circuit is, however, not suitable. This can occur either from electric reasons or due to difficult implementation. Then it is possible to use as the short-circuit a quarter-wave disconnected segment of the line. In this case the impedance is of

$$Z_1 = Z_0' j \tan \gamma l \tag{2.276}$$

where γ is the propagation constant, l is the line length and Z_0' is the characteristic impedance of this quarter-wave segment. It is obvious that for achieving as large width of the band as possible it is necessary to use the line with as low Z_0' as possible i.e. with the largest width. This, however, brings a further increase of parasitic elements.

Instead of segments of the microstrip line, segments of radial lines are frequently used. The parasitic elements are reduced in this way and, in addition, with respect to features of this line the band width is enlarged. The dimensions can be calculated from relationships presented in Chapter 4.

Damping elements are more complex than reflectionless terminations. They consist of several resistive elements selected and arranged in such a way that they could be adjusted for the given impedance with providing the attenuation of the preceding signal as required. The connection of three resistors into a π-element or into a T-element is most frequently used (Fig. 2.69). From the above mentioned requirements it is easy to derive that when neglecting parasitic elements the following relationships hold for values of resistance of

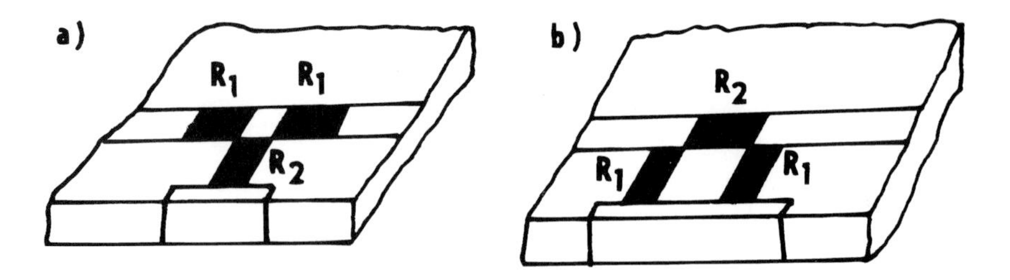

Fig. 2.69 The connection of three resistors in to a) π-element or in to b) T-element.

T-elements:

$$\frac{R_1}{Z_0} = -\frac{\sqrt{G}-1}{\sqrt{G}+1} \qquad \frac{R_2}{Z_0} = \frac{2\sqrt{G}}{1-G} \tag{2.277}$$

and for resistances of π-elements

$$\frac{R_1}{Z_0} = -\frac{\sqrt{G}+1}{\sqrt{G}-1} \qquad \frac{R_2}{Z_0} = -\frac{G-1}{2\sqrt{G}} \tag{2.278}$$

where G is the gain required (lower than 1).

With increasing frequency, all the parasitic effects start to be implemented, connected with MIC resistances and with microwave losses. In addition, similar effects as in the T-junction occur, as mentioned in the Chapter during with the line discontinuities.

REFERENCES 2.1

Bryant T.G., Weiss J.A. (1968) Parameters of Microstrip Transmission Lines and of Coupled Pairs of Microstrip Lines, *IEEE Trans. Microwave Theory Tech., MTT-16*, 12, pp. 1021–7.

Cohn S.B. (1969) Slotline on a Dielectric Substrate, *IEEE Trans. Microwave Theory Tech., MTT-17*, 10, pp. 768–78.

Garg R., Gupta K.C. (1976) Expression for Wavelength and Impedance of Slotline, *IEEE Trans. Microwave Theory Tech., MTT-24*, 8, p. 532.

Ghione G., Naldi C.U. (1987) Coplanar Waveguides for MMIC Applications: Effect of Upper Shielding, Conductor Backing, Finite-Extent Ground Planes, and Line-to-Line Coupling, *IEEE Trans. Microwave Theory Tech., MTT-35*, 3, pp. 260–7.

Green H.E., (1965) The Numerical Solution of Some Important Transmission Line Problems, *IEEE Trans. Microwave Theory Tech., MTT-13*, pp. 676–92.

Gupta K.C., Garg R., Bahl I.J. (1979) *Microstrip Lines and Slotlines*, Dedham, Massachusetts: Artech House.

Hammerstadt E., Jensen O. (1980) Accurate Models for Microstrip Computer-Aided Design, *IEEE MTT-S Int. Symp. Dig.*, Washington, D.C., pp. 407–9.

Harrington R.F. (1968) *Field computation by moment methods*, Macmillan, New York.

Hoefer W.J.R. (1985) The Transmission-Line Matrix Method - Theory and Applications, *IEEE Trans. Microwave Theory Tech., MTT-33*, 10, pp. 882–93.

Hoffman R.K. (1983) *Integrierte Mikrowellenschaltungen*, Springer-Verlag, Berlin Heidelberg New York Tokyo.

Jansen R.H. (1978) High-Speed Computation of Single and Coupled Microstrip Parameters Including Dispersion, High-Order Modes, Loss and Finite Strip Thickness, *IEEE Trans. Microwave Theory Tech., MTT-26*, 1, pp. 78–82.

Jansen R.H. (1979) Unified User Oriented Computation of Shielded, Covered and Open Planar Microwave and Millimetre Wave Transmission Line Characteristics, *Microwave, Optics and Acoustics*, 3, 1, pp. 14–22.

Jansen R.H. (1983) Kirschning M., Arguments and an Accurate Model for the Power-Current Formulation of Microstrip Characteristic Impedance, *AEÜ*, Band 37, Heft 3/4, pp. 108–12.

Jansen R.H. (1985) The Spectral-Domain Approach for Microwave Integrated Circuits, *IEEE Trans. Microwave Theory Tech., MTT-33*, 10, pp. 1043–56.

Johns P.B., Beurle R.L. (1971) Numerical Solution of 2-dimensional Scattering Problems Using a Transmission-Line Matrix, *Proc. Inst. Elect. Eng.*, 118, 9, pp. 1203–8.

Kirschning M., Jansen R.H. (1982) Accurate Model for Effective Dielectric Constant of Microstrip with Validity up to Millimetre-wave Frequencies, *Electronic Letters*, 18, 6, pp. 272–3.

Kirschning M., Jansen R.H. (1984) Accurate Wide-Range Design Equations for the Frequency-Dependent Characteristic of Parallel Coupled Microstrip Lines, *IEEE Trans. Microwave Theory Tech., MTT-32*, 1, pp. 83–90.

Kitazawa T., Hayashi Y., Suzuki M. (1976) A Coplanar Waveguide with Thick Metal-Coating, *IEEE Trans. Microwave Theory Tech., MTT-24*, 9, pp. 604–8.

Kowalski G., Pregla R. (1971) Dispersion characteristics of shielded microstrips with finite thickness, *AEÜ*, V.2S, No. 4, pp. 193–6.

March S. (1981) Microstrip Packing: Watch The Last Step, *Microwaves*, 12, pp. 83, 84, 87, 88, 92, 94.

Mirshekar-Syahkal D., Davies J.B. (1982) An Accurate, Unified Solution to Various Fin-Line Structures, of Phase Constant, Characteristic Impedance, and Attenuation, *IEEE Trans. Microwave Theory Tech., MTT-30*, 11, pp. 1854–61.

Pantic Z., Mittra R. (1986) Quasi-TEM Analysis of Microwave Transmission Lines by the Finite-Element Method, *IEEE Trans. Microwave Theory Tech., MTT-34*, 11, pp. 1096–103.

Pramanick P., Bhartia P. (1985) Accurate Analysis Equations and Synthesis Technique for Unilateral Finlines, *IEEE Trans. Microwave Theory Tech., MTT-33*, 1, pp. 24–30.

Pramanick P., Bhartia P. (1985) Computer-Aided Design Models for Millimeter-Wave Finlines and Suspended-Substrate Microstrip Lines, *IEEE Trans. Microwave Theory Tech., MTT-33*, 12, pp. 1429–35.

Pramanick P., Bhartia P. (1986) A New Model for the Apparent Characteristic Impedance of Finned Waveguide and Finlines, *IEEE Trans. Microwave Theory Tech., MTT-34*, 12, pp. 1437–41.

Pramanick P., Mansour R.R., MacPhie R.H. (1987) Computer Aided Design Models for Unilateral Finlines with Finite Metalization Thickness and Arbitrarily Located Slot Width, *IEEE MTT-S Int. Microwave. Symp.* Dig., Las Vegas, pp. 703–6.

Pucel R.A., Massé D.J., Hartwig C.P. (1968) Losses in Microstrip, *IEEE Trans. Microwave Theory Tech., MTT-16*, 6, pp. 342-350, correction *MTT-16*, 6, p. 1064.

Schmidt L.P., Itoh T. (1980) Spectral Domain Analysis of Dominant and Higher Modes in Finite, *IEEE Trans. Microwave Theory Tech., MTT-28*, 9, pp. 981–5.

Silvester P. (1968) TEM Properties of Microstrip Transmission Lines, *Proc. IEE, Vol. 115*, pp. 42–9.

Schneider M.V. (1969) Microstrip Lines for Microwave Integrated Circuits, *B.S.T.J., Vol. 48*, pp. 1421–44.

Schneider M.V. (1969a) Dielectric Loss in Integrated Microwave Circuits, *The Bell System Technical Journal, 48*, pp. 2325–32.

Spielman B.E. (1977) Dissipation Loss Effects in Isolated and Coupled Transmission Lines, *IEEE Trans. Microwave Theory Tech., MTT-25*, 8, pp. 648–55.

Tomar R.S., Bhartia P. (1986) New Accurate Design Data for a Suspended Microstrip Line, *Int. Journal of Inf. and MM Waves, 7, 9*, pp. 1281–95.

Tomar R.S., Bhartia P. (1987) New Quasi-Static Models for the Computer-Aided Design of Suspended and Inverted Microstrip Lines, *IEEE Trans. Microwave Theory Tech., MTT-35*, 4, pp. 453–457, correction *MTT-35*, 11, p. 1076.

Tomar R.S., Bhartia P. (1987a) Modeling the Dispersion in a Suspended Microstrip Line, *IEEE-MTT-S Int. Microwave Symp. Dig.*, Las Vegas.

Wheeler H.A. (1977) Transmission-line Properties of a Strip on a Dielectric Sheet on a Plane, *IEEE Trans. Microwave Theory Tech., MTT-25*, 8, pp. 631–47.

Worm S.B., Pregla R. (Feb. 1984) Hybrid. mode analysis of arbitrarily shaped planar microwave structures by method of lines, *IEEE Trans. Microwave Theory Tech.,MTT-32*, pp. 191–6.

REFERENCES 2.2

Alexopoulos N.G., Jackson D.R., Katehi P.B. (1985) Criteria for Nearly Omnidirectional Radiation Patterns for Printed *Antenas, AP-33*, 2.

Altschuler H.M., Oliner A.A. (1960) Discontinuities in the Center Conductor of Symetric Strip Transmission Line, *IEEE Trans. Microwave Theory Tech., MTT-8*, 5, pp. 328–39.

Benedek P., Silvester P. (1972) Equivalent Capacitances for Microstrip Gaps and Steps, *IEEE Trans. Microwave Theory Tech., MTT-20*, 11, pp. 729–33.

Chadha R., Gupta R.C. (1982) Compensation of Discontinuities in Planar Transmission Lines, *IEEE MTT-S Int. Microwave Symp. Dig.*, Dallas, pp. 308–10.

Collin R.E. (1960) Field Theory of Guided Waves, McGraw-Hill Book Company, London.

Douville R., James D. (1973) Experimental Study of Symmetric Mictrostrip Bends and Their Compensation, *IEEE Trans. Microwave Theory Tech., MTT-26*, 3, pp. 175–82.

Easter B. The Equivalent Circuit of Some Microstrip Discontinuities, *IEEE Trans. Microwave Theory Tech., MTT-23*, 8, pp. 655–60.

Gopinath A., Thompson A.F., Stephenson I.M. (1976) Equivalent Circuit Parameters of Microstrip Step Change in Width and Cross Junctions, *IEEE Trans. Microwave Theory Tech., MTT-24*, 3, pp. 142–4.

Gupta K.C., Garg R., Bahl I.J. (1979) Microstrip Lines and Slotlines, Artech House.

Gupta K.C. (1979) Design of Parallel Coupled Line Filter with Discontinuity Compensation in Microstrip, *Microwave Journal*, pp. 39–57.

Gupta K.C., Garg R., Chadha R. (1981) Computer Aided Design of Microwave Circuits, Artech House Inc..

Hammerstadt E.O. (1975) Equations for Microstrip Circuit Design, *5th Europe Microwave Conf.*, Hamburg, pp. 268–71.

Helard M., Citerne J., Picon O., Hanna V. (1985) Theoretical and Experimental Investigation of Fin-Line Discontinuities, *IEEE Trans. Microwave Theory Tech., MTT-33*, 10, pp. 994–1003.

Jansen R.H. (1985) The Spectral Domain Approach for Microwave Integrated Circuits, *IEEE Trans. Microwave Theory Tech., MTT-33*, 10, pp. 1043–56.

Jansen R.H. (1984) Hybrid Mode Analysis of End Effects of Planar Microwave and Millimeter Wave Transmission Lines, *Proc. Inst. Elect. Eng.*, 128, pp. 77–86.

Jansen R.H., Koster N. (1975) Accurate Results on the End Effect of Single and Coupled Microstrip Lines for Use in Microwave Circuit Design, *Electron. & Commun. (AET)*, 29, pp. 241–7.

Jansen R.H. (1978) High-Speed Computation of Single and Coupled Microstrip Parameters Including Dispersion, High-order Modes, Loss and Finite Strip Thickness, *IEEE Trans. Microwave Theory Tech., MTT-26*, 1, pp. 78–82.

Jansen R.H., Koster N. (1982) New Aspects Concerning the Definition of Microstrip Characteristic Impedance as a Function of Frequency, *IEEE MTT-S Int. Microwave Symp. Dig.*, Dallas, 1982, pp. 305–7.

Katehi P.B., Alexopoulos N.G. (1985) Frequency-Dependent Characteristics of Microstrip Discontinuities in Millimeter-Wave Integrated Circuits, *IEEE Trans. Microwave Theory Tech., MTT-33*, 10, pp. 1029–35.

Kirschning M., Jansen R.H., Koster N. (1981) Accurate Model for Open End Effect of Microstrip Lines, *Electronic Letters*, 3, 27, pp. 123–5.

Kirschning M., Jansen R.H. (1984) Accurate Wide-Range Design Equations for the Frequency-Dependent Characteristic of Parallel Coupled Microstrip Lines, *IEEE Trans. Microwave Theory Tech., MTT-32*, 1, pp. 83–90.

Knorr J.B., Saenz J. (1973) End Effect in Shorted Slot, *IEEE Trans. Microwave Theory Tech., MTT-21*, 9, pp. 579–80.

Knorr J.B. (1981) Equivalent Reactance of a Shorting Septum in a Finline: Theory and Experiment, *IEEE Trans. Microwave Theory Tech., MTT-29*, 11, pp. 1196–202.

Koster N., Jansen R.H. (1986) The Microstrip Step Discontinuity: A Revised Description, *IEEE Trans. Microwave Theory Tech., MTT-34*, 2, pp. 213–23.

Kvasil B. (1957) Teoretické základy techniky centimetrových vĺn, SNTL Praha.

Miklin S.C., Smolitsky K.L. (1967) *Approximate for Solution of Differential and Integral Equations*, American Elsevier, New York.

Pic E., Hoefer W. (1981) Experimental Characterization of Fin Line Discontinuities, *IEEE MTT-S Int. Microwave Symp. Dig.*, New York, pp. 108–10.

Schiavon G., Tognolati P., Sorrentino R. (1988) Fulwave Analysis of Coupled-Finline Discontinuities, *IEEE MTT-S Int. Microwave Symp. Dig.*, New York, pp. 725–8.

Silvester P., Benedek P. (1973) Microstrip Discontinuity Capacitances for Right-Angle Bends, T-junctions and Crossings, *IEEE Trans. Microwave Theory Tech., MTT-21*, 5, pp. 341–6.

Silvester P., Benedek P. (1972) Equivalent Capacitance of Microstrip Open Circuits, *IEEE Trans. Microwave Theory Tech., MTT-20*, 8, pp. 511–6.

Thompson A.F., Gopinath A. (1975) Calculation of Microstrip Discontinuity Inductances, *IEEE Trans. Microwave Theory Tech., MTT-23*, 8, pp. 648–54.

Webb K.J., Mittra R. (1985) Solution of the Finline Step-Discontinuity Problem Using the Generalized Variational Technique, *IEEE Trans. Microwave Theory Tech., MTT-33*, 10, pp. 1004–10.

Zangh X., K. Kenneth (1988) Time-Domain Finite Difference Approach to the Calculation of the Frequency-Dependent Characteristics of

Microstrip Discontinuities, *IEEE Trans. Microwave Theory Tech.*, *MTT-36*, 12, pp. 1775–87.

REFERENCES 2.3

Cabana D. (1983) A new Transmission Line Approach for Design Spiral Microstrip Inductors for Microwave Integrated Circuits, *IEEE MTT-S Int. Microwave Symp. Dig.*, Boston, pp. 245–7.

Caverly R. (1986) Characteristic Impedance of Integrated Circuit Band Wires, *IEEE Trans. Microwave Theory Tech.*, *MTT-34*, 9, pp. 982–4.

Gupta K.C., Garg R., Chadha R. (1981) Computer Aided Design of Microwave Circuits, Artech House.

Parisot M. (1984) Highly Accurate Design of Spiral Inductors for MMIC's with Small Size and High Cut-off Frequency Characteristics, *IEEE MTT-S Int. Microwave Symp. Dig.*, San Francisco, pp. 106–10.

Pengelly R.S., Rickard D.C. (1977) Design, Measurement and Application of Lumped Elements up to Y-band, *Proc. 7th European Microwave Conf.*, Copenhagen, pp. 460–4.

Volman V.I. (1982) Spravočnik po rasčetu i konstruirovanii SVČ poloskovych ustrojstv, Radio i svjaz, Moskva.

Wolff I., Knoppik N. (1974) Rectangular and Circular Microstrip Disc Capacitors and Resonators, *IEEE Trans. Microwave Theory Tech.*, *MTT-22*, 10, pp. 857–64.

3

Modelling of active semiconductor circuit elements

P. Bezoušek

3.1 SCHOTTKY-BARRIER DIODES

These are typical high-frequency devices where the active region is given by the transition metal — semiconductor. The current is almost exclusively provided by electrons and according to the known thermoion theory the static $I - V$ characteristic of this diode is given by

$$I_{jR} = I_S e^{\left(\frac{U_j}{nkT}\right)} - 1 \tag{3.1}$$

where

$$I_S = A^* S T^2 e^{\frac{-e\phi B}{kT}}$$

I_{jR} is the current through the transition, U_j is the voltage at the transition, n is ideality factor, I_S is the saturated current through the transition, A^* is the Richardson's constant, Φ_B is the height of the potential barrier from the metal to the semiconductor and S is the transition area.

The ideality factor $n = 1$ for the ideal diode, whereas in real transitions it has a value of 1.1 to 10 (point-contact diode) depending on the transition quality. It is associated with the thickness and nature of the intermediate layer, which is formed between the metal and semiconductor in the production process. The ideality factor behaviour is weakly dependent on the voltage and in accordance with experiments it was found that with increasing voltage U_j it increases approximately according to the relationship

$$n = n_0 \left(1 + \frac{n_1 e U_j}{kT}\right) \tag{3.2}$$

where $n_1 = 6 \times 10^{-3}$ to 4×10^{-2}. The height of the potential barrier is also not quite constant, however, it is a weakly increasing function of the voltage at the transition. This fact results, besides others, in reducing the detection sensitivity of the Schottky-barrier diode down to one third of the theoretical value and thus, in the calculation of the voltage sensitivity of the detector, it is always necessary to consider the actual, experimentally determined curvature of the voltampere characteristic of the diode in the vicinity of zero voltage.

In the case of the voltage at the transition variable with time, in the Schottky-barrier diodes, similarly as in varactors, changes of the charge of ionized donors in the transition region will be manifested. For the dependence of the magnitude of this charge on the voltage at a steep transition the following relationships were derived (see part 3.2)

$$\begin{aligned} Q_j &= S\sqrt{2\varepsilon\left(\phi_b - U_j + \frac{kT}{e}\right)eN_D} \quad \text{for } U_j \leq \Phi_B \\ Q_j &= 0 \qquad \text{for } U_j > \Phi_B \end{aligned} \tag{3.3}$$

where k is the Boltzmann's constant, N_D is the density of donors in the semiconductor, ε is the dielectric permittivity of the semiconductor, Φ_B is the barrier potential (height of the barrier from the semiconductor into the metal). Thus, at voltage changes U_j, the capacitive current also passes through the transition

$$I_{jC} = \frac{\partial Q_D}{\partial t} = \frac{\partial Q_D}{\partial U_j}\frac{\partial U_j}{\partial t} = C_j\frac{\partial U_j}{\partial t} \tag{3.4}$$

The capacitance of the transition of the Schottky diode C_j was introduced, associated with charge of ionized donors. In contrast to the p–n transition the diffusion capacitance is quite missing, since the current is transferred only by majority carriers. There is, however, a fact that the electrons injected from the semiconductor into the metal have a very high energy (hot electrons) in comparison with electrons that are in thermal equilibrium with the metal lattice, however, the characteristic relaxation period of their energy equilibration is shorter than 1 ps, so that in the whole microwave band the effect of these electrons on the transition impedance can be neglected.

In the high-frequency region it is also necessary to consider effects associated with a finite time of flight of electrons through the transition. Two groups of electrons are differentiated from this standpoint. The first group includes electrons emitted from the boundary of empty and non-emptied regions of the semiconductors with an energy lower than the barrier height. These electrons are reflected back and the time of their return τ_R to a first approximation is independent of their initial energy so that

$$\tau_R = \frac{\pi}{e}\sqrt{\frac{m\varepsilon}{N_D}} \tag{3.5}$$

where m is the effective mass of the electron. For common microwave diodes $\tau_R = 2$ to 10 ps. These electrons do not contribute to the direct current, however, they result in an additional admittance of the diode Y_R, which exerts a distinct maximum at a frequency $f_R \approx 1.3/\tau_R$:

$$\Re\{Y_R\} = \frac{1}{2}\frac{I_S e^{\frac{e\phi_B}{kT}} - I_{jR}}{\phi_B - U_j}\frac{\omega\tau_R}{1-(\omega\tau_R)^2}\left[\frac{1-\cos\pi(1-\omega\tau_R)}{1-\omega\tau_R}\right.$$
$$\left. - \frac{1-\cos\pi(1+\omega\tau_R)}{1+\omega\tau_R}\right] \tag{3.6a}$$

$$\Im\{Y_R\} = \frac{I_S e^{\frac{e\phi_B}{kT}} - I_{jR}}{\phi_B - U_j}\frac{\omega\tau_R}{2}\left[1 - \frac{\sin(1-\omega\tau_R)\pi}{1-\omega\tau_R} + \frac{\sin(1+\omega\tau_R)\pi}{1+\omega\tau_R}\right] \tag{3.6b}$$

The second group includes electrons that have sufficient energy to pass through the transition. The longest time of the transition of electrons passing from the semiconductor into the metal is one half of the time of return τ_R, however, for most electrons it is even essentially shorter. This also holds for electrons passing from the metal into the semiconductor (they form the current I_C) which are accelerated in the transition region. In the microwave frequency region the effect of the second group of electrons can be neglected.

The non-emptied part of the semiconductor and both contacts represent a certain resistance R_S, which is in series with the transition

$$R_S = R_{sc} + R_{se} \tag{3.7}$$

where R_{sc} is the series resistance of contacts and R_{se} is the resistance of the non-emptied part of the semiconductor. The resistance R_{se} depends on the voltage U_j, since with a change of the voltage U_j the length of the emptied region is changed. This dependence can be, however, neglected.

The current I_{jR} is subject to permanent fluctuations. One of the basic reasons for these fluctuations is the discontinuous nature of this current, which consists of two statistically independent components:

a) current of electrons from the semiconductor into the metal

$$I_{jR_1} = I_S e^{\frac{eU}{nkT}}$$

b) current of electrons from the metal into the semiconductor

$$I_{jR_2} = I_S \tag{3.8}$$

Both components are a source of the mutually uncorrelated shot noise and thus, the total spectral density of the shot noise is

$$< \bar{i}_S^2 >= 2e\left(I_{jR} + I_S\right) \tag{3.9}$$

The electrons that do not have a sufficient energy for overcoming the barrier also contribute to the current fluctuations in the transition region. The spectral density of the noise current $\langle i_R^2 \rangle$ formed by these electrons was derived e.g. in Trippe *et al.* (1986). It is almost independent of the voltage U_j and in the vicinity of the zero frequency it is about proportional to the square of the frequency achieving a maximum value in the vicinity of the frequency f_R. This source is particularly manifested at a low bias and at $f > 10\,\text{GHz}$.

A further noise source is the series resistance $R_S = R_{sc} + R_{se}$ generating the thermal noise, which is characterized by the series noise voltage source with a spectral density

$$< \bar{u}_T^2 >= 4kTR_S \tag{3.10}$$

At a large current, in the non emptied part of the semiconductor, electrons are generated with their energy essentially exceeding the thermal energy $3/2kT$ and this effect increases thermal fluctuations expressed by (3.10). According to Jelenski *et al.* (1986) it can be expressed by a complementary noise source with a spectral density of

$$< \bar{u}_H^2 >= \frac{8}{3}\frac{\tau_\varepsilon}{N_D}\frac{1}{Sd} - I^2 R_{Se}^2 \tag{3.11}$$

where τ_ε is the relaxation time of the equilibration of the electron energy in the seminconductor, d is the length of the non-emptied region of the semiconductor and I is the current in the diode.

A further source of the noise in semiconductors is the fluctuation of the number of charge carriers resulting from their capture in trapping centres. Each trapping centre is characterized by the relaxation time τ_T. In the presence of only one type of trap with one relaxation period τ_T, for the spectral density of the fluctuation of the electron number it is possible to write

$$< \Delta\bar{n}^2 >= \alpha N_T V \frac{\tau_T}{1+\omega^2\tau_T^2} \tag{3.12}$$

where α is a constant, N_T is the density of traps and V is the semiconductor volume. This corresponds to an equivalent source of the noise voltage u_T in series with the resistance R_{Se}:

$$< \bar{u}_T^2 >= \frac{I^2}{N_D^2 V^2} S_n R_{Se} \tag{3.13}$$

In the transition region, these fluctuations can be similarly expressed with the help of the additional noise current i_T as follows

$$< \bar{i}_T^2 >= 4e\, e^{\frac{eU_j}{nkT}} \left(\frac{f_c}{f}\right)^\alpha \tag{3.14}$$

The flicker noise should also be included in the calculation of noise sources. It occurs in a lower frequency band and its spectral density decreases with the frequency approximately in accordance with the $1/f$ law. It is a matter of general consideration that it results from the presence of traps with a light spectrum of relaxation periods. No satisfactory explanation of this phenomenon has, however, yet been presented and thus, in the calculation the experimental relationship for an equivalent noise current is used

$$< \bar{i}_f^2 >= 4e\, e^{\frac{eU}{nkT}} \left(\frac{f_c}{f}\right)^\alpha \tag{3.15}$$

where $\alpha = 0.8$ to 1.2; f_c is the cut-off frequency of the flicker noise, which means the frequency at which the flicker noise is identical with the white noise, in this case the shot noise.

For an equivalent circuit of the diode see Fig. 3.1. In addition to the elements already described it also includes the assembling capacitance C_m and series inductance L_s characterizing the inlet

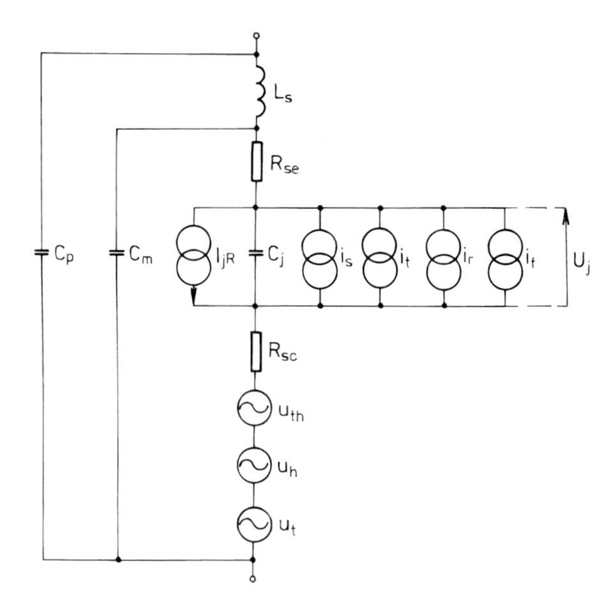

Fig. 3.1 Equivalent circuit of the diode.

line to the diode and capacitance of the capsule C_p (absent when the diode is only in the chip form, without capsule).

In the case of a weak signal the diagram is simplified in such a way that the current source I_{jR} is replaced by a differential resistance of the transition

$$R_j = \frac{1}{\frac{\partial I_{jR}}{\partial U_j}}$$

The model described, particularly the noise source, is however, considerably tedious and thus, the noise characteristics of this diode are more frequently characterized by the effective noise temperature T_e. This is a temperature, which should have a resistance corresponding to the diode resistance for the generation of the same noise power as that of this diode. This is an integral parameter instructively expressing the diode quality. On the other hand, it is impossible to determine either with the help of it or directly the parameters of a circuit with this diode, since it is defined assuming a weak signal in a steady state. In the calculation of T_e the diagram in Fig. 3.1 is considered. The current source I_{jR} is replaced by the resistance R_j and the elements C_m, L_s and C_p are neglected, since they have no effect on noise characteristics of the diode. A usual method is employed for the calculation of the total voltage noise source u_s in the series with the diode

$$< \bar{u}_n^2 >= \left(i_s^2 + i_f^2 + i_T^2 + i_R^2\right) \frac{R_j^2}{|1 + jC_v R_j C_j|^2} + u_{TH}^2 + u_T^2 + u_H^2 \tag{3.16}$$

The effective noise temperature is of

$$T_e = \frac{u_n^2}{4kR_d} \tag{3.17}$$

where

$$R_d = \Re\left\{R_s + \frac{R_j}{1 + j\omega C_j R_j}\right\}$$

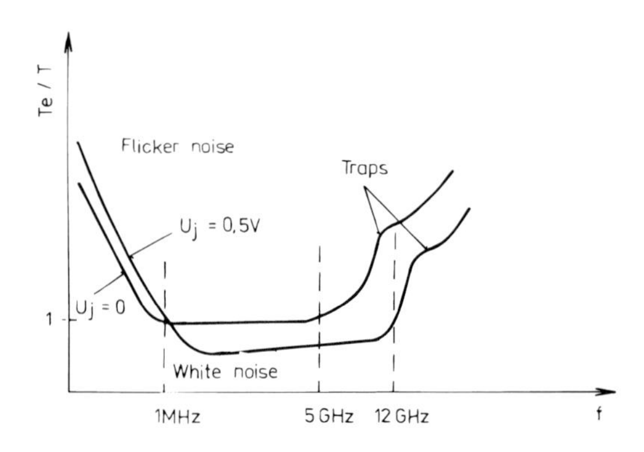

Fig. 3.2 Typical dependence of T_e of a microwave Schottky-barrier diode on the frequency and bias.

Figure 3.2 shows a typical dependence of T_e of a microwave Schottky-barrier diode on the frequency and bias. In the lower part of the frequency spectrum the flicker noise is prevalent and in the high-frequency region fluctuations resulting from electrons reflected from the barrier. In the region of intermediate frequencies the noise is about constant. This wide minimum is, however, dependent on the bias, as can be seen from the second part of the Figure. At a low current $T_e \approx T$. Then it first decreases due to a drop of the shot noise with a subsequent increase as a result of increasing the effect of the thermal noise of the resistance R_s. At a high current, T_e disproportionally increases due to the noise of hot electrons and isolated trapping centres. For low temperatures ($T < 100\,\mathrm{K}$) it is impossible to calculate the effective noise temperature on the basis of the relationships presented, since the other types of electron emission become dominant, for example the tunnel effect, and thus T_e is increased with respect to the physical temperature.

In practice the Schottky-barrier diodes are divided according to their use into mixing and detecting ones, according to the method of the production into the point and flat ones, according to the potential barrier height into the ZBS (zero bias Schottky), LBS (low barrier Schottky) and NBS (normal barrier Schottky) ones. Table 3.1 presents a summary of typical characteristics of particular types of Schottky-barrier diodes and their field of use.

Table 3.1 Schottky diodes

Parameter	Mixer		Detector	
	LBS	NBS	point	ZBS
barrier height Φ [V]	$0.3 \div 0.4$	$0.5 \div 0.7$	–	$0.1 \div 0.15$
coefficient of ideality	< 1.1	< 1.1	$2 \div 3$	< 1.2
effective noise temp. $t_e = e/T_0$	$0.6 \div 0.8$	$0.6 \div 0.7$	$2 \div 5$	–
pinch-off voltage U_B [V]	$5 \div 10$	> 10	$2 \div 3$	$3 \div 5$
cut-off frequency f_c [GHz]	$10 \div 15$	$15 \div 40$	90	$20 \div 40$
I_S [μA]	$(10 \div 30)10^{-3}$	$(1 \div 10)10^{-3}$	$10 \div 100$	$0.1 \div 5$

3.2 VARACTOR DIODES

The varactor is a high-frequency diode that uses the dependence of the transition capacitance on the voltage. From the standpoint of the diode structure this can be either a Schottky-barrier diode polarized in the back direction or a diode with a p–n transition.

The theory of the p–n transition is well known. Its static volt-ampere characteristic is given by the relationship:

$$I_{jR} = I_S \left[e^{\frac{eU_j}{kT}} - 1 \right] \tag{3.18}$$

At a low current, the equivalent circuit of the varactor is nearly identical with that of the Schottky-barrier diode as shown in Fig. 3.1. The essential difference is that at a high current the diffusion capacitance starts to be implemented resulting from the accumulated charge of minority carriers injected through the transition and stepwise recombined with a mean effective life time τ.

The dependence of the transition capacitance C_j connected with the charge of ionized admixture on the voltage can be affected by the profile of the admixture concentrations in the transition region. For a known dependence of the concentration of donors $N_D(x)$ and acceptors $N_A(x)$ on the distance from the transition x, the dependence C_j on U_j can be calculated by the integration of the Laplace's equation for the potential in the one-dimensional case:

$$\frac{d^2\phi}{dx^2} = \frac{e\,(N_D - N_A + n_p - n_e)}{\varepsilon} \tag{3.19}$$

$$\phi\,(-\infty) - \phi\,(+\infty) = U_j + \phi_B$$

where Φ is the potential, Φ_B is the height of the potential barrier, n_p, n_e are concentrations of electrons in the conductive band and holes in the valency band, respectively, ε is the dielectric permittivity.

A great part of the transition is actually completely empty and the remaining free electrons and holes are situated due to the diffusion only in the vicinity of the boundary between the empty and non-empty parts. Thus, in Equation (3.19) it is possible to consider $n_p = n_e = 0$ except for the cases when $U_j \approx \Phi_B$.

The following two types of concentration profiles are of the greatest importance in practice:

a) The diode with an abrupt transition with a profile shown in Fig. 3.3. The concentration of admixtures, i.e. of donors in this case (n type material) is homogeneous in a certain region. At point $x = L_N$ the concentration strongly increases by several orders of

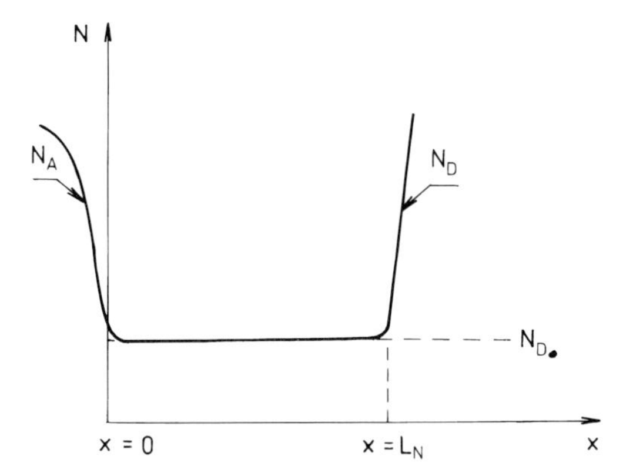

Fig. 3.3 Concentration profile of diode with an abrupt transition (N_D — donors concentration, N_A — acceptors concentration).

magnitude since at this point there is a beginning of the region n^+. At point $x = 0$, there is a diffused abrupt transition (region p^+). The dependence of the charge Q_j and capacitance C_j of the transition on the voltage U_j on the transition can be approximated by the following relationships:

$$\begin{aligned} Q_i(U_j) &= Q_j(U_p) + C_j(U_p)(U_p - U_j) \\ C_j(U_j) &= C_j(U_p) \end{aligned} \quad \text{for } U_j < U_p \tag{3.20a}$$

$$\begin{aligned} Q_j(U_j) &= S\sqrt{2\varepsilon e N_D\left(\phi_B - U_j + \frac{kT}{e}\right)} \\ C_j(U_j) &= \frac{\partial Q_i}{\partial U_j} = \frac{S}{2}\sqrt{\frac{2\varepsilon e N_{Do}}{\phi_B - U_j + \frac{kT}{e}}} \end{aligned} \quad \text{for } U_p \leq U_j \leq U_B \tag{3.20b}$$

$$\begin{aligned} Q_i(U_j) &= 0 \\ C_j(U_j) &= 0 \end{aligned} \quad \text{for } U_j > \phi_B,\ U_p = \phi_B \tag{3.20c}$$

where U_p is the voltage, for which the spatial charge just achieves the contact $x = L_n$ (punch-through voltage).

b) The diode with a hyperabrupt junction. The concentration profile of this diode is shown in Fig. 3.4. It can be seen that in the

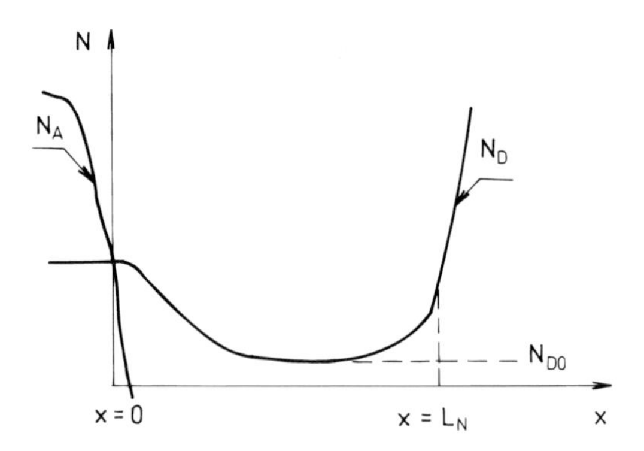

Fig. 3.4 Concentration profile of diode with a hyperabrupt junction.

vicinity of the transition there is a considerable concentration of admixtures in both parts. The concentration of donors in the direction from the transition, however, strongly decreases till its stabilization at a rather low value N_{Do}. The result of this profile is that the capacitance varies considerably with a change of the voltage on the diode, even at higher voltage values, whereas in the abrupt transition the largest change of the capacitance is concentrated about zero or at positive voltages. These diodes are assigned for tuning oscillators. The basic requirement is the linearity of tuning, from which it is possible to establish the requirement for the course of the capacitance and concentration profile. The frequency f of the oscillator tuned by a varactor in the simplest case is given by the relationship:

$$f = \frac{1}{2\pi\sqrt{L_0\left(C_0 + C_j\right)}} \tag{3.21}$$

where L_o and C_o are the equivalent inductance and capacitance of the resonance circuit, respectively, C_j is the capacitance of the varactor transition.

The following equation should hold according to the tuning linearity requirement

$$\frac{df}{dU_j} = const = K \tag{3.22}$$

After substituting f from (3.21) the following result is obtained as

$$C_j\left(U_j\right) = \frac{1}{K^2 4\pi^2 L_0^2 \left(U_j + U_{j0}\right)^2} - C_0 \tag{3.23}$$

Processes associated with the accumulation of the charge in minority carriers in the transition region are analysed in Section 3.3 since they represent the principle of the function of the p-i-n diode. It is shown that at a high current the transition capacitance is strongly increased and thus, the relationships for the passage of the current through the transition are considerably modified. One of the consequences is that the varactor with the p-n transition cannot be used as a non-linear resistor in the band of microwave frequencies. The strong non-linearity of the transition capacitance is, however, used in efficient frequency multipliers and in generators of narrow pulses. In this case the effect of the rapid diode recovery is used i.e. a very rapid transition of the diode from the conductive to the non-conductive condition.

The following types of varactors are most frequently used:

The parametric diode for parametric amplifiers, reactance mixers and multipliers. In this diode the voltage dependence is used of the capacitance of the transition C_j coupled with the charge of ionized admixtures. The most important parameters of these diodes are as follows:

— the transition capacitance at a zero voltage, $C_j(0)$
— the cut-off frequency at a zero voltage

$$f_c\,(0) = \frac{1}{2\pi C_j\,(0)\,R_s\,(0)}.$$

— the coefficient of the capacitance change

$$\gamma = \frac{C_{j_{max}} - C_{j_{min}}}{2\left(C_{j_{max}} + C_{j_{min}}\right)}$$

— the breakdown voltage U_{BR}
— the dynamic coefficient of the diode quality

$$Q_d = \frac{S_1}{\omega R_s}$$

where S_1 is the maximum amplitude of the first harmonic component of the varactor elastance $S = \dfrac{1}{C_j}$ during the excitation by a harmonic signal. This diode is typically produced with an abrupt transition.

The equivalent diagram in Fig. 3.1 and relationships (3.18), (3.20) can be used for it.

The tuning varactor as a tuning element in resonance circuits, most frequently in oscillators. It is produced with either an abrupt or hyperabrupt transition. The diodes with the abrupt transmission have a better quality coefficient, those with the hyperabrupt transition are characterized by a higher ratio between the maximum and minimum capacitances and more uniform dependence of the frequency on the voltage. The equivalent diagram in Fig. 3.1 and relationships (3.18), (3.20) can be used. The most important parameters of the tunning varactor are as follows:
— the capacitance at a zero voltage, $C_j(0)$
— the ratio $C_j(0)/C_j(U_{BR})$
— the breakdown voltage U_{BR}
— the quality coefficient

$$Q\left(O_v\right) = \frac{1}{2\pi R_s C_j\left(0\right)}$$

The diode with rapid recovery for frequency multipliers with a high multiplying ratio (as high as 200x) and for generators of narrow pulses (min 50 ps). In this diode, the non-linearity of the diffusion capacitance is used and thus, the equivalent diagram in Fig. 3.1 and relationships (3.18), (3.20) cannot be used for it. The circuits with these diodes are analysed in the time domain. The principle of the activity is obvious from the description of the p–i–n diode Part 3.3. The diode with the rapid recovery has the following most important parameters:
— the capacitance at a zero voltage $C(0)$
— the off-period t_r
— the life time of minority carriers τ_{ef}.

3.3 p–i–n DIODES

The name of this diode is derived from its structure formed by the following three regions: p type semiconductor, intrinsic region i and

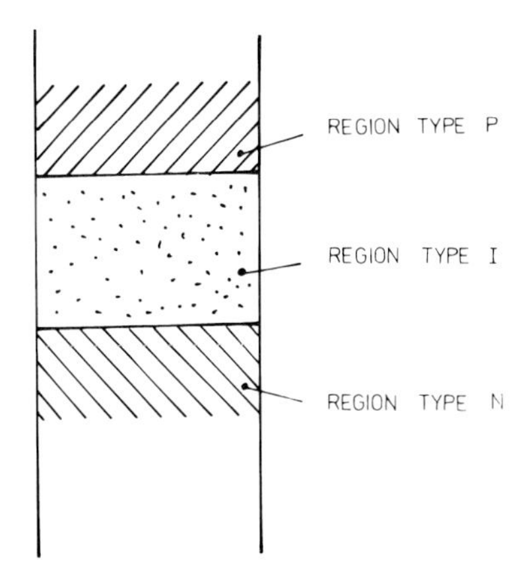

Fig. 3.5 p-i-n diode.

n type semiconductor (Fig. 3.5). This diode is used for controlling and switching RF energy by means of the accumulated charge of carriers injected through both transitions from the regions p and n into the layer i. These carriers form a high-conductivity plasma and they are stepwise recombined with a life time τ_{ef}. The conductivity of the layer i can be controlled within wide ranges by the current passing through the diode.

The p–i–n diode polarized in the forward direction by the current I_0 has the accumulated charge in its layer i

$$Q_{p,n} = I_0 \tau_{ef}$$

Thus, at mobilities of electrons and holes μ_n, μ_p the layer i has a conductivity:

$$G_i = \frac{I_0 \tau_{ef}}{d^2} (\mu_n + \mu_p) \tag{3.24}$$

where d is the length of the layer i.

The interfaces between particular layers actually represent two transitions with characteristics similar to the p–n transitions. The differential admittance of these transitions in the case of a large accumulated charge $Q_{p,n}$ is in accordance with Watson (1969)

$$Y_j = Y_0 \sqrt{1 + j\omega\tau_{ef}} \tag{3.25}$$

where Y_o is the low-frequency differential conductivity of the transition

$$Y_0 = \frac{\partial I_j}{\partial U_j}$$

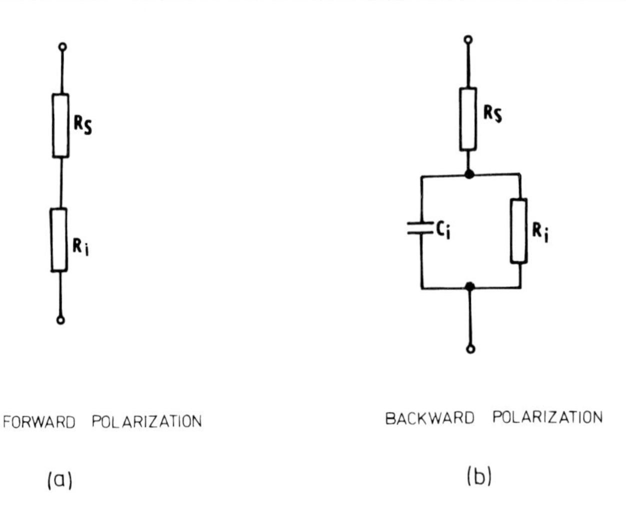

Fig. 3.6 Equivalent circuit of p-i-n diode (a) forward polarization (b) backward polarization.

For an equivalent circuit of the p–i–n diode in this regime see Fig. 3.6a. The criterion of the applicability of this model from the standpoint of the RF signal value can be for example the magnitude of the charge ΔQ exhausted in the course of a negative half-period of the RF signal with respect to the total charge:

$$\frac{\Delta Q}{Q} = \frac{I_{vf}}{I_0}\frac{2}{\omega\tau_{ef}}. \tag{3.26}$$

where I_{vf} is the amplitude of the RF signal.

It is possible to check that for $\omega\tau_{ef} \gg 2$ there is a possibility that $I_{RF} \gg I_o$ and in spite of this the mentioned model will yield a sufficiently good approximation. Thus, in the forward direction the p–i–n diode is manifested as a resistance $R_F \approx R_s + R_i$, also at a high RF power, which can be controlled within a wide range by the value of the direct current I_o. From the standpoint of the RF energy transfer it is manifested as a linear element.

For an equivalent diagram of the diode in the back direction see Fig. 3.6b. The layer i is represented by a low capacitance

$$C_i = \frac{\varepsilon S}{d}$$

and resistance (3.27)

$$R_i = \frac{\rho_i d}{S}$$

where S is the diode area, d is the length of the layer i, ρ_i is the specific resistance of the empty layer i.

At a sufficiently negative direct bias the non-linearity of the diode is also not manifested, since the capacitance of the ideal p–i–n diode is independent of the voltage in the back direction. This holds as far as the amplitude of the RF voltage of the signal on the diode U_{RF} satisfies to the condition:

$$\begin{aligned} U_{RF} &\leq U_0 \\ U_{RF} &\leq U_{BR} - U_0 \end{aligned} \tag{3.28}$$

where U_{BR} is the diode breakdown voltage.

When the diode operates without bias, then at a larger amplitude of the RF voltage, carriers are injected into the layer i for the time of the positive half-period. In this way the mean concentration of carriers in the layer i is increased and the resistance R_j is decreased. Leenov, 1964 found that this resistance can be expressed with the help of diode parameters and RF current amplitude as follows:

$$R_i = \frac{d}{\frac{e}{kT} - I_{vf}\sqrt{\frac{D}{\omega}}} \tag{3.29}$$

where D is the diffusion constant. For sufficiently short diodes this effect leads to an independent switching on of the diode. As can be seen, the resistance decreases with increasing power, which is used in RF power limiters.

The above mentioned diagrams characterize the p–i–n diode only under cut-off conditions. For the description of the behaviour of the diode in switching on and in pulsed control of the RF energy, it is necessary to pass into the time domain. Detailed solution of the courses of the voltage and current in the diode are very complicated. Only the phenomenologically most important effects can be mentioned occurring in this function. Let us consider the diode to be connected in a control circuit according to Fig. 3.7. At time t_0 the current starts to pass through the diode and the layer i starts to be filled. This process takes a certain time referred to as an on-period t_{on}. At the end of this time the layer i is homogeneously filled with

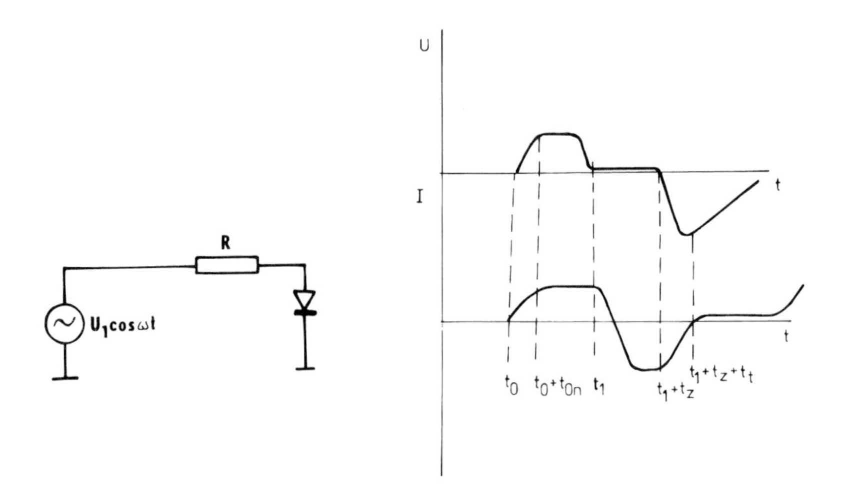

Fig. 3.7 Time dependences of U and I on the p-i-n diode.

charge carriers and the RF impedance of the diode achieves a low value. Immediately after overswitching the controlling voltage to the back direction at time t_1 the layer i starts to be emptied, however the diode impedance is sufficiently low. The voltage on the diode is also maintained at a low value and remains in the forward direction. This time is termed the recovery time t_z and it terminates at a moment of disappearing the charge-neutral region at the centre of the layer i. From this time a rapid decrease of the current passing through the diode and an increase of the voltage and impedance occur. This phase is named the off-period t_t and it is usually very short. By its order of magnitude it corresponds to the time necessary for the flight of carriers through one half of the layer i at a saturated velocity v_s (from the centre to the layer i boundary):

$$t_t = \frac{d}{2v_s} \tag{3.30}$$

The high velocity of this last phase is utilized in diodes with rapid recovery (Section 3.2) for the generation of abrupt edges, narrower pulses and for the multiplication of the frequency. For this purpose, a diode with a very short layer i is, however, necessary, which is rather similar to the varactor. In p–i–n diodes assigned for controlling the RF energy, parasitic effects are described and their duration is typically suppressed by choosing a suitable connection of the control circuit.

Table 3.2 p-i-n diode's basic properties

Type of diode	C_j [pF]	$U_{\rm BR}$ [V]	W_I [μm]	$t_{\rm on}$ [μs]	τ [μs]
power switch	$0.2 \div 0.5$	$200 \div 2000$	$100 \div 300$	$1 \div 10$	$5 \div 10$
speed switch	$0.1 \div 0.5$	$20 \div 200$	$10 \div 100$	< 1	$1 \div 2$
limiter	$0.1 \div 0.5$	$10 \div 50$	$0.5 \div 5$	< 1	< 1

The typical p–i–n diodes are divided into three groups depending on their applications: high-power switching diodes, snap-off diodes and limiters. Table 3.2 summarizes typical features of particular types.

3.4 BIPOLAR TRANSISTORS

The structure and function of the bipolar transistor are well known from its low-frequency applications. The high-frequency properties of this transistor are usually characterized by the limit frequency

$$f_T = \frac{1}{2\pi\tau_{ec}} \tag{3.31}$$

where

$$\tau_{ec} = \tau_b + \tau_c + \tau_e + \tau_{rc}$$

τ_b, τ_c and τ_e are times of the flight through the base, region of the spatial charge of the collector and emitter, respectively, and τ_{rc} is the collector time constant.

In the high-frequency region the course of the current gain of the transistor h_{21e} in the connection with the earthed emitter approximately follows the relationship

$$h_{21e} \sim \left(\frac{f_T}{f}\right)^2 \tag{3.32}$$

The transition limit frequency is thus simultaneously the current gain limit frequency. The further important limit frequency is the maximum oscillation frequency $f_{\max}$, which is approximately given by the relationship

$$f_{max} = \frac{1}{4\pi\sqrt{R_b' C_c \tau_{ec}}} = \sqrt{\frac{f_T}{8\pi R_b' C_c}} \tag{3.33}$$

where R_b' is the equivalent resistance of the base and C_c is the capacitance between the collector and base.

Thus, to achieve a high operation frequency, it is first of all necessary to minimize the base resistance, parasitic capacitive elements and delay in the transistor. This can be achieved of the transistor (comb shaping of the base and emitter contact). A more detailed description of the structure and technology of microwave transistors can be found for example in Section 4.3.1.

For a simple non-linear equivalent diagram of the bipolar transistor see Fig. 3.8. Particular current sources i_{be} and i_{bc} according

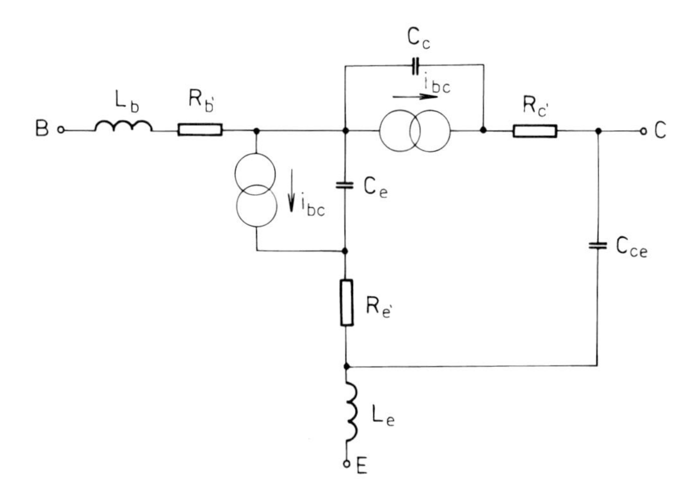

Fig. 3.8 A simple non-linear equivalent circuit of the bipolar transistor.

to Leenov (1964), respectively, represent on the one hand currents through relevant diodes and, on the other hand, currents into the third electrode associated with the concentration of minority carriers in the base (so called α-generators) and they can be expressed by the relationships

$$\begin{aligned} i_{be} &= i_{e0}\left(e^{\frac{eU_{be}}{kT}} - 1\right) + \alpha_i i_{c0}\left(e^{\frac{eU_c}{kT}} - 1\right) \\ i_{bc} &= \alpha_0 i_{e0}\left(e^{\frac{eU_{be}}{kT}} - 1\right) + i_c 0\left(e^{\frac{eU_{bc}}{kT}} - 1\right) \end{aligned} \tag{3.34}$$

where i_{c0}, i_{e0} are residual currents of diodes base-collector and base-emitter, respectively; α_0, α_i are the current amplifying coefficients with earthed base in the direct and inversion direction, respectively, U_{be}, U_{bc} are voltages on the diodes base-emitter and base-collector, respectively.

The amplifying coefficient is a function of the frequency. It is most frequently expressed by the following relationships

$$\begin{aligned} \alpha_0 &= \frac{\alpha_{00}}{1 + j\frac{\omega}{\omega_0}} \\ \alpha_i &= \frac{\alpha_{i0}}{1 + j\frac{\omega}{\omega_0}} \end{aligned} \tag{3.35}$$

where $\omega_0 = 2\pi f_T / \alpha_{00}$ is the breakpoint frequency.

The capacitors C_c and C_e represent the base-collector and base-emitter capacitances, respectively. They can be approximated as a sum the capacitance of the spatial charge and diffuse capacitance of these transitions. The resistance R_b', R_c' and R_e' are series resistances of particular electrodes. A more perfect model of microwave parameters of the bipolar transistor is, however, actually the distributed model where the regions of the base and collector are distributed into several equal sections formed by current sources and capacitance again (Fig. 3.9). The work with this model is, how-

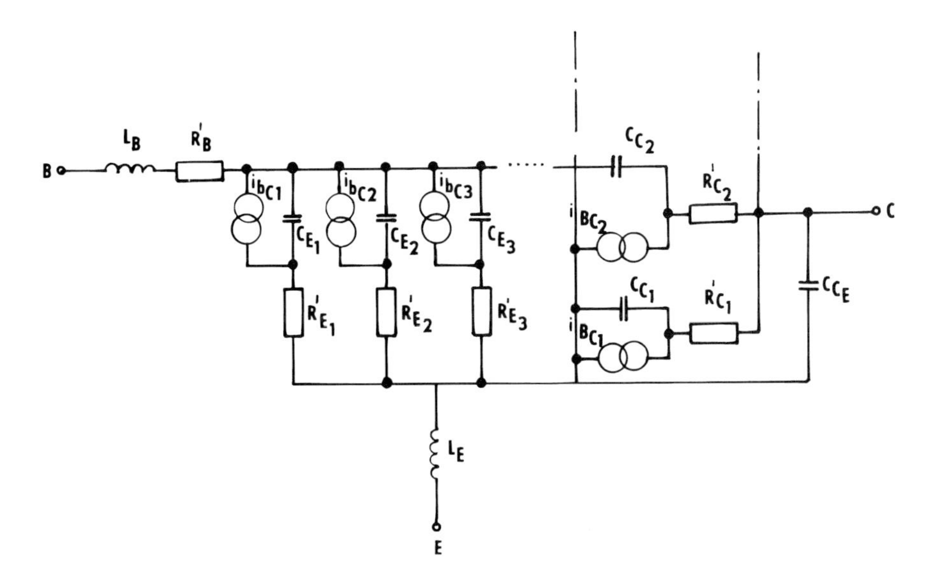

Fig. 3.9 A more complicated equivalent circuit of the bipolar transistor.

ever, very complicated. It is particularly difficult to establish values of individual elements of these equivalent circuits and thus, their use is restricted only to theoretical considerations. When designing circuits with bipolar transistors, experimentally obtained scattering parameters S are usually used in the microwave region.

The minimum noise figure of a bipolar transistor is calculated according to an approximate formula

$$F_{min} = 1 + k\left(1 + \sqrt{\frac{2}{K}}\right) \tag{3.36}$$

where

$$K = \frac{I_E}{kT} R_b' \left(\frac{f}{f_T}\right)^2$$

and I_E is the operation current of the emitter. It can be seen that the noise figure critically depends on the limit frequency f_T, base resistance R_b' and emitter current I_E.

Silicon microwave bipolar transistors of the npn type achieve $f_T < 8\,\text{GHz}$; $f_{\max} < 12\,\text{GHz}$. The increases of the limit frequency are mainly prevented by a considerable diffusion capacitance of the transition base-emitter resulting from the back injection of holes into the emitter. It is eliminated when forming a heterotransition between the materials with a divided width of the forbidden band instead of the usual p-n transition. It is schematically shown in Fig. 3.10. The material with a wider forbidden band w_z serves as an emitter. The discontinuity in the course of the energy levels in the transition between the two materials acts as a high-efficiency potential barrier for holes passing from the base to the emitter and thus, it essentially reduces their concentration in the emitter region, thus also reducing the diffusion capacitance of this transition. The transistors have the HBT junction (heterojunction bipolar transistor) and they achieve $f_{\max}$ as high as $70\,\text{GHz}$.

Bipolar transistors, in comparison with field controlled transistors, are principally more advantageous for power applications and for applications calling for oscillators with a low flicker noise for example. The HBT transistors shift the applicability limit of this type to the high-frequency region of the microwave spectrum.

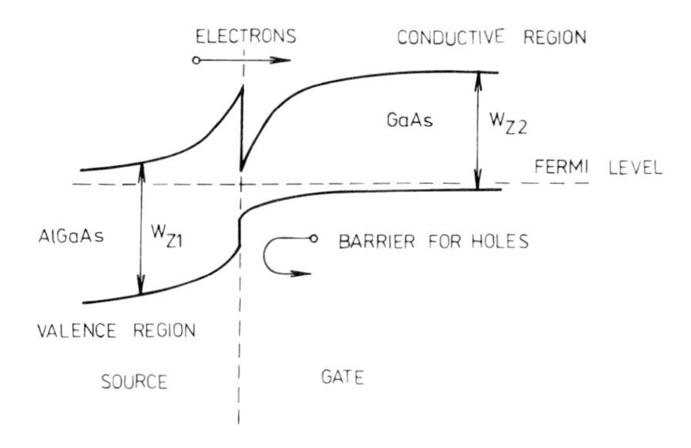

Fig. 3.10 Energy level diagram.

3.5 MESFETs

The MESFET (MEtal-Semiconductor Field Effect Transistor) is a typical microwave device. This is a transistor controlled by the field, where the controlling electrode is formed by the metal-semiconductor transition. The main advantage of this structure, whose profile and equivalent circuit are shown in Fig. 3.11 is the simplicity of its shape. The transistor system can have very small geometric dimensions which is the basic condition for good high-frequency characteristics of this active semiconductor element. The MESFET is exclusively produced on the basis of semiconductor compounds GaAs or possibly InP with a high mobility of electrons and high saturated velocity.

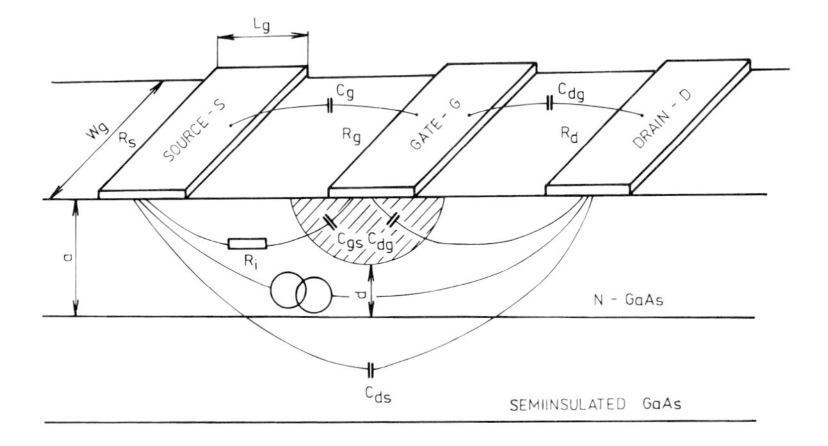

Fig. 3.11 Profile and equivalent circuit of MESFET.

The MESFET function can be most properly explained on a simplified model, for example according to Watson (1969). Under the controlling electrode — barrier gate G — an empty region is formed (marked by shading in Fig. 3.11) similarly as in the transition in the Schottky's diode or varactor. The current through the channel I_d between the emitter and collector can pass only through the non-empty region of a height d. In the saturated regime, i.e. when the electrons at the narrowest site of the channel have the saturated velocity v_s ($10^5\,\mathrm{ms}^{-1}$)

$$I_d = eN_d w_g d v_s \tag{3.37}$$

where N_d is the donor density in the channel, w_g is the channel width.

In an abrupt transition approximation (see Section 3.2) the depth of the non-empty region will be given by the relationship

$$d = a\left(1 - \sqrt{\frac{U_{B_i} - U_g}{U_{B_i} - U_p}}\right) \tag{3.38}$$

where a is the total thickness of the channel, U_g is the voltage between the barrier gate and emitter, U_{Bi} is the built-in potential and U_p is the voltage of the channel locking. Then, for the channel current I_d, static steepness g_{mo} and gate capacitance C_{gs}

$$I_d = I_m\left(1 - \sqrt{\frac{U_{B_i} - U_g}{U_{B_i} - U_p}}\right) \tag{3.39a}$$

$$g_{mo} = \frac{\partial I_d}{\partial U_g} = \varepsilon\frac{w_g v_s}{d} = I_m\, 2\sqrt{(U_{Bi} - U_p)(U_{Bi} - U_g)} =$$

$$\frac{I_m}{2(U_{Bi} - U_p)}\frac{1}{\left(1 - \frac{I_d}{I_m}\right)} \tag{3.39b}$$

$$C_{gs} = \frac{w_g}{a - d} = C_{gso}\frac{1}{\sqrt{1 - \frac{U_g}{U_{Bi}}}} \tag{3.39c}$$

where $I_m = eN_d w_g a v_s$ is the maximum current in the channel, L_g is the length of the barrier gate and C_{gso} is the gate capacitance at a zero voltage U_g. In this approximation, the time of the passage of electrons under the barrier gates τ_t is given by a simple dependence on basic parameters of the transistor

$$\tau_t = \frac{L_g}{v_s} = \frac{C_{gs}}{g_{mo}} \tag{3.40}$$

However, certain assumptions are not actually valid. First of all, the region under the gate is not homogeneous due to the voltage U_d on the collector and thus, equation (3.39c) for the capacitance C_{gs} derived under the assumption $U_d = 0$ is not precisely valid. The further consequence of this is that equation (3.38) does not precisely hold for the thickness of the non-empty region, however, this thickness is also weakly dependent on the voltages U_d and U_p. Thus, the relationships (3.39) and (3.40) should be understood only as basic equations making it possible to explain the MESFET function.

In applications of MESFET type transistors their classification according to the value of the threshold voltage U_p is of importance:

1) $U_p < 0$ D type transistor (depletion mode)
2) $U_p \geq 0$ E type transistor (enhancement mode)

The D type transistor has an open channel at a zero voltage of the gate and it usually operates with a negative bias of the gate. The E type transistor has a closed channel at a zero voltage of the gate and it usually operates at a positive bias of the gate. For a summary see Table 3.3.

A complete equivalent circuit of MESFET is shown in Fig. 3.12. The role of circuit elements of the chip diagram is obvious from Fig. 3.11. The elements beyond the chip are parasitic inductances and capacitances of the capsule C_{gp}, C_{dgp}, C_{dp}; connection of the transistor to the capsule L_g, L_d, L_s; and connection of the capsule into the circuit L_{gp}, L_{dp}, L_{sp}.

For a current of the channel I_d and capacitance of the gate as the main non-linear element of the transistor the following partially empirical relationships are valid:

$$I_d(t) = I_m \left[1 - \left(\frac{U_{Bi} - U_g(t - \tau_t)}{U_{Bi} - U_p}\right)^{\alpha}\right] \qquad \text{for } U_p \leq U_g \leq U_{Bi}$$

$$I_d(t) = I_m \qquad \text{for } U_g > U_{Bi}$$

$$I_d(t) = 0 \qquad \text{for } U_g < U_p$$

Table 3.3 Typical parameters and applications of MESFETs

Application	Type	Parameters							
		$F_{\min}$ [dB]	G_a [dB]	f [GHz]	I_{dss} [mA]	U_{DS} [V]	U_{P} [V]	g_m [mS]	η_{add} [%]
Low-noise amplifiers	HEMT MESFET-D	1 1.4	9 9.5	18 12	15 30	2 3	−0.7 −0.8	25 40	
Power amplifiers	MESFET-D	$P_{1\,\text{dB}} = 20\,\text{dBm}$ $G = 5\,\text{dB}$		18	150	10	−3	200	23
	MESFET-D	$P_{1\,\text{dB}} = 33\,\text{dBm}$ $G = 7.5\,\text{dB}$		10	100	10	−2	600	31
General using	MESFET-D	2.5	8.5	12	150	3	−3	50	28
Logical functions SDFL, BFL DCFL	 MESFET-D MESFET-E				 5 0.5	 2 1.5	 −1 ±0.1	 10 0.9	

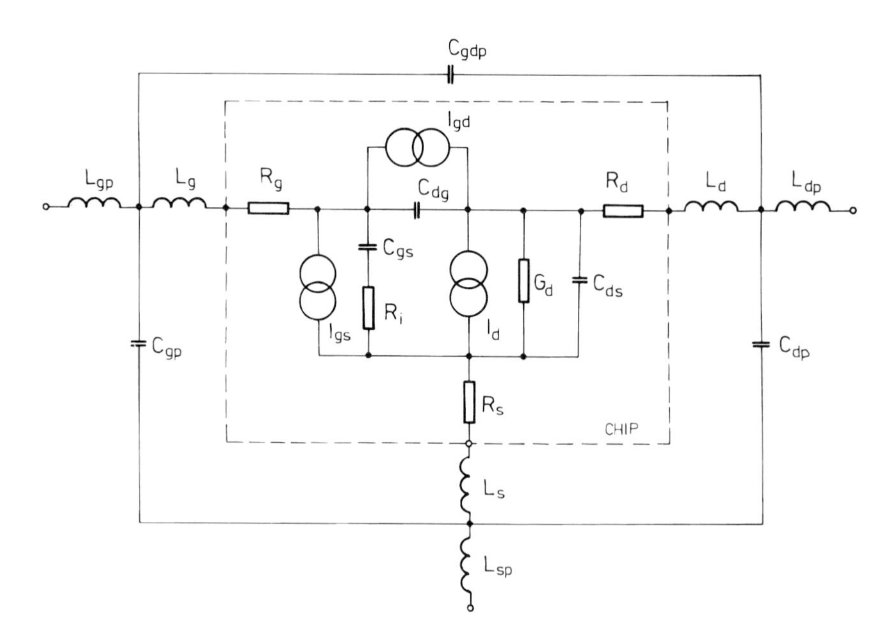

Fig. 3.12 Complete equivalent circuit of MESFET.

$$C_{gs} = C_{gs}(U_p)\frac{2}{\pi}\arctan\left(\frac{U_{Bi}-U_p}{U_{Bi}-U_g}\right)^{\alpha} \qquad \text{for } U_g < U_a$$

$$C_{gs} = \frac{U_{gs}-U_a}{0.23}\left\{\frac{C_0}{\left(1-\frac{U_g}{U_{bi}}\right)^{\alpha}}+\right. \tag{3.41}$$

$$\left. C_{gs}(U_p)\left[A-\frac{2}{\pi}\arctan\left(\frac{U_{Bi}-U_g}{0.15}\right)^{\alpha}\right]\right\} \qquad for\ U_a \le U_g \le U_b$$

$$C_{gs} = C_0\left(1-\frac{U_g}{U_{Bi}}\right)^{\alpha} + C_{gs}(U_p) \qquad for\ U_g > U_b$$

$$C_0 = C_{gs}(0) - C_{gs}(U_p)$$

where $C_0 = C_{gs}(0) - C_{gs}(U_p)$, α is a dimensionless coefficient ($\alpha = 0.3 - 0.7$), $U_a = U_p - 0.08$ [V], and $U_b = U_p + 0.15$ [V].

The change of the current I_d of the collector follows the change of the gate voltage U_g only after a certain time necessary for the new stabilization of the boundary under the gate between the empty and non-empty regions. This time is about the same as the time of the transition of electrons under the gate τ_t. This is also expressed in the relationships for I_d and C_{gd}.

In the model described, further non-linear dependences were neglected, such as the dependences of C_{gd} and C_d on voltages U_g and U_d and a dependence of C_{gs} on U_d.

The current of Schottky diodes between the gate and emitter and between the gate and collector can also be unambiguously expressed

$$\begin{aligned}
I_{gs} &= 0 && \text{for } U_{BR} \le U_g \le U_{Bi} \\
I_{gs} &= \frac{U_g - U_{Bi}}{R_F} && \text{for } U_g > U_{Bi} \\
I_{gs} &= \frac{U_g - U_{BR}}{R_R} && \text{for } U_g < U_{BR} \\
I_{gd} &= 0 && \text{for } U_{BR} \le U_{dg} \le U_{Bi} \\
I_{gd} &= \frac{U_{dg} - U_{Bi}}{R_F} && \text{for } U_{dg} > U_{Bi} \\
I_{dg} &= \frac{U_{dg} - U_{BR}}{R_R} && \text{for } U_{dg} < U_{BR}
\end{aligned} \tag{3.42}$$

where U_{BR} is the voltage of the gate breakdown, R_F is the resistance of the gate diode in the forward direction and R_R is the gate diode resistance in the reverse direction. In this way, it is possible to approximate the voltampere characteristic of the Schottky's diode formed by the gate particularly with respect to the fact that the gate current is mostly negligible in comparison with the other currents in the transistor and when it becomes comparable, then series resistances are already prevalent in these diodes over the resistance of transitions.

In the case of a small signal the sources I_{gs} and I_{gd} can be neglected. The linearization of relationships (3.41) for small RF amplitudes results in a new equivalent circuit of the transistor chip (Fig. 3.13) which also includes noise sources. C_{gso} and the static

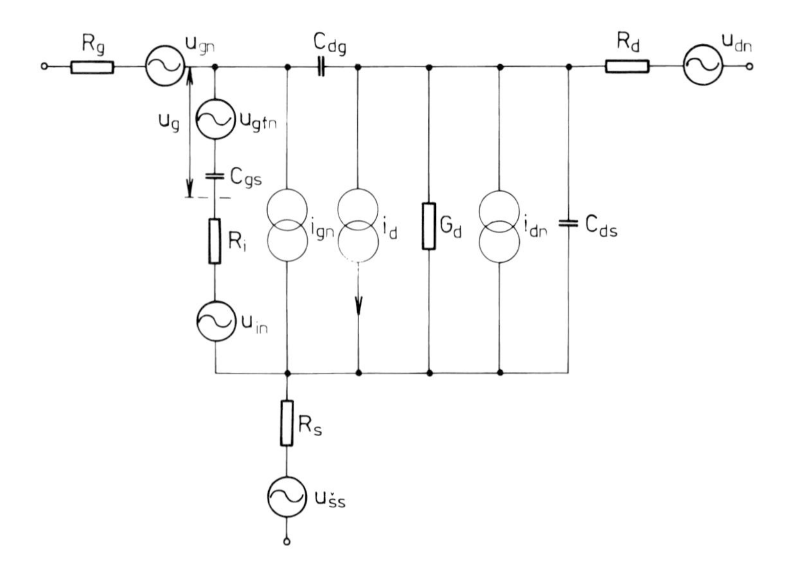

Fig. 3.13 Equivalent circuit of the transistor chip for small RF amplitudes.

steepness g_{mo} is calculated from relationships (3.41) for the operation point given by direct components U_{g0} and U_{d0} of the voltage U_g and U_d so that

$$C_{gs} = C_{gs}(U_{go}) = C_{gso}\left(1 - \frac{U_{go}}{U_{Bi}}\right)^{\alpha}$$

$$g_{mo} = \left(\frac{\partial I_d}{\partial U_g}\right)_{U_{go}, U_{do}} = \frac{\alpha I_m}{U_{Bi} - U_p}\left(\frac{U_{Bi} - U_g}{U_{Bi} - U_p}\right)^{\alpha - 1} \qquad (3.43)$$

The relationship for the steepness g_m depending on angular frequency ω as presented in Fig. 3.13 is derived from the Taylor's expansion of the function $I_d(t)$ given by relationship (3.41) assuming that τ_t is very small with respect to the period of the RF signal (which is fairly adhered to in the microwave region since $\tau_t \approx 1 - 5\,\text{ps}$).

The small-signal nature of the transistor is characterized by limit frequencies.

The transit limit frequency f_T

$$f_T = \frac{g_{mo}}{2\pi C_{gs}} = \frac{\omega_T}{2\pi} \doteq \frac{1}{2\pi\tau_t} \tag{3.44}$$

is tightly related to the time of flight of electrons under the gate τ_t, (see 3.40) and it determines the noise characteristics of the transistor and the bandwidth.

The maximum frequency of the gain attainable $f_{\max}$ is a frequency, for which the available gain decreases to unity, so that

$$f_{mag} = \frac{g_{mo}}{2\pi C_{gs}} \left\{ 4C_g G_d \frac{(\tau_i + \tau_s + \tau_g)}{C_{gs}} + \frac{2C_{dg}}{C_{gs}} \left[C_{dg} + g_{mo} (\tau_i + \tau_s + 2\tau_g) \right] \right\}^{-1} \tag{3.45}$$

In the high-frequency region, the course of the maximum available gain approximately follows the relationship

$$G_{mag} \doteq \left(\frac{f_{mag}}{f} \right)^2 \tag{3.46}$$

Maximum oscillation frequency

$$f_{max} \doteq \frac{g_{mo}}{2\pi C_{gs}} \frac{1}{\sqrt{4G_d (R_i + R_s + R_g) + 4C_{dg} g_{mo} \tau_g}} \doteq \sqrt{\frac{f_T}{8\pi\tau_g C_{dg}}} \tag{3.47}$$

where $R' = R_g + (R_i + R_s + R_g)(C_d C_{gs}/(C_{dg} C_{\omega_o})$ is simultaneously a frequency for which the unilateral gain equals unity. It can be seen that the final approximate relationships for $f_{\max}$ essentially agrees with the expression for $f_{\max}$ of a bipolar transistor (compare with 3.33).

As far as the source noise is concerned u_{gn}, u_{in}, u_{sn}, u_{dn} represent the thermal noise of relevant resistors and for their magnitude in the frequency band Δf it is possible to write:

$$\begin{aligned} <\bar{u}_{gn}^2> &= 4kTR_g\Delta f \\ <\bar{u}_{in}^2> &= 4kTR_i\Delta f \\ <\bar{u}_{sn}^2> &= 4kTR_s\Delta f \\ <\bar{u}_{dn}^2> &= 4kTR_d\Delta f \end{aligned} \tag{3.48}$$

The sources i_{dn} and i_{gn} are already connected with the transistor function. The reason for this are fluctuations of the charge density in the transistor channel resulting from collisions between electrons and phonons, i.e. a mechanism known for the generation of the thermal noise in a linear resistance. In the case of the transistor, in the channel the electrons are drifted with a saturated velocity v_s. In spite of the fact that, as already shown by Van der Ziel (1962, 1963), the noise of a channel in the RF region can also be expressed as a thermal noise of the conductivity g_m at an effective temperature rather higher than the temperature of the material itself:

$$<\bar{i}_{gn}^2>= 4kTg_mP\Delta f \tag{3.49}$$

where P is a dimensionless factor ($P = 1 - 3$) depending on a particular transistor and operation point used.

The noise of the barrier gate represented by the source i_{gs} results from fluctuations of the charge under the gate and thus, it is considerably correlated with the channel noise:

$$\begin{aligned} <\bar{i}_{gn}^2> &= 4kTR\frac{\omega^2C_{gs}^2}{g_{mo}} \\ \frac{<\bar{i}_{gn}\bar{i}_{dn}^*>}{|i_{sn}|\,|i_{dn}|} &= jC \end{aligned} \tag{3.50}$$

where R, C are dimensionless coefficients ($R = 0.1 - 0.4$; $C = 0.5 - 0.9$) depending on the operation point of the transistor.

In the low-frequency region further sources of the noise are present. This is particularly the flicker noise and the noise resulting from random captures of electrons at capture centres under the gate with a defined relaxation capture time τ. In the equivalent circuit in Fig. 3.13 their effect is expressed by a voltage source u_{gfn} for which

$$< \bar{u}_{gfn}^2 >= 4kTR_i \left[\frac{f_c}{f} + \sum_{r=1}^{N} \frac{\frac{\tau_r}{\tau_{00}}}{1 + (2\pi f \tau_r)^2} \right] \tag{3.51}$$

where f_c is the limit frequency of the flicker noise, τ_r is the relaxation period of the r-th trap and τ_{oo} is a constant with the dimension of time. The value f_c ranges between 1 and 10 MHz for a transistor based on GaAs. Similarly as the Schottky barrier diode of this material, the MESFET is also unsuitable for low-frequency applications, for mixers with a low intermediate frequency as well as for oscillators in the case of requirements of a low FM noise.

The characteristics of the MESFET as a low-noise amplifier are given by four real noise parameters, which can be expressed according to the equivalent diagram in Fig. 3.13 as follows:

$$\begin{aligned} F_{min} &= 1 + 2\sqrt{P + R - 2C\sqrt{PR}}\frac{f}{f_T} \\ &\quad \cdot \sqrt{g_{mo}\left(R_g + R_i + R_s\right) + \frac{PR\left(1 - C^2\right)}{P + R - 2C\sqrt{PR}}} \\ R_n^{-1} &= g_{mo}\left(\frac{f_T}{f}\right)^2 \sqrt{P + R - 2C\sqrt{PR}} \\ G_{opt} &= \Re\left\{Y_{opt}\right\} \\ B_{opt} &= \Im\left\{Y_{opt}\right\} \end{aligned} \tag{3.52}$$

where $F_{\min}$ is the minimum noise figure, R_n in the noise resistance and G_{opt}, B_{opt} are components of the optimum noise admittance of the source, given by

$$Y_{opt} = \frac{1}{C_v C_{gs}} \left[\sqrt{\frac{g_{mo}\left(R_s + R_g + R_i\right) + \frac{PR(1-C^2)}{P+R-2C\sqrt{PR}}}{P + R - 2C\sqrt{PR}}} - j\frac{P - C\sqrt{PR}}{P + R - 2C\sqrt{PR}} \right] \tag{3.53}$$

The noise figure of the transistor for a general admittance Y at the input is then calculated from

$$F = F_{min} + R\frac{(G - G_{opt})^2 + (B - B_{opt})^2}{G} \tag{3.54}$$

where $G = \text{Re}\{Y\}, B = \text{Im}\{Y\}$.

It can be seen that the minimum noise figure is connected with the limit frequency f_T and thus also with the capacitance of the barrier gate and with its length L_g – see (3.44) and (3.40). Its dependence on the magnitude of parasitic resistances R_s, R_i, R_g and on the quantity P is also of importance. This quality is a function of the transistor current. It has a distinct minimum at low current values, which is directly reflected by the noise number, which achieves its minimum value for currents of about 10 to 15% of the current at a zero voltage on the barrier gate.

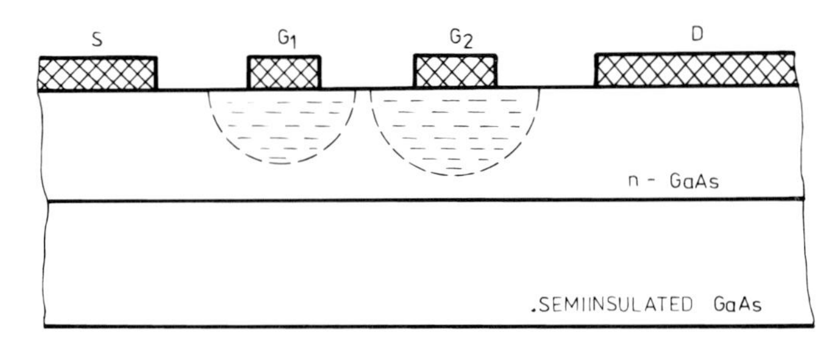

Fig. 3.14 MESFET with two bases.

The frequently used modification of the MESFET type transistor according to Fig. 3.11 is the MESFET with two bases Fig. 3.14. For the equivalent circuit of this transistor, which is model by two MESFET transistors in cascade connection, see Fig. 3.15. The two-base transistors are used as mixers, switches and amplifiers with a controllable gain.

In designing hybrid circuits, it is necessary to know the MESFET parameters expressed by scattering and noise parameters measured in the arrangement assumed for the use of the transistors. In this way, nonaccuracies are eliminated resulting from different parasitic reactances, which are considerably important particularly in the microwave frequency region. These measurements are, however, very tedious and thus, most important producers of transistors present

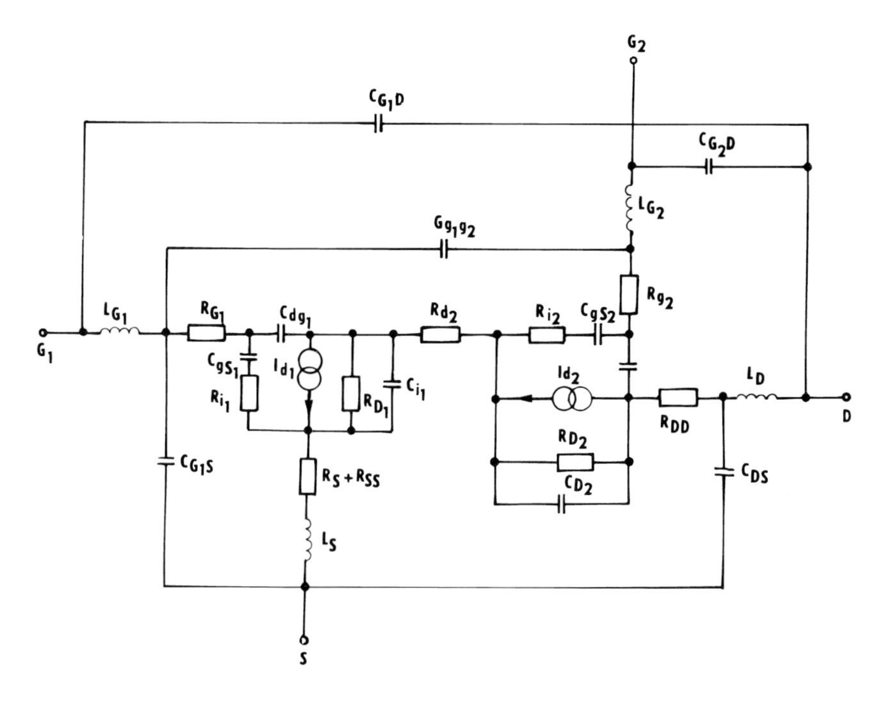

Fig. 3.15 Equivalent circuit of two-base MESFET.

these data as typical parameters of their products, which can be considered when designing circuits.

When designing monolithic circuits, it is advantageous to use the equivalent circuits in Fig. 3.12 and Fig. 3.13. The values of elements of these circuits should be determined for a particular technology and structure of the transistor based on measuring many chips. For orientation purposes, approximate relationships are presented below, which can also serve for determining the most important elements of the circuits on the basis of geometric dimensions of the transistor structure:

$$R_g = 1.3\frac{w_g}{n^2 L_g} \quad [\Omega]$$

$$R_d = \frac{1}{w_g} \quad [\Omega]$$

$$\tau_t = \frac{C_{gs}}{g_{mo}}$$

$$R_i = \frac{6}{w_g} \quad [\Omega]$$

$$
\begin{aligned}
Cgso &= 1.23 w_g L_g^{-1} \ \text{[pF]} \\
\alpha &= 0.5 \\
R_s &= \frac{2}{w_g} \ [\Omega] \\
C_{dg} &= 0.1 w_g \ \text{[pF]} \\
R_{ds} &= \frac{200}{w_g} \ [\Omega] \\
C_{ds} &= 0.2 w_g \ \text{[pF]}
\end{aligned}
\tag{3.55}
$$

where w_g is the gate width [mm], L_g is the gate length [μm] and n is number of sections of the gate.

The MESFETs transistors are classed, depending on their use to low-noise, power and digital ones. In low-noise transistors the minimum noise number is of the highest importance, which can be achieved by reducing the gate length l_g. In contrast to this, the power transistors have a very wide gate, which is divided into a series of parallel sections to minimize the gate resistance R_g. The other resistances R_s and R_d should also be very low for this transistor. Thus, in power transistors, high-doped regions n^{++} are frequently implanted under emitter and collector electrodes. Since the high frequency matching of power transistors is usually difficult with respect to very low input and output impedances, certain producers supply transistors with their internal matching i.e. with matching circuits incorporated inside the transistor capsule. It is, however, necessary to remind that this matching is possible only in a limited frequency band. Both transistor types mentioned are produced only in the form of the D-MESFET with a voltage of channel locking of 2 to 5 V.

The transistors for digital circuits are produced in both versions. The D-MESFET transistors occur more frequently, since their production is more properly managed and rather shorter switching times are achieved with them. The E-MESFET types are, however, more advantageous for logic circuits. First of all, the level of the input and output voltages are similar in these transistors (both voltages are positive), which considerably simplifies the construction of logic circuits, since the complicated shift of levels is avoided. These transistors also exert a lower power loss and thus, they can be integrated to form larger sets. Their production is, however, more complicated with respect to the required accuracy of adhering to the thickness and channel doping, which considerably reduces the yield.

In Table 3.3, data concerning different types of MESFET are summarized. Transistors of the type HEMT are also included described in the following section.

3.6 HEMTs

The HEMT is a transistor controlled by the field with a heterotransition, where the channel is formed by a very thin semiconductor layer with a high concentration of electrons and high mobility. This layer is formed due to special characteristic of the heterotransition between materials with different widths of the forbidden band.

Several institutions were involved in the development of the transistor based on this principle and thus, several very similar structures were obtained with a number of different names such as HEMT (High Electron Mobility Transistor), MODEFET (MOdulation Doped FET), TEGFET (Two-dimensional Electron Gas FET), SDHT (Selective Doped Heterostructure Transistor), and SISFET (Semiconductor Insulator Semiconductor FET). In this work, the uniform name HEMT is used.

The function of this transistor is very similar to that of the MESFET and the same can also be about their equivalent circuits (compare for example Figs. 3.18 and 3.13). The main characteristics of these transistors are summarized in Table 3.3. From them it is possible to see that the HEMTs achieve an essentially better noise number, particularly due to a higher steepness g_m at a low gate capacitance C_{gs}, which favourably affects the limit frequency

$$f_T = \frac{g_m}{2\pi C_{gs}}$$

In Fig. 3.16, there is a schematic diagram of the energy level in the vicinity of an abrupt heterotransition between the materials AlGaAs and GaAs. These materials have a divided width of the forbidden band and thus, at their interface there is a discontinuity of

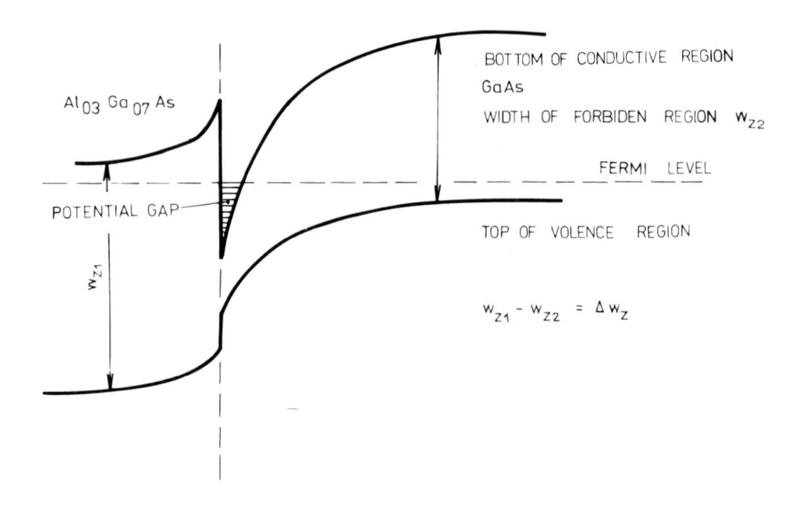

Fig. 3.16 Energy level diagram of HEMT.

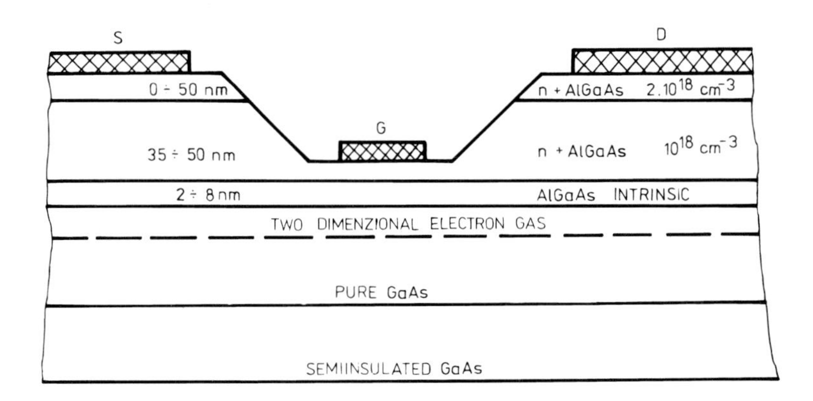

Fig. 3.17 HEMT structure.

boundaries of energy zones. The value of the step ΔE_c equals the difference between electron affinities in the two materials. The electrons from n-AlGaAs diffuse into GaAs till the equilibration of the Fermi's levels in the two materials. Thus, a very narrow potential well is formed in a close vicinity of the boundary of both materials, filled with electrons on the side of GaAs. Here, the electrons form quasi-two dimensional electron gas. Since very pure undoped GaAs material is used for these purposes, these electrons have a very high mobility as well as high saturated velocity achieved at electric fields, which are essentially lower than the normal critical field in GaAs.

The HEMT structure is shown in a schematic diagram in Fig. 3.17. At a zero bias of the gate the empty region under the gate is extended up to the boundary between the two materials. During a drop of the voltage on the gate the electron density in the quasi-two dimensional layer rapidly decreases, which means that the channel is narrowed. Due to this, the transistor shows a very high steepness. At a positive gate voltage, the conductivity of the material n-AlGaAs starts to be implemented, which leads to a decrease of the steepness, which is not observed for the MESFET.

The capacitance of the HEMT is slightly dependent on the voltage for the negative gate bias. The dependence results only from scattering capacitances along sides of the gate due to changes in the volume of the empty region in direction of sides. These changes can, however, be important in transistors with a very short gate ($1 \geq 0.25\mu$m) with respect to a very low capacitance of the gate itself.

An approximately linear equivalent circuit of the HEMT is shown in Fig. 3.18. The circuit is no different from that of MESFET except for replacing R_s by a combination RCR.

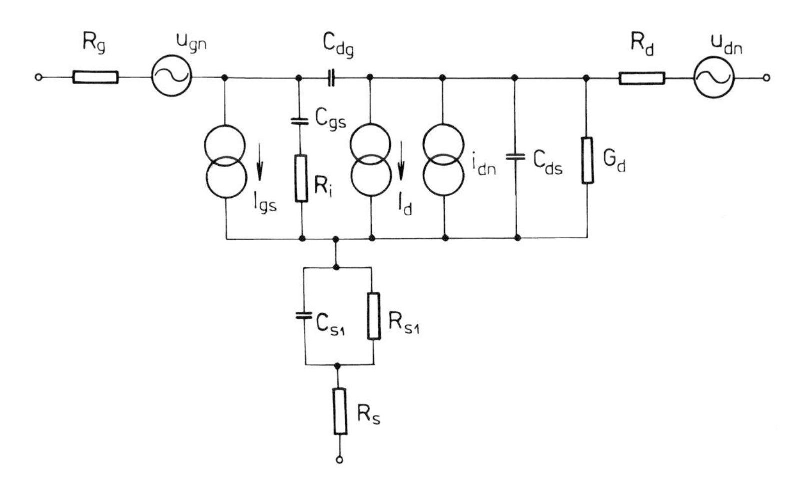

Fig. 3.18 Linear equivalent circuit of the HEMT.

The calculation of resistances R_s, R_i and R_d is very difficult in the case of the HEMT due to the complex nature of the whole structure. It is of essential importance that the resistances are larger than those in the MESFET, since the channel is very thin. Particular layers form a distributed RC structure, which results in a considerable drop of the value of these resistances at a higher frequency.

REFERENCES

Jelenski A. *et al.* (1986) Broad Band Noise Measurements and Noise Measurement of Metal-Semiconductor Junctions, *IEEE Trans. Microwave Theory and Tech., MTT-34*, 11, pp. 1193–201.

Leenov D. (1964) The Silicon PIN Diode as Microwave Protector at Megawatt Levels, *IEEE Trans. Electr. Dev.*, 11, pp. 53–61.

Trippe M. *et al.* (1986) Transit Time effect in the noise of Schottky Barrier diodes. *IEEE Trans. Microwave Theory and Tech., MTT-34*, 11, pp. 1183–92.

Van der Ziel (1962) Thermal moise in FET's. *Proc. IRE, Vol. 50*, pp. 1808–12.

Van der Ziel (1963) Gate noise in field effect transistors at moderate frequencies, *Proc. IRE, Vol. 51*, pp. 461–7.

Watson H.A. (1969) Microwave Semiconductor Devices and Their Circuit Applications, McGraw Hill Book Company.

4

Basic circuits

I. Kneppo and J. Fabian

4.1 METHODS OF THE MIC SYNTHESIS

By the MIC synthesis we mean a complex of mathematical procedures with the aim to choose and arrange one system from theoretical models of microwave elements, a microwave integrated circuit, in order that an electric circuit could be obtained, able to perform the electric function required (amplification, oscillation, mixing, multiplication) in the desired frequency region of the microwave spectrum of electromagnetic waves with the preliminarily determined characteristics or with optimum characteristics established on the basis of a preliminarily chosen criterion. The key point of the synthesis is arranging and solving the system of mathematical equations expressing relationship between characteristics of elements and circuit quantities on the one hand and external quantities affecting the given circuit by means of its input and output lines on the other hand. The synthesis also includes checking of the MIC proposed according to the stability criterion for the given operation regime, estimation of the effect of relationship between the elements on electric characteristics and stability of the whole MIC and optimization of the MIC characteristics.

The procedure and selection of the synthesis method is to a considerable extent affected by the type of microwave elements used in designing the given MIC circuit.

For example in the MIC synthesis with active elements, it is necessary to consider the problem of the circuit stability, whereas when introducing lossless elements, in spite of the fact that these microwave elements cannot be technically implemented, the theoretical consideration in the analysis or synthesis of the MIC are usually essentially simplified.

A further standpoint, which should be considered in the MIC synthesis is the type of the mathematical function used for expressing the basic electric characteristic of the given microwave element. When a linear dependence is valid between the basic electric quantities, then the given element is linear. The starting point in arranging equations for input and output quantities of the microwave integrated circuit synthesized from linear electric elements is the principle of superposition. A non-linear dependence between the voltage and current of a non-linear electric element included into the MIC is the reason for the non-linear behaviour of the whole circuit. This circuit is typically characterized by a number of functions, which are not known in the linear circuit, as e.g. the production of harmonic components of the basic harmonic voltage or current signals, mixing of frequency components of signals, limiting of the output power level or parametric oscillations. The microwave circuit is considered as non-linear already in the case when it includes even only one non-linear element. In this case, the MIC features depend not only on the shape of the volt-ampere characteristics of the non-linear element but also on the site of the circuit where the non-linear element is included. In general, the solution of problems of the analysis or synthesis of the non-linear MIC results in a requirement for solving a system of non-linear equations which cannot be always solved in a closed analytical form. It is almost unavoidable to use proper approximation and iteration procedures, which are currently common parts of miscellaneous computing programs.

Depending on the relationship between geometric dimensions of the element and operation wavelength, the microwave elements are either lumped elements or elements with distributed parameters. When the largest dimension of a given element is smaller than the operation wavelength in the transmission structure of the given MIC by an order of magnitude, then this elements falls into the first group. In contrast to this, a larger dimension means that the phase of the signal at different sites of the elements is not identical and thus, the circuit quantities are functions of spatial co-ordinates. In general, the integrated circuit can include, in addition to elements with distributed parameters, many lumped elements. The latter ones, due to a high level of current technologies employed in the MIC production exert acceptable electric characteristics in a wide frequency region of microwaves. The lumped microwave element is characterized by an equivalent scheme, which also includes lumped elements, however, with ideal characteristics. The principle of methods of the MIC synthesis with concentrated elements is usually based on the topology of branch nodes (Simonyi, 1963; Desoer and Kuh, 1969) and the electric characteristics of the circuit are characterized by the impedance, admittance or hybrid matrix.

The distributed elements are given in the MIC by segments of transmission lines, planar parts, directional couplers, circulators, resonators and filters. In this case, the methods of the microwave circuit synthesis are based on a multiport representation and matrix formalism. The scattering and transmission matrices are most frequently used (Collin, 1966; Gupta *et al.*, 1981). With respect to the fact that in general, MIC can consist of elements of both dimension modes, in the circuit synthesis it is suitable to represent all the elements by a uniform method, either as lumped or as distributed equivalents. The distributed elements can be replaced by the lumped ones only approximately, whereas the multipole lumped elements can be precisely transformed to the multiport circuits (Monaco and Tiberio, 1970). This is also the reason for the fact the MIC is most frequently synthesized as a multiport network.

4.1.1 The matrix representation

With respect to the multiparameter nature of the microwave integrated circuits, in the theoretical design it is of advantage to use symbols and mathematical methods of the matrix algebra. Each circuit element and last also the resulting integrated circuit is from this standpoint an n-port, whose characteristics (assuming that this is a linear circuit) can be expressed by a matrix of complex coefficients. At microwave frequencies, it is of advantage to use the formalism of scattering matrices. The definition base of complex scattering coefficients are complex amplitudes a_k and a_k of the incident and emerging wave, respectively in the k-th port of the given n-port (Fig. 4.1). The amplitude b_k depends on reflections in the k-th port and or contribution from incident waves in the other $n - l$ ports of the n-port, so that

$$b_k = S_{k1}a_1 + S_{k2}a_2 + \cdots + S_{kk}a_k + \cdots + S_{kn}a_n \tag{4.1}$$

where $S_{ki}(i = 1, 2, \ldots, n)$ are scattering coefficients characterizing electric properties of the k-th port of the given microwave linear n-

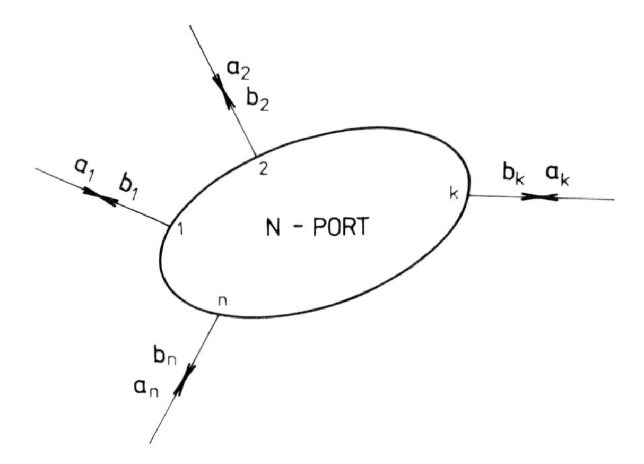

Fig. 4.1 Definition of the n-port.

port. The complete system of linear equations of the n-port in the matrix form is

$$\mathbf{b} = \mathbf{S}\mathbf{a} \tag{4.2}$$

where

$$\mathbf{b} = \begin{bmatrix} b_1 \\ b_2 \\ \vdots \\ b_n \end{bmatrix}, \qquad \mathbf{a} = \begin{bmatrix} a_1 \\ a_2 \\ \vdots \\ a_n \end{bmatrix}, \qquad \mathbf{S} = \begin{bmatrix} S_{11} & S_{12} & \dots & S_{1n} \\ S_{21} & S_{22} & \dots & S_{2n} \\ \vdots & \vdots & \dots & \vdots \\ S_{n1} & S_{n2} & \dots & S_{nn} \end{bmatrix}$$

As can be seen, the scattering matrix $\mathbf{S}$ is a square-type $n \times n$ matrix. Its elements, scattering coefficients S_{ki} $(k, i = 1, 2, \dots, n)$ can be determined by measuring the ratio between the amplitudes of the emerging wave in the k-th port and incident wave in the i-th port assuming a zero amplitude of the other incident waves (this will be achieved by terminating these ports with a normalized impedance).

The average power flowing into the port k is given by

$$P_k = \frac{1}{2}\left(a_k a_k^* - b_k b_k^*\right) \tag{4.3}$$

This shows that P_k equals the power carried into the port k by the incident wave, $\frac{1}{2}a_k a_k^*$, less the power reflected back, $\frac{1}{2}b_k b_k^*$.

It is useful to remind the following important characteristics of S-matrices:

(1) for a reciprocal n-port the S-matrix is symmetrical,

$$\mathbf{S} = \mathbf{S}^T$$

where the superscript T indicates the transpose of a matrix

(2) for a lossless n-port

$$\sum_{k=1}^{n} S_{ki} S_{ki}^* = 1$$

for all $i = 1, 2, \ldots, n$. This relation is a direct consequence of the power conservation property of a lossless electromagnetic system

(3) for these lossless n-ports, the power conservation condition yields an orthogonality constraint given as follows

$$\sum_{k=1}^{n} S_{ks} S_{ks}^* = 0$$

for all $s, r = 1, 2, \ldots, n$; $s \neq r$.

Based on complex amplitudes a_k and b_k ($k = 1, 2, \ldots, n$) of incident and emerging waves, respectively, the other system of complex parameters is also defined, the transmission or T- coefficients. Transmission coefficients are defined in terms of wave variable normalized with respect to impedances at various ports exactly in the same way as for S-coefficients. Thus, for a two-port we define a set of T-coefficients (Kerns and Beatty, 1969) as

$$\begin{bmatrix} b_1 \\ a_1 \end{bmatrix} = \begin{bmatrix} T_{11} & T_{12} \\ T_{21} & T_{22} \end{bmatrix} \begin{bmatrix} a_2 \\ b_2 \end{bmatrix} \tag{4.4}$$

With respect to the fact that the T-coefficients are not commonly presented in catalogues of microwave parts and that they are not even measured by common microwave measuring techniques, transformation relationships can be used when designing the MIC (Gupta *et al.*, 1981)

$$\begin{aligned} T_{11} &= \frac{(-S_{11}S_{22} + S_{12}S_{21})}{S_{21}} \\ T_{12} &= \frac{S_{11}}{S_{21}} \\ T_{21} &= -\frac{S_{22}}{S_{21}} \\ T_{22} &= \frac{1}{S_{21}} \end{aligned} \tag{4.5}$$

or back

$$S_{11} = \frac{T_{12}}{T_{22}}$$
$$S_{12} = T_{11} - \left(\frac{T_{12}}{T_{22}}\right)$$
$$S_{21} = \frac{1}{T_{22}} \tag{4.6}$$
$$S_{22} = -\frac{T_{21}}{T_{22}}$$

In a more general case, when the microwave circuit is the n-port, in defining the T-matrix it is first necessary to divide all the ports of the given n-port into two parts:

(1) the n-port input interface, where the waves $\mathbf{a}_1=[a_1, a_2, \ldots, a_r]^T$ and $\mathbf{b}_1 = [b_1, b_2, \ldots, b_r]^T$ are present, and

(2) the n-port output interface, where waves $\mathbf{a}_2 = [a_r + 1, a_r + 2, \ldots, a_r + s]^T$ and $\mathbf{b}_2 = [b_r + 1, b_r + 2, \ldots, b_r + s]^T$ occur (Fig. 4.2). It is obvious that $r+s = n$. By definition, the transmission matrix $\mathbf{T}$

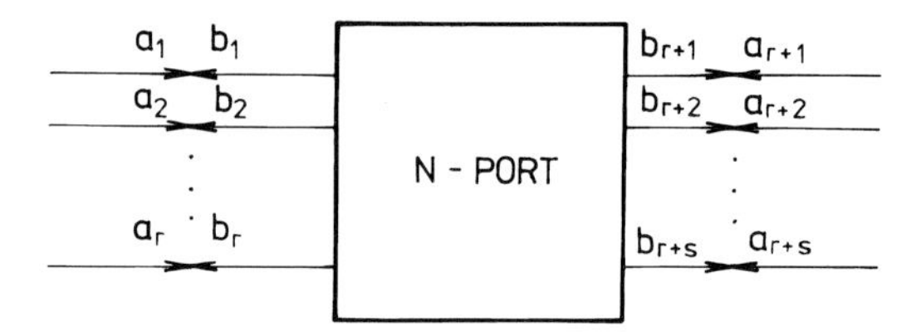

Fig. 4.2 Definition of the input and output ports of the microwave multiport.

of the n-port relates the vector $\mathbf{b}_1$ and $\mathbf{a}_1$ of the input-interface waves to the vectors $\mathbf{a}_2$ and $\mathbf{b}_2$ of the output-interface waves according to the matrix expression

$$\begin{bmatrix} \mathbf{b}_1 \\ \mathbf{a}_1 \end{bmatrix} = \mathbf{T} \begin{bmatrix} \mathbf{a}_2 \\ \mathbf{b}_2 \end{bmatrix} = \begin{bmatrix} \mathbf{T}_1 & \mathbf{T}_2 \\ \mathbf{T}_3 & \mathbf{T}_4 \end{bmatrix} = \begin{bmatrix} \mathbf{a}_2 \\ \mathbf{b}_2 \end{bmatrix} \tag{4.7}$$

where submatrices $\mathbf{T}_1$, $\mathbf{T}_2$, $\mathbf{T}_3$ and $\mathbf{T}_4$ are all $r \times s$ matrices. The transformation relations are as follows

$$\mathbf{T}_1 = \mathbf{S}_2 - \mathbf{S}_1\mathbf{S}_3^{-1}\mathbf{S}_4$$
$$\mathbf{T}_2 = \mathbf{S}_1\mathbf{S}_3^{-1}$$
$$\mathbf{T}_3 = -\mathbf{S}_3^{-1}\mathbf{S}_4 \tag{4.8}$$
$$\mathbf{T}_4 = \mathbf{S}_3^{-1}$$

where $\mathbf{S}_1$, $\mathbf{S}_2$, $\mathbf{S}_3$ and $\mathbf{S}_4$ represent the blocks of the S-matrix and relate the vectors of the emergent and incident waves as

$$\begin{bmatrix} \mathbf{b}_1 \\ \mathbf{b}_2 \end{bmatrix} = \begin{bmatrix} \mathbf{S}_1 & \mathbf{S}_2 \\ \mathbf{S}_3 & \mathbf{S}_4 \end{bmatrix} \begin{bmatrix} \mathbf{a}_1 \\ \mathbf{a}_2 \end{bmatrix} \tag{4.9}$$

In the back transformation the following relationships are valid

$$\begin{aligned} \mathbf{S}_1 &= \mathbf{T}_2\mathbf{T}_4^{-1} \\ \mathbf{S}_2 &= \mathbf{T}_1 - \mathbf{T}_2\mathbf{T}_4^{-1}\mathbf{T}_3 \\ \mathbf{S}_3 &= \mathbf{T}_4^{-1} \\ \mathbf{S}_4 &= -\mathbf{T}_4^{-1}\mathbf{T}_3 \end{aligned} \tag{4.10}$$

It is necessary to note that transformation equations (4.8) and (4.10) are applicable only when $r = s$. In the opposite case, when the number of input ports is not the same as the number of output ports, the submatrices $\mathbf{S}_3$ and $\mathbf{S}_4$ are not square matrices and thus, relevant inversion matrices $\mathbf{S}_3^{-1}$ and $\mathbf{T}_4^{-1}$ do not exist.

In a special case, when $r < s$, the transformation problem can be solved by using the inversion matrix $\mathbf{g} = \mathbf{S}^{-1}$ (Kneppo and Fabian, 1991). Now, the matrix relation connecting the incident and emergent waves of the multiport may be written in the following form

$$\begin{bmatrix} \mathbf{a}_1 \\ \mathbf{a}_2 \end{bmatrix} = \begin{bmatrix} \mathbf{g}_1 & \mathbf{g}_2 \\ \mathbf{g}_3 & \mathbf{g}_4 \end{bmatrix} \begin{bmatrix} \mathbf{b}_1 \\ \mathbf{b}_2 \end{bmatrix} \tag{4.11}$$

where the submatrices, $\mathbf{g}_1$, $\mathbf{g}_4$ are square matrices $r \times r$, $s \times s$, respectively and submatrices $\mathbf{g}_2$, $\mathbf{g}_3$ are rectangular matrices $r \times s$, $s \times r$, respectively of the matrix $\mathbf{g}$. The formulae for submatrices of transmission coefficients of the given n-port are

$$\begin{aligned} \mathbf{T}_1 &= \mathbf{R}^{-1}\mathbf{S}_2 \\ \mathbf{T}_2 &= \mathbf{R}^{-1}\mathbf{S}_1\mathbf{g}_2 \\ \mathbf{T}_3 &= \mathbf{Q}^{-1}\mathbf{g}_1\mathbf{S}_2 \\ \mathbf{T}_4 &= \mathbf{Q}^{-1}\mathbf{g}_2 \end{aligned} \tag{4.12}$$

where $\mathbf{R} = \mathbf{I} - \mathbf{S}_1\mathbf{g}_1$ and $\mathbf{Q} = \mathbf{I} - \mathbf{g}_1\mathbf{S}_1$. The elements of submatrices $\mathbf{g}_1$ and $\mathbf{g}_2$ can be obtained either by the inversion of the matrix

S as whole or block by block. In the second case, the following relationships can be used

$$\mathbf{g}_1 = \left(\mathbf{S}_1 - \mathbf{S}_2\mathbf{S}_4^{-1}\mathbf{S}_3\right)^{-1}$$
$$\mathbf{g}_2 = -\left(\mathbf{S}_1 - \mathbf{S}_2\mathbf{S}_4^{-1}\mathbf{S}_3\right)^{-1}\mathbf{S}_2\mathbf{S}_4$$

For sake of completeness it should be noted that the transformation equations (4.12) present an unambiguous solution only under an assumption of the regularity of matrices **Q** and **R**.

4.1.2 Network matrix decomposition

Each MIC consists of several electric elements, which are electrically connected in such a way that the whole circuit is able to fulfil the electric function required. In general, it can be decomposed into two basic blocks, Fig. 4.3

(1) block E of electric elements also including possibly present sources and Monaco and Tiberio (1974); Pospíšil (1986)

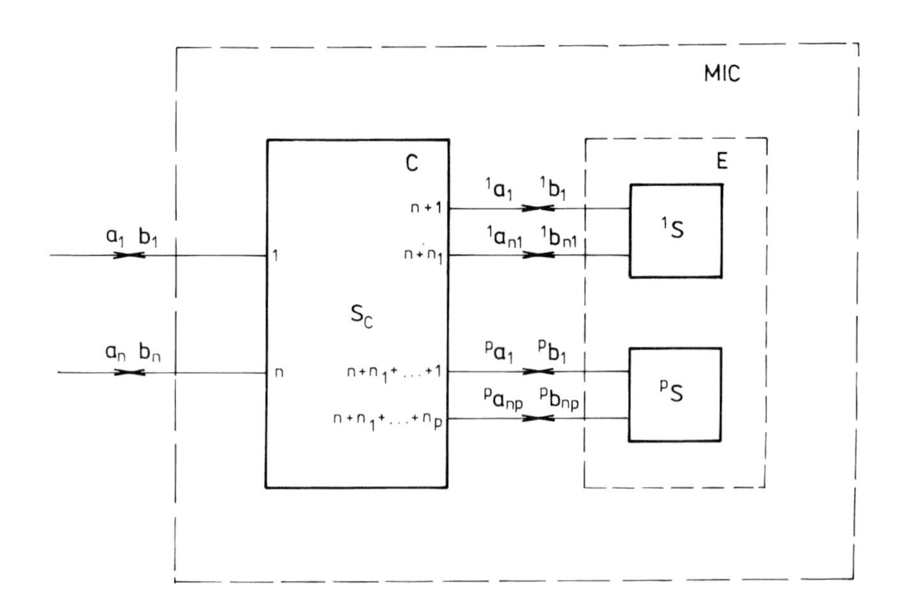

Fig. 4.3 Decomposition of the microwave circuit.

(2) block C of interconnections containing internal connections between elements and external connections between elements and external circuits.

Each of p elements of block E is an n_k-port, $k = 1, 2, \ldots, p$, with a scattering matrix ${}^k\mathbf{S}$, so that for their input waves

$$ {}^k\mathbf{b} = \left[{}^{\mathrm{k}}b_1, {}^{\mathrm{k}}b_2, \ldots, {}^{\mathrm{k}}b_{n_{\mathrm{k}}}\right]^{\mathrm{T}} $$

and output waves

$$ {}^{\mathrm{k}}\mathbf{a} = \left[{}^{\mathrm{k}}a_1, {}^{\mathrm{k}}a_2, \ldots, {}^{\mathrm{k}}a_{n_{\mathrm{k}}}\right]^{\mathrm{T}} $$

the relationship

$$ {}^{\mathrm{k}}\mathbf{b} = {}^{\mathrm{k}}\mathbf{S}{}^{\mathrm{k}}\mathbf{a} + {}^{\mathrm{k}}\mathbf{c} \tag{4.13} $$

holds; when taking into account internal sources in the elementary n_{k}-ports expressed by a vector of output waves

$$ {}^{\mathrm{k}}\mathbf{c} = \left[{}^{\mathrm{k}}c_1, {}^{\mathrm{k}}c_2, \ldots, {}^{\mathrm{k}}c_{n_{\mathrm{k}}}\right]^{\mathrm{T}} $$

for the k-th n_{k}-port. Symbol T stands for the transposition of the matrix.

The vector $\mathbf{b}_E$ of emergent waves of block E is

$$ \mathbf{b}_{\mathrm{E}} = \mathbf{S}_{\mathrm{E}}\mathbf{a}_{\mathrm{E}} + \mathbf{c}_{\mathrm{E}} \tag{4.14} $$

where the scattering matrix $\mathbf{S}_{\mathrm{E}}$ is a diagonal block matrix

$$ \mathbf{S}_{\mathrm{E}} = diag\left\{{}^1\mathbf{S}, {}^2\mathbf{S}, \ldots, {}^{\mathrm{p}}\mathbf{S}\right\} $$

with $\mathbf{a}_{\mathrm{E}}$, $\mathbf{b}_{\mathrm{E}}$ and $\mathbf{c}_{\mathrm{E}}$ column vectors

$$ \mathbf{a}_{\mathrm{E}} = \begin{bmatrix} {}^1\mathbf{a} \\ {}^2\mathbf{a} \\ \vdots \\ {}^{\mathrm{p}}\mathbf{a} \end{bmatrix} \qquad \mathbf{b}_{\mathrm{E}} = \begin{bmatrix} {}^1\mathbf{b} \\ {}^2\mathbf{b} \\ \vdots \\ {}^{\mathrm{p}}\mathbf{b} \end{bmatrix} \qquad \mathbf{c}_{\mathrm{E}} = \begin{bmatrix} {}^1\mathbf{c} \\ {}^2\mathbf{c} \\ \vdots \\ {}^{\mathrm{p}}\mathbf{c} \end{bmatrix} $$

As a whole, the matrix $\mathbf{S}_{\mathrm{E}}$ has its dimension $m \times m$, while

$$ m = \sum_{k=1}^{P} n_k \tag{4.15} $$

The electric characteristics of the block C are characterized by the scattering matrix $\mathbf{S}_{\mathrm{C}}$. This block includes direct short connections between n external inlets of the MIC and m internal inlets of electric elements associated in the block E, so that the order of the matrix $\mathbf{S}_{\mathrm{C}}$ is $(n+m) \times (n+m)$. When all these short connections are mutually matched by their impedances and reflectionless, then the elements ${}^{\mathrm{C}}S_{ij}$ of the matrix $\mathbf{S}_{\mathrm{C}}$ are only (1) units in the i-th line and j-th column of the matrix corresponding to the connection between the i-th inlet, $i = 1, 2, \ldots, n, n+1, n+2, \ldots, n+m$ and j-th inlet $j = 1, 2, \ldots, n, n+1, n+2, \ldots, n+m$ where $i <> j$ and (2) zeros for $i = j$. Thus, the matrix record of the function of the connecting block is as follows

$$\mathbf{b}_{\mathrm{C}} = \mathbf{S}_{\mathrm{C}} \mathbf{a}_{\mathrm{C}} \tag{4.16}$$

where

$$\mathbf{a}_{\mathrm{C}} = \begin{bmatrix} \mathbf{a} \\ \mathbf{b}_{\mathrm{E}} \end{bmatrix} \qquad \mathbf{b}_{\mathrm{C}} = \begin{bmatrix} \mathbf{b} \\ \mathbf{a}_{\mathrm{E}} \end{bmatrix} \qquad \mathbf{S}_{\mathrm{C}} = \begin{bmatrix} {}^{\mathrm{C}}\mathbf{S}_1 & {}^{\mathrm{C}}\mathbf{S}_2 \\ {}^{\mathrm{C}}\mathbf{S}_3 & {}^{\mathrm{C}}\mathbf{S}_4 \end{bmatrix}$$

having in mind that $\mathbf{a}$ and $\mathbf{b}$ are $n \times 1$ column vectors $\mathbf{a} = [a_1, a_2, \ldots, a_n]^T$ and $\mathbf{b} = [b_1, b_2, \ldots, b_n]^{\mathrm{T}}$ of input and output waves in the external ports of the circuit (Fig. 4.3), respectively.

A more detailed inspection of the MIC topology shows that the submatrix ${}^{\mathrm{C}}\mathbf{S}_1$ expresses the mutual interconnection between n external MIC inlets is typically a zero matrix of the order $n \times n$, since the non-zero value of a certain element of this matrix would indicate that the corresponding pair of external inlets is interconnected and that these inlets do not lead to any internal inlet of a certain elementary n_k-port. Submatrices ${}^{\mathrm{C}}\mathbf{S}_2$ and ${}^{\mathrm{C}}\mathbf{S}_3$ of the orders $n \times m$ and $m \times n$, respectively, represent matrices of mutual connections between external MIC inlets and certain internal inlets of a certain n_k-port. For example, in the case of an integrated amplifier, this would be an input connected to the gate of the first transistor and the output connected with the collector of the last transistor of the given amplifying chain. With respect to the fact that short direct connections are electrically reciprocal elements it holds that

$${}^{\mathrm{C}}\mathbf{S}_3 = {}^{\mathrm{C}}\mathbf{S}_2^{\mathrm{T}} \tag{4.17}$$

Last, the $m \times m$ submatrix ${}^{\mathrm{C}}\mathbf{S}_4$ represents a topology of internal connections between the n_k-ports themselves. It is assumed that each internal inlet is connected either (1) with a certain other internal inlet (then no internal inlet is free in such a way that it could be connected with an external inlet) and the given MIC is a closed system (for example an oscillator with self-excitation) or (2) a certain internal inlet is connected with a certain external inlet. In the first case the submatrices ${}^{\mathrm{C}}\mathbf{S}_2$ and ${}^{\mathrm{C}}\mathbf{S}_3$ are null matrices and the submatrix ${}^{\mathrm{C}}\mathbf{S}_4$ is a permutation matrix. In the second case, a non-zero element — unit — occurs in a corresponding line and column of the submatrices ${}^{\mathrm{C}}\mathbf{S}_2$ and ${}^{\mathrm{C}}\mathbf{S}_3$ and vice versa in the corresponding line and column of the submatrix ${}^{\mathrm{C}}\mathbf{S}_4$ there will be a zero. As a definite consequence, the whole matrix ${}^{\mathrm{C}}\mathbf{S}$ will be a permutation $(n+m)\times(n+m)$ matrix symmetrical about the main diagonal with zeros in the main diagonal and with an even number of lines and columns. Here it should be remembered that in this concept of the matrix ${}^{\mathrm{C}}\mathbf{S}$ interconnecting nodes are not included into the connecting block C and that these multiple interconnections are considered as further elementary n_k-ports falling into the block E.

The resulting MIC consisting of the blocks E and C is externally an n-port, whose electric properties can be expressed by the scattering $n \times n$ matrix $\mathbf{S}$. As a matter of fact, $\mathbf{S}$ is a function (1) of all the parameters of elementary n_k-ports mutually represented by the scattering matrix $\mathbf{S}_{\mathrm{E}}$, and (2) of all the interconnections represented by the matrix $\mathbf{S}_{\mathrm{C}}$ so that

$$\mathbf{S} = f\left(\mathbf{S}_{\mathrm{E}}, \mathbf{S}_{\mathrm{C}}\right) \tag{4.18}$$

The matrix $\mathbf{S}$ bounds column vectors of external input and output waves according to the relationship

$$\mathbf{b} = \mathbf{S}\mathbf{a} \tag{4.19}$$

When using active elements in the MIC design, then the resulting circuit is also an active n-port and the vector $\mathbf{c} = [c_1, c_2, \ldots, c_n]^{\mathrm{T}}$ of its source waves is also a function of matrices $\mathbf{S}_{\mathrm{E}}$ and $\mathbf{S}_{\mathrm{C}}$ so that

$$\mathbf{c} = g\left(\mathbf{S}_{\mathrm{E}}, \mathbf{S}_{\mathrm{C}}, \mathbf{c}_{\mathrm{E}}\right) \tag{4.20}$$

The source waves are added to emergent waves of the given MIC, so that, as a final consequence, on the right hand of (4.20), there is also the column vector $\mathbf{c}$ and (4.19) is converted to

$$\mathbf{b} = \mathbf{S}\mathbf{a} + \mathbf{c} \tag{4.21}$$

In deriving the total scattering matrix $\mathbf{S}$ of a microwave integrated circuit, the starting point is (4.16), which will be changed to the system of two matrix equations after inserting for $\mathbf{b}_\mathrm{E}$ from (4.14)

$$\mathbf{b} = {}^\mathrm{C}\mathbf{S}_2\mathbf{S}_\mathrm{E}\mathbf{a}_\mathrm{E} + {}^\mathrm{C}\mathbf{S}_2\mathbf{c}_\mathrm{E}$$

$$\mathbf{a}_\mathrm{E} = {}^\mathrm{C}\mathbf{S}_3\mathbf{a} + {}^\mathrm{C}\mathbf{S}_4\mathbf{S}_\mathrm{E}\mathbf{a}_\mathrm{E} + {}^\mathrm{C}\mathbf{S}_4\mathbf{c}_\mathrm{E}$$

By its solution with respect to $\mathbf{b}$, assuming the regularity of the matrix $(\mathbf{I}_\mathrm{m} - \mathbf{S}_\mathrm{E}{}^\mathrm{C}\mathbf{S}_4)$, the following result is obtained:

$$\mathbf{b} = {}^\mathrm{C}\mathbf{S}_2\left(\mathbf{I}_\mathrm{m} - \mathbf{S}_\mathrm{E}\,{}^\mathrm{C}\mathbf{S}_4\right)^{-1}\mathbf{S}_\mathrm{E}\,{}^\mathrm{C}\mathbf{S}_3\mathbf{a} + {}^\mathrm{C}\mathbf{S}_2\left(\mathbf{I}_\mathrm{m} - \mathbf{S}_\mathrm{E}\,{}^\mathrm{C}\mathbf{S}_4\right)^{-1}\mathbf{c}_\mathrm{E} \tag{4.22}$$

where $\mathbf{I}_\mathrm{m}$ is the unit matrix $m \times m$. Comparing (4.22) and (4.21) it follows that when they should be identically valid for arbitrary input and output waves of the microwave integrated circuit, then

$$\mathbf{S} = {}^\mathrm{C}\mathbf{S}_2\left(\mathbf{I}_\mathrm{m} - \mathbf{S}_\mathrm{E}\,{}^\mathrm{C}\mathbf{S}_4\right)^{-1}\mathbf{S}_\mathrm{E}\,{}^\mathrm{C}\mathbf{S}_3 \tag{4.23}$$

and

$$\mathbf{c} = {}^\mathrm{C}\mathbf{S}_2\left(\mathbf{I}_\mathrm{m} - \mathbf{S}_\mathrm{E}\,{}^\mathrm{C}\mathbf{S}_4\right)^{-1}\mathbf{c}_\mathrm{E} \tag{4.24}$$

Equations (4.23) and (4.24) are the results sought for, since they determine the scattering matrix and the vector of source waves of the MIC consisting of p elementary n_k-ports. As a matter of fact, the number of scattering coefficients of the resulting matrix $\mathbf{S}$ is always lower than the number of scattering coefficients of all the electric elements used in arranging the given MIC and thus, (4.23) and (4.24) yield unambiguous results.

By the concept the MIC synthesis, a problem is considered in the sense of the above mentioned matrix formalism, where (4.18) and (4.20) are solved with respect to the matrix of circuit elements

$$\mathbf{S}_\mathrm{E} = f_1\left(\mathbf{S}, \mathbf{S}_\mathrm{C}\right)$$

and the column vector of source waves

$$\mathbf{c}_\mathrm{E} = g_1\left(\mathbf{S}, \mathbf{S}_\mathrm{C}, \mathbf{c}\right)$$

on the basis of preliminarily determined target values of scattering coefficients of the matrix $\mathbf{S}$ of the resulting MIC and preliminarily known values of source waves $\mathbf{c}_\mathrm{E}$ of elementary n_k-ports.

By a formal solution of the system of (4.16) and (4.21), it is possible to obtain the following result for $\mathbf{b}_{\rm E}$

$$\mathbf{b}_{\rm E} = \left({}^{\rm C}\mathbf{S}_3\mathbf{S}^{-1}\,{}^{\rm C}\mathbf{S}_2 + {}^{\rm C}\mathbf{S}_4\right)^{-1}\mathbf{a}_{\rm E} + \left({}^{\rm C}\mathbf{S}_3\mathbf{S}^{-1}\,{}^{\rm C}\mathbf{S}_2 + {}^{\rm C}\mathbf{S}_4\right)^{-1}\,{}^{\rm C}\mathbf{S}_3\mathbf{S}^{-1}\mathbf{c}$$

from which it follows after a comparison with (4.14) that

$$\mathbf{S}_{\rm E} = \left({}^{\rm C}\mathbf{S}_3\mathbf{S}^{-1}\,{}^{\rm C}\mathbf{S}_2 + {}^{\rm C}\mathbf{S}_4\right)^{-1} \tag{4.25}$$

and

$$\mathbf{c}_{\rm E} = \mathbf{S}_{\rm E}\,{}^{\rm C}\mathbf{S}_3\mathbf{S}^{-1}\mathbf{c} \tag{4.26}$$

With the help of (4.25) and (4.26), and assuming the invertibility of matrices $({}^{\rm C}\mathbf{S}_3\mathbf{S}^{-1}\,{}^{\rm C}\mathbf{S}_2 + {}^{\rm C}\mathbf{S}_4)$ and $\mathbf{S}$, the matrix $\mathbf{S}_{\rm E}$ and vector $\mathbf{c}_{\rm E}$ can be formally calculated, however, this result is not unambiguous. Thus, the formally calculated values cannot be considered as the only characteristics of elementary n_k-ports synthesizing the given MIC. Certain cases can occur, where they even cannot be implemented by any real microwave element. The calculated matrix $\mathbf{S}_{\rm E}$ also will not necessarily be the expected diagonal matrix. The reason for this is that the number of equations in (4.25) is lower than the number of unknown S-coefficients of all the electric elements of the given MIC. In this case, the starting point of the MIC synthesis can be stepwise introducing limiting conditions, which lead to reducing the degree of freedom of solving the problem. So, for example the diagonal form of the matrix $\mathbf{S}_{\rm E}$ is first rewritten and further values or definition intervals of scattering coefficients of certain elements in the MIC proposed are *a priori* chosen and last a criterion is introduced of the optimality of a certain important characteristic of the circuit proposed (for example maximum power amplification or minimum noise number). In most cases, the only useful way will, however, be a method of 'trial and error', which inheres in repeating analyses of one MIC proposed with different stepwise varying values of parameters of electric elements until the desired characteristics of the MIC proposed are achieved. In the MIC synthesis, when choosing a certain number of electric elements with preliminarily known S-coefficients and when determining the interconnection between these elements, then the whole problem of the synthesis is reduced only to searching for the remaining parameters and interconnections. In this case, the function (4.18) can be expressed in the form

$$\mathbf{S} = f\left(\mathbf{S}_{\rm EG}, \mathbf{S}_{\rm EU}, \mathbf{S}_{\rm CG}, \mathbf{S}_{\rm CU}\right) \tag{4.27}$$

where the submatrices $\mathbf{S}_{\mathrm{EG}}$, $\mathbf{S}_{\mathrm{CG}}$ contain *a priori* given parameters and submatrices $\mathbf{S}_{\mathrm{EU}}$, $\mathbf{S}_{\mathrm{CU}}$ searched for, unknown parameters of electric elements and they are solved for

$$\mathbf{S}_{\mathrm{EU}} = f_3\left(\mathbf{S}, \mathbf{S}_{\mathrm{EG}}, \mathbf{S}_{\mathrm{CG}}\right) \tag{4.28}$$

or

$$\mathbf{S}_{\mathrm{CU}} = f_4\left(\mathbf{S}, \mathbf{S}_{\mathrm{EG}}, \mathbf{S}_{\mathrm{CG}}\right) \tag{4.29}$$

The necessary condition for arranging the system of equations (4.28) or (4.29) is that the number of equations in the system should be the same as the number of unknown S-coefficients or number of interconnections to be found.

4.1.3 Synthesis of n-port on the base of elementary two-ports

Many useful and widely used MICs can be obtained when the elementary n_k-ports in block E (Fig. 4.3) are exclusively the two-ports ($n_k = 2, k = 1, 2, \ldots, p$). By meaningful arrangement of the interconnection in block C typical connections are obtained referred to as cascade, series, parallel, series-parallel or parallel-series. The synthesis of this MIC is essentially simplified when choosing, depending on the type of the interconnection, a suitable matrix representation of complex coefficients characterizing the elementary two-ports, i.e. T-, Z-, Y-, H- or G-matrices, respectively (Weinberg, 1966). In this procedure, it is typically necessary to transform the initial scattering matrices of elementary two ports to a matrix of the required type and last, after calculating the resulting matrix of the synthesized MIC to return by the back transformation to the base of scattering coefficients, again. It can happen that the advantage acquired by simplifying the mathematical relationships between the transformed matrices of elementary two-ports will be, on the other hand, abolished by the additional effort and work necessary for the re-calculation of matrix coefficients, when disregarding errors originating from rounding during these calculations.

Cascading of elementary two-ports the block C forms an interconnection between the output port of the k-th, $k = 1, 2, \ldots, p-1$, elementary two-port and input port of the k+1-th elementary two-port, the input of the first elementary two-port being simultaneously

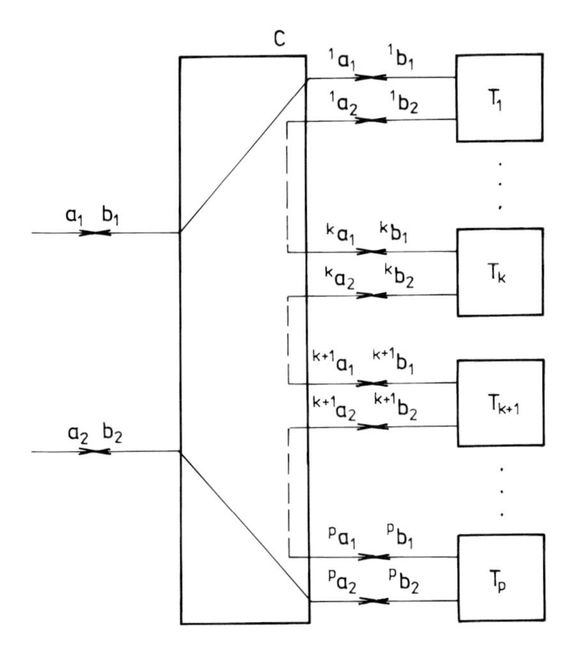

Fig. 4.4 Cascading of the two-ports.

the input MIC port and the output port of the p-th elementary two-port being the output MIC port (Fig. 4.4).

Electric characteristics of the microwave two-port formed by the cascade connection of p elementary two-ports can be most simply expressed by multiplying transmission matrices $\mathbf{T}_1$, $\mathbf{T}_2$, ... , $\mathbf{T}_p$ of elementary two-ports, so that the result is the transmission matrix

$$\mathbf{T} = \mathbf{T}_1\mathbf{T}_2 \ldots \mathbf{T}_\mathrm{p} \qquad (4.30)$$

With respect to the fact that the most frequently used characteristics of microwave parts are not the transmission but the scattering coefficients, the S-matrices of elementary two-ports should be first transformed according to (4.5) and last the resulting **T**-matrix (4.30) should be transformed back to the S-matrix of the whole two-port with the help of (4.6).

In a more general case, where the MIC synthesized is an n-port, which was obtained by a cascade connection of elementary n_k-ports, $k = 1, 2, \ldots, p$, the resulting T-matrix is also given by the product (4.30), however, in this case the elementary T_k-matrices are matrices of the order $2r_k \times 2s_k$ where r_k is the number of input ports and s_k

number of output ports of the elementary n_k-port and it holds that $n_k = r_k + s_k$. Of course, when cascading the elementary n_k-ports, a condition has to be further adhered to that the number of the output ports of the k-th elementary n_k-port equals the number of input ports of the subsequent $k+1$-th n_{k+1}-port. Transformation equations (4.8), (4.10) and (4.12) can be useful in the conversion of S-coefficients to T-coefficients and vice versa.

The calculation of the resulting T-matrix of the microwave multiport by multiplying T_k-matrices of elementary multiports is a simple task from the mathematical standpoint. On the other hand, problems can occur in the calculation resulting from high values of T-coefficients, particularly in the transformation of certain types of microwave parts (idealized and symmetrical microwave devices, for example).

The direct calculation of the scattering matrix of a cascade of microwave $2r$-ports by a method of the generalized composite scattering matrix, which is based on a definition of the star matrix multiplication (Chu and Itoh, 1986; Redheffer, 1962) leads to relationships which are as a matter of fact mathematically more complicated, however, the results usually exert a higher precision.

Under an assumption that the scattering matrix ${}^{\mathrm{A}}\mathbf{S}$ of the first elementary multiport in the cascade is assumed by four submatrices and thus

$$ {}^{\mathrm{A}}\mathbf{S} = \begin{bmatrix} {}^{\mathrm{A}}\mathbf{S}_1 & {}^{\mathrm{A}}\mathbf{S}_2 \\ {}^{\mathrm{A}}\mathbf{S}_3 & {}^{\mathrm{A}}\mathbf{S}_4 \end{bmatrix} \tag{4.31} $$

similarly as the corresponding matrix

$$ {}^{\mathrm{B}}\mathbf{S} = \begin{bmatrix} {}^{\mathrm{B}}\mathbf{S}_1 & {}^{\mathrm{B}}\mathbf{S}_2 \\ {}^{\mathrm{B}}\mathbf{S}_3 & {}^{\mathrm{B}}\mathbf{S}_4 \end{bmatrix} \tag{4.32} $$

of the second multiport, the scattering matrix is assumed as a result of the star multiplication ${}^{\mathrm{A}}\mathbf{S} * {}^{\mathrm{B}}\mathbf{S} = \mathbf{S}$

$$ \mathbf{S} = \begin{bmatrix} \mathbf{S}_1 & \mathbf{S}_2 \\ \mathbf{S}_3 & \mathbf{S}_4 \end{bmatrix} = \begin{bmatrix} {}^{\mathrm{A}}\mathbf{S}_1 + {}^{\mathrm{A}}\mathbf{S}_2\,{}^{\mathrm{B}}\mathbf{S}_1\mathbf{U}_1\,{}^{\mathrm{A}}\mathbf{S}_3 & {}^{\mathrm{A}}\mathbf{S}_2\mathbf{U}_2\,{}^{\mathrm{B}}\mathbf{S}_2 \\ {}^{\mathrm{B}}\mathbf{S}_3\mathbf{U}_1\,{}^{\mathrm{A}}\mathbf{S}_3 & {}^{\mathrm{B}}\mathbf{S}_4 + {}^{\mathrm{B}}\mathbf{S}_3\,{}^{\mathrm{A}}\mathbf{S}_4\mathbf{U}_2\,{}^{\mathrm{B}}\mathbf{S}_2 \end{bmatrix} $$

$$ \text{where } \mathbf{U}_1 = (\mathbf{I} - {}^{\mathrm{A}}\mathbf{S}_4{}^{\mathrm{A}}\mathbf{S}_1)^{-1} \text{ and } \mathbf{U}_2 = (\mathbf{I} - {}^{\mathrm{A}}\mathbf{S}_1{}^{\mathrm{B}}\mathbf{S}_4)^{-1} \tag{4.33} $$

In spite of the fact that the method of the generalized composite scattering matrix is explicitly applicable in cascading the $2r$-ports, i.e. multiports with an even number of ports, it can also be used for an odd number of ports. In this case, it is necessary to introduce dummy ports, by which the given elementary multiport is complemented to the $2r$-port with the same number of input as well as output ports (Overfelt and White, 1989).

4.1.4 Nonlinear MICs

The synthesis of a nonlinear MIC is different from that of a linear MIC because the transfer characteristic between the input and output is a function of the signal level. This has considerable consequences with essential changing of the method by which it is necessary mathematically to describe the non-linear MIC in the synthesis. For all the non-linear microwave elements in the MIC it is first necessary to have available large-signal models describing their transfer function over the desired range of operating levels. Second, the synthesis must account for any new frequency components which are generated. Third, information on the bias sources and drive level must be accounted for in the synthesis. Finally, the solution may not be unique, stable, or even absent.

Mathematical methods, which are currently being most frequently used in the analysis and synthesis of non-linear MICs are principally as follows:

(1) time-domain methods
(2) methods of harmonic balance
(3) Volterra series
(4) describing functions.

Each of these methods has its advantages and disadvantages and thus, in their applications it is necessary to carefully consider which of them is most suitable for the solution of the given circuit problem.

Time-domain methods are based on (1) arranging, usually by applying Kirchhoff laws, a system of differential state equations

$$g\left(\frac{dX}{dt}, X, t\right) = 0 \tag{4.34}$$

characterizing the given MIC and (2) solution of these equations in the time domain. In (4.34)

$$X = [X_1; X_2, ..., X_n]$$

is a set of voltages and currents, typically at different nodes and different time instants. The time derivative of these quantities is a function of time, voltages and currents in the circuit. Equation (4.34) imposes topological constraints on the MIC. Each microwave element in the circuit must then be described by its constitutive relation. For non-linear resistors

$$p\left(v_R, i_r\right) = w_R$$

where w_R is the power dissipated in the resistor. Non-linear capacitors are described by a pair of equations

$$C\left(v_c, q_c\right) = 0$$
$$i_c = \frac{dq_c}{dt}$$

where q_c is the charge on the capacitor, and non-linear inductors by a similar pair, namely

$$L\left(i_L, \phi_L\right) = 0$$
$$v_L = \frac{d\phi_L}{dt}$$

where ϕ_L is the (magnetic) flux across the inductor.

With respect to the fact that, even in the case of a simple MIC, the solution of this system of equations would be rather tedious and time-consuming, numerical integration methods are used (Chua and Lin, 1975; Jastrzebski and Sobhy, 1984; Skelboe, 1980; Skelboe, 1982; Sobhy and Jastrzebski, 1985). In the direct integration, the non-linear differential equations are converted to non-linear algebraic equations by means of the discretization of the time variable. The resulting algebraic equations are solved by iteration, usually by Newton's method. In spite of the fact that this procedure is rather general, it has certain restrictions, particularly in its application for the MIC. For example one of the problems is the choice of a suitable time interval for the numerical integration. Microwave circuits typically have widely separated time constants and a resulting set of equations having widely varying time-constants is difficult to solve. The consequence is that a small time interval of integration must be chosen and a large number of iterations my be required to reach the steady-state, leading to excessive computation time.

A related time-domain method involves the use of associated discrete models, representing the constitutive equations with which the components are mathematically modelled as finite difference equations. The resulting system of non-linear algebraic equations is then solved by the iteration for each time sample.

For strictly periodic excitation, shooting methods are often used to bypass the transient response altogether. An attempt to achieve this is made by correctly choosing the initial conditions so that transients are not excited (Aprille and Trick, 1972a; 1972b; Colon and Trick, 1973; Director, 1971, Director and Current, 1976; Nakhla and Branin, 1977; Norenkov *et al.*, 1987). Shooting methods are attractive for problems that have small periods. Unlike the direct integration methods, the circuit equations are only integrated over one period.

Today there are various popular and widely used time-domain computer software tools such as SPICE (Nagel and Pederson, 1973), MICROWAVE SPICE (Eesof, 1985), CIRCEC (Baucells *et al.*, 1988), and ANAMIC (Jastrzebski and Sobhy, 1984), for instance.

Method of harmonic balance (El-Rabaie *et al.*, 1988, Nakhla and Vlach, 1976; Rizzoli *et al.*, 1988) is based on the distribution of a given non-linear MIC into two subsystems: linear (LS) and non-linear (NS) ones (Fig. 4.5). The subsystems obtained in this way are

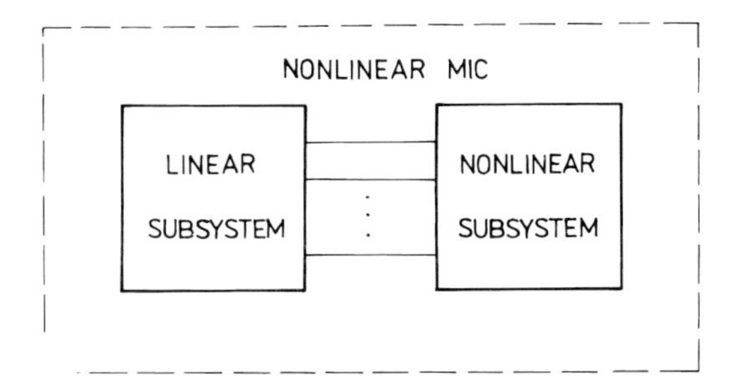

Fig. 4.5 Division of the non-linear circuit into linear and non-linear parts.

separately solved by using suitable mathematical methods: time-domain for a non-linear subsystem, and frequency-domain for a linear one. The solution of a complete system is obtained by combining the particular subsystem solutions taking into account the existing mutual links. Once the system is partitioned, an iteration process is used to obtain a solution. This process relies on achieving convergence of the solution through comparison of the spectral components using Fourier transform.

There are many modifications of the harmonic balance method and software means for the computer solution of non-linear MIC (Camacho-Peñalosa, 1983; Filicori and Naldi, 1983; Gilmore *et al.*, 1985; Gilmore, 1986; Gilmore and Steer, 1991; Steer *et al.*, 1991; Hicks and Khan, 1982a, 1982b; Hwang and Itoh, 1986; Kerr, 1975; Kerr, 1979). In general, the efficiency of the harmonic equilibrium method depends on (1) the type of the non-linear circuit to be solved and method of its division into the linear and non-linear subsystems, (2) mathematical methods employed for solving the subsystems, (3) number of harmonic frequency components considered in the analysis and (4) choice of initial conditions of the iteration solution.

The linear subsystem forms a multiport, where n-ports represent connections with the non-linear subsystems and m-ports serve for coupling external sources to the system. This multiport can be characterized by an arbitrary set of linear parameters (impedance, admittance, hybrid or scattering) depending on which of them are available for the given MIC configuration, which can possibly be obtained by using transformation relationships. When using scattering coefficients

$$\mathbf{S} = \begin{bmatrix} S_{11} & S_{12} & \cdots & S_{1,n+m} \\ S_{21} & S_{22} & \cdots & S_{2,n+m} \\ \vdots & \vdots & \vdots & \vdots \\ S_{n+m,1} & S_{n+m,2} & \cdots & S_{n+m,n+m} \end{bmatrix}$$

then the terminal variables are the incident $\mathbf{a} = [a_1, a_2, \ldots, a_{n+m}]$ and emergent $\mathbf{b} = [b_1, b_2, \ldots, b_{n+m}]$ waves in the ports of the given circuit. This gives rise to a system of linear equations, as follows

$$\mathbf{b} = \mathbf{S}\mathbf{a} \tag{4.35}$$

The linear circuit matrix $\mathbf{S}$ is calculated at each frequency component present in the circuit.

The non-linear subsystem is typically represented by a non-linear set of equations

$$i_j(t) = f_j\left[v_1(t), v_2(t), \ldots, v_j(t), \ldots, v_n(t)\right] \tag{4.36}$$

where f_j is a certain non-linear function and v_j and i_j are the voltage and current in the j-th port, respectively. The equation (4.35) is stated in the frequency domain, and (4.36) in the time domain but simultaneous solution requires that they be expressed in the same domain, usually in the frequency domain. Time-to-frequency conversion is achieved using the discrete Fourier transform or fast Fourier transform.

At the beginning of the solution, it is first necessary to estimate the initial values of the variables v_j and i_j. Iteration between (4.35) and (4.36) is performed using the Fourier transform to obtain frequency components for the time samples obtained from (4.36) until a self-consistent set of variables is attained. In this case, the self-consistence means that these variables adhere to the current continuity equations. The continuity equation for current states that the currents

$$i_N = \left[{}^N i_1, {}^N i_2, \ldots, {}^N i_n\right]$$

from the non-linear subsystem must equal currents

$$i_L = \left[{}^L i_1, {}^L i_2, \ldots, {}^L i_n\right]$$

of the linear subsystem. This corresponds to zero error function formed as

$$E\left(i_L, i_N\right) = \sum_p \sum_j \left|{}^N i_j - {}^L i_j\right|^2$$

The iteration process continues till the value of the error function achieves the preliminarily determined limit. At this moment, the linear and non-linear parts present self-consistent results, due to identical values of currents on both sides of the interconnection in n interface ports. Thus, the quantities i_j and v_j are determined and they can serve as a basis for the calculation of voltages or currents, respectively, in any required point of the circuit by using Ohm's law.

The method of the harmonic balance is an efficient tool of the synthesis of non-linear analogue microwave circuits operating in the periodic or quasi-periodic steady state regime. It can be used to derive the continuous-wave response of numerous non-linear MICs including amplifiers, mixers, and oscillators. On the other hand, the advantage is the fact that when solving certain circuits, it is necessary to consider many frequency components, which makes the solution more complicated with prolonging the time of the calculations.

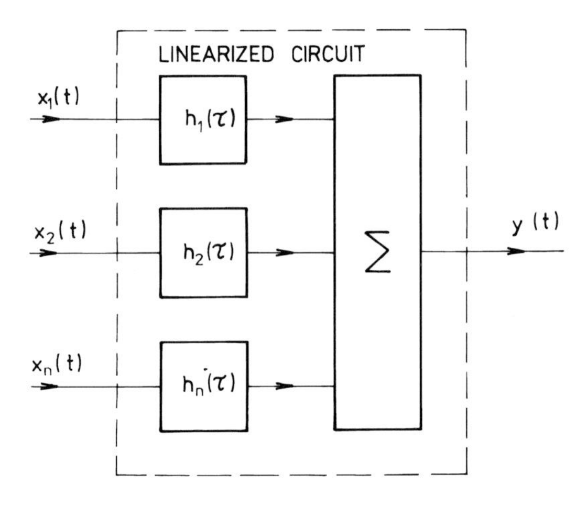

Fig. 4.6 Linearization in the describing function method.

Volterra series method (Obregon, 1985; Weiner and Spina, 1980) is based on a mathematical model of the non-linear circuit (Fig. 4.6) according to which the relationship between the output signal $y(t)$ and input signal $x(t)$ is given by a set of functions

$$y\,(t) = \sum_{n=1}^{\infty} y_n\,(t) \tag{4.37}$$

where $y_n(t)$ are Volterra integrals

$$y_n\,(t) = \int_{-\infty}^{\infty} \cdots \int_{-\infty}^{\infty} h_n\,(\tau_1, \ldots, \tau_n)\, x\,(t - \tau_1) \ldots x\,(t - \tau_n)\; \mathrm{d}\tau_1 \ldots \mathrm{d}\tau_n$$

The n-th order kernel $h_n(\tau_i, \ldots, \tau_n)$ represents the n-th order impulse response. One can see that for a linear system $n = 1$, and we have known convolution integral

$$y\,(t) = \int_{-\infty}^{\infty} h\,(\tau)\, x\,(t - \tau)\; \mathrm{d}\tau$$

giving time response of the linear circuit. In the frequency domain the n-th order output signal is given by

$$Y_n(f) = \int_{-\infty}^{\infty} \cdots \int_{-\infty}^{\infty} H_n(f_1, \ldots, f_n)\, \delta(f - f_1, \ldots, f - f_n) \cdot \prod_{i=1}^{n} X(f_i)\, \mathrm{d}f_i$$

where

$$\int_{-\infty}^{\infty} \cdots \int_{-\infty}^{\infty} \delta(f - f_1, \ldots f - f_n)\, \mathrm{d}f \ldots \mathrm{d}f_n = 1$$

and Fourier transforms of the n-th order kernel and input signal are

$$H_n(f_1, \ldots, f_n) = \int_{-\infty}^{\infty} h_n(\tau_1, \ldots, \tau_n) \prod_{i=1}^{n} \mathrm{e}^{-j2\pi f_i \tau_i}\, \mathrm{d}\tau_i$$

and

$$X(f) = \int_{-\infty}^{\infty} x(t)\, \mathrm{e}^{-j2\pi f t}\, \mathrm{d}t$$

respectively. The complete frequency domain output signal from the non-linear circuit is given as follows

$$Y(f) = \sum_{n=1}^{\infty} Y_n(f)$$

If the input signal is represented by a set of harmonic signals of different frequencies:

$$x(t) = \sum_{k=1}^{p} A_k \mathrm{e}^{j2\pi f_k t}$$

where A_k is a complex amplitude of the k-th frequency component of the input signal and f_k is its frequency, then n-th order frequency domain output signal is given by

$$\sum_{k_1=1}^{p} \cdots \sum_{k_n=1}^{p} \prod_{i=1}^{n} A_{k_i} \mathrm{e}^{j\omega_{k_i} t} \int_{-\infty}^{\infty} \cdots \int_{-\infty}^{\infty} h_n(\tau_1, \ldots, \tau_n) \prod_{i=1}^{n} \mathrm{e}^{-j\omega_{k_i} \tau_i}\, \mathrm{d}\tau_i$$

where $\omega = 2\pi f$. Using (4.37) we obtain

$$y_n(t) = \sum_{k_1=1}^{p} \sum_{k_n=1}^{p} \left(\prod_{i=1}^{n} A_{k_i} \right) H_n(f_{k_1}, \ldots, f_{k_n}) \, \mathrm{e}^{j(\omega_{k_1} + \cdots + \omega_{k_n})t}$$

The method of Volterra series is a suitable approach to solving the non-linear MIC when requiring a complete knowledge of the output signal $y(t)$. It is, however, necessary to remember that the range and complex character of the calculations strongly increases with the order n. The method is very suitable in the calculation of intermodulation products in amplifiers (Hu *et al.*, 1986; Lambranianou and Aitchison, 1977; Minasian, 1980) and in designing frequency multipliers and mixers (Minasian, 1981).

Describing-function method is an approximation method. Its concept is based on attempts to extend methods used in solving linear circuits to the non-linear circuits even on account of certain restrictions and imprecisions. The principle of the procedure is replacing of the non-linear operator, characterizing the given non-linear circuit, by an equivalent (in a certain defined sense) by the linear operator, or possibly by a set of linear operators. The non-linear manifestations of the circuit are simulated by variations of the gain of the describing function depending on the input signal magnitude. The shape deformation of signals resulting from nonlinearities is disregarded during this.

The procedure of forming the describing function depends on (1) the linearization operator used, (2) shape of the input signal (harmonic, periodic, random) and (3) error function representing the residual differences between linear and non-linear outputs.

For example in the case of a non-linear circuit without memory elements the relationship between the output signal

$$y_N(t) = \left[{}^N y_1(t), {}^N y_2(t), \ldots, {}^N y_m(t) \right]$$

and input signal

$$x(t) = [x_1(t), x_2(t), \ldots, x_n(t)]$$

is given by the non-linear operator F as

$$y_N(t) = F[x(t)] \tag{4.38}$$

It is obvious that

$$^{N}y_i(t) = f_i[x_1(t), x_2(t), \ldots, x_n(t)]$$

where f_i, $i = 1, 2, \ldots, m$ is a non-linear function of many variables. The describing function, which should be determined for (4.38) is represented by the linear function

$$y_L(t) = \beta x(t) \tag{4.39}$$

where the matrix $m \times n$

$$\boldsymbol{\beta} = \begin{bmatrix} \beta_{11} & \beta_{12} & \ldots & \beta_{1n} \\ \vdots & \vdots & \vdots & \vdots \\ \beta_{m1} & \beta_{m2} & \ldots & \beta_{mn} \end{bmatrix}$$

is a matrix of linear coefficients, so that

$$^{L}y_i(t) = \beta_{i_1} x_1(t) + \beta_{i_2} x_2(t) + \cdots + \beta_{i_n} x_n(t) \tag{4.40}$$

The optimum value of coefficients β_{ij}, $i = 1, 2, \ldots, m$; $j = 1, 2, \ldots, n$ is determined by minimizing the mean quadratic error

$$E_i = \left\langle \left(^{L}y_i - {^{N}y_i} \right)^2 \right\rangle = \frac{1}{T} \int_{t_0}^{t_0+T} \left[^{L}y_i(t) - {^{N}y_i(t)} \right]^2 \mathrm{d}t \tag{4.41}$$

where T is a period of the lowest frequency components of the input signal. Minimizing (4.41) gives

$$\frac{\partial E_i}{\partial \beta_{ij}} = 0$$

so that

$$\int_{t_0}^{t_0+T} \left[^{L}y_i(t) - {^{N}y_i(t)} \right] x_j(t)\, \mathrm{d}t = 0 \tag{4.42}$$

Integrating (4.42) gives rise to a system of n equations linear with respect to the set of coefficients $\boldsymbol{\beta}_i = [\beta_{i1}, \beta_{i2}, \ldots, \beta_{in}]^T$, i.e.

$$\mathbf{A}\boldsymbol{\beta}_i - \mathbf{B}_i = \mathbf{0} \tag{4.43}$$

where

$$\mathbf{A} = \begin{bmatrix} < x_1 x_1 > & < x_1 x_2 > & \dots & < x_1 x_n > \\ < x_2 x_1 > & < x_2 x_2 > & \dots & < x_2 x_n > \\ \vdots & \vdots & \vdots & \vdots \\ < x_n x_1 > & < x_n x_2 > & \dots & < x_n x_n > \end{bmatrix}$$

$\mathbf{B}_i = [< x_1 f_i >< x_2 f_i > \cdots < x_n f_i >]^T$, and $\mathbf{0}$ is a zero column vector. Solving matrix equation (4.43) with respect to unknown vector $\boldsymbol{\beta}_i$ gives

$$\boldsymbol{\beta}_i = \mathbf{A}^{-1}\mathbf{B}_i \tag{4.44}$$

providing regularity of the matrix $\mathbf{A}$.

A further example can be a dynamic non-linear circuit with a response

$$y_N(t) = \int_{-\infty}^{t} h[x(\tau)] \, d\tau$$

where the input signal $x(t)$ consists of several random or deterministic signals

$$x(t) = \sum_{i=1}^{n} x_i(t) \tag{4.45}$$

The equivalent linear circuit consists of a set of linear filters. The responOnse of this set of filters to the signal is according to 4.45 obviously

$$y_L(t) = \sum_{i=1}^{n} \int_{0}^{\infty} h_i(\tau) x_i(t-\tau) \, d\tau \tag{4.46}$$

The describing function method makes it possible to propose MIC with low levels of non-linearities. When it is used even in the case where the non-linear elements in the circuit exert strongly non-linear characteristics, then the result can encounter considerable errors. The disadvantage is that the method is neither to evaluate nor to correct the error. On the other hand, the describing function method can be, however, useful in predicting the autostability of the MIC proposed.

4.2 BASIC LINEAR CIRCUITS

P. Bezoušek, F. Hrníčko and M. Pavel

4.2.1 Non-reflecting termination

The termination of a microstrip line is most frequently used. At lower frequencies (up to 2 GHz) an inserted resistive chip for flat bonding is used as the non-reflecting termination (Fig. 4.7). The solution

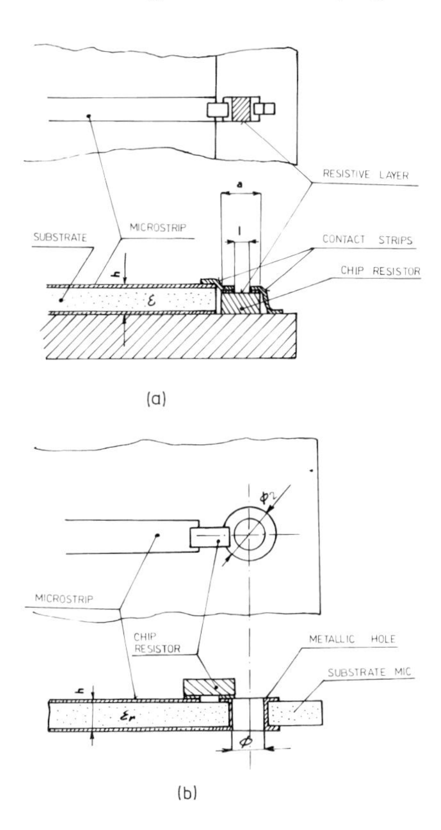

Fig. 4.7 Non-reflecting termination.

with a metal-coated opening is suitable since the substrate with the circuit is self-supporting. This terminating resistance has, however, more considerable parasitic components due to the inductance of the metal-coated opening. Its proper function (VSWR = 1.2) is conditioned by adhering to the following dimension requirements:

$$\begin{aligned} l &< 0.05\lambda_g \\ a &\leq 3l \\ w &\leq 0.1\lambda_g \\ h &\leq 0.05\lambda_g \\ \Phi &\leq h. \end{aligned} \tag{4.47}$$

A quality non-reflecting termination of the microstrip line on ceramic substrate (VSWR < 1.1) can be obtained by a direct deposition of the resistive layer on the substrate, preferably by the thin-layer technique. For a concentrated non-reflecting end obtained by this technique see Fig. 4.8. The wide-band end in Fig. 4.8a has

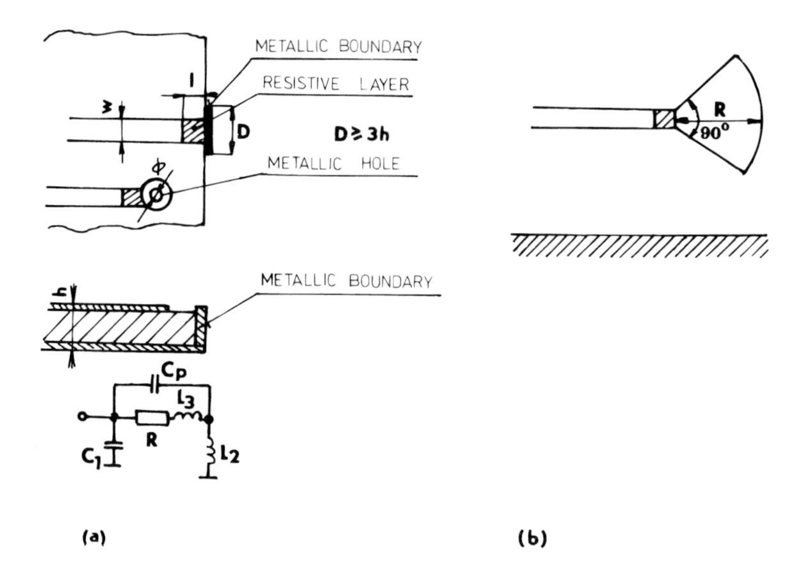

Fig. 4.8 Band-pass (a) and low-pas (b) non-reflecting termination.

a direct short circuit to the earthing side of the plate formed either on the metal-coated side of the plate or by the opening metal-coating. The solution with the opening makes it possible to prepare this termination at any site, which is of importance in more complex circuits, however, in higher frequency bands the results are not as good as those with the termination with a metal-coated edge. On the other hand, in the case of this solution, in addition to topologic problems, there is frequently a technology problem connected with requirements for the formation of precise openings on edges of the substrate. To minimize parasitic elements C_1, C_p and L_2, L_3, it is necessary to choose the ratio $l/w = 0.35$ and the substrate thickness $h \leq 0.05\lambda_g$.

Better parameters in a narrower frequency band can be achieved with a termination according to Fig. 4.8b, where the short-circuit is replaced by a radial quarter-wave stub. This stub represents a short circuit at a side of the resistor in a band of about 20% in the vicinity of the resonant frequency f_0. The stub length R can be calculated by solving the transcendent equation

$$\chi(kR) = \Theta\left(k\frac{w}{2}\right) - \frac{\pi}{2} \tag{4.48}$$

where

$$k = 2\pi f_0\sqrt{\varepsilon_r\varepsilon_0\mu_0}$$

$$\Theta(X) = \arctan\left[\frac{N_0(X)}{J_0(X)}\right]$$

$$\chi(X) = \arctan\left[-\frac{J_1(X)}{N_1(X)}\right]$$

In a particular case it is possible to optimize the stub length experimentally.

The best parameters in a rather wide frequency band can be achieved with a termination in a distributed form according to Fig. 4.9. The resistive layer is shaped to form a narrow wedge. The

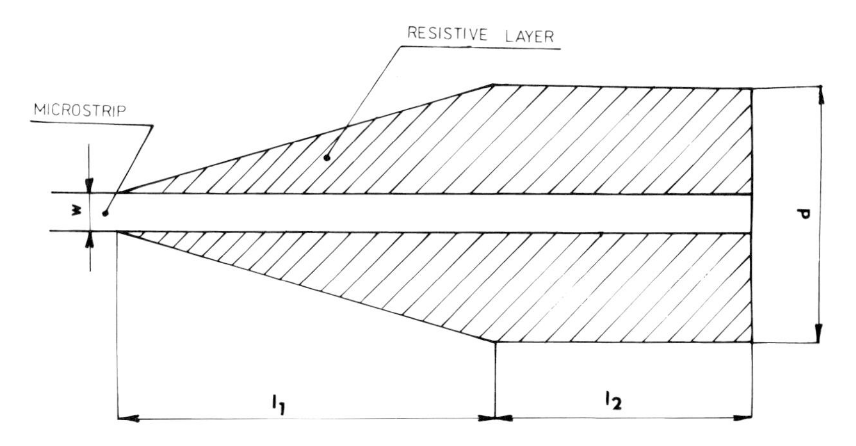

Fig. 4.9 Termination in a distributed form.

dimensions are determined from empirical relationships mentioned with this figure $\lambda_{0\,max}$ is the wavelength in the air at the lowest frequency used and $\lambda_{0\,min}$ is the wave length at the highest frequency.

4.2.2 Resonators

In the MIC, a resonator formed by half-wave segment of a microstrip line or of the other type of line is most frequently used. When designing its length, it is, however, also necessary to consider the correction for the terminal capacitances. In the case of a very weak coupling, the length of the half-wave resonator with two open ends can be approximately calculated from the relationship

$$l = \frac{\lambda_{g0}}{2} - 2\Delta l \tag{4.49}$$

where

$$\lambda_{g0} = \frac{\lambda_0}{\sqrt{\varepsilon_{ef}}}$$

and

$$\Delta l = 0.412h \frac{\varepsilon_{ef} + 0.3}{\varepsilon_{ef} - 0.258} \left(\frac{\frac{w}{h} + 0.264}{\frac{w}{h} + 0.8} \right)$$

where λ_0 is the wavelength in the vacuum, corresponding to the resonant frequency of the resonator.

The resonator shown in Fig. 4.10 is less frequently used. Its resonant frequency

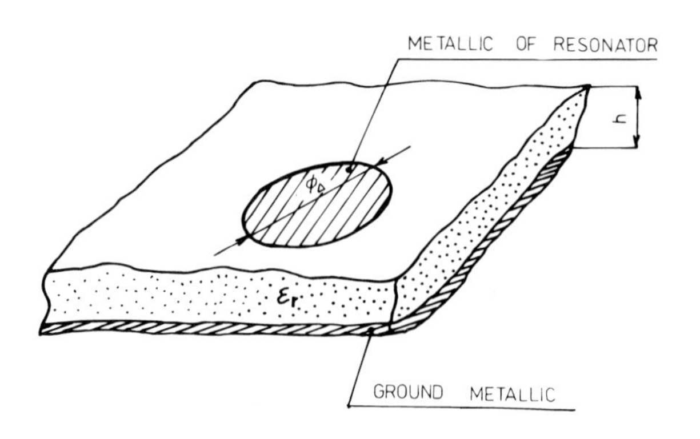

Fig. 4.10 Resonator in the MIC.

$$f_0 = \frac{x_1 c}{\pi\sqrt{\varepsilon_r}D'} \tag{4.50}$$

where

$$D' = D + kh, \ k = \begin{cases} 5.06 & \text{for } \varepsilon_r = 1 \\ 3.4 & \text{for } \varepsilon_r = 2.65 \\ 2.4 & \text{for } \varepsilon_r = 9.6 \end{cases}$$

and x_1 is the first root of the Bessel function of the first kind of the zero order and c is the velocity of light in the vacuum.

4.2.3 Matching transformers

The matching transformer is a lossless reciprocal two-port serving for non-reflecting matching of a generator with an internal real impedance Z_0 to the load with a real impedance $Z_L \neq Z_0$. In an ideal case, the elements of the scattering matrix of this four-port adhere to the following relationships:

$$\begin{aligned} S_{11} &= S_{22} = 0 \\ |S_{12}| &= |S_{21}| = 1 \end{aligned} \tag{4.51}$$

In the MIC technique, this circuit is most frequently formed by a cascade of quarter-wave line segments with different impedance values. The number of steps and impedances of particular segments are chosen depending on the bandwidth required, maximum VSWR and its frequency response. The maximum flat response of the ripple and the ripple according to Chebyshev are most frequently used. For these cases, the number of steps n is established as follows:

a) for a maximum flat response

$$n = INT\left\{0.5 + \frac{\log\left(\frac{\rho_{max}}{k}\right)}{\log\left[\cos\left(\frac{\pi BW}{4}\right)\right]}\right\} \tag{4.52}$$

b) for the Chebyshev response

$$n = INT\left\{0.5 + \frac{\arg\cosh\left[\frac{k^2}{\rho_{max}^2}\left(\frac{1-\rho_{max}^2}{1-k^2}\right)\right]}{\arg\cosh\left[\frac{1}{\sin\left(\frac{\pi BW}{4}\right)}2\right]}\right\}$$

where

$$\rho_{max} = \frac{PSV_{max} - 1}{PSV_{max} + 1}$$

$$BW = 2\frac{f_2 - f_1}{f_2 + f_1}$$

$$k = \frac{R - 1}{R + 1}$$

$$R = \frac{Z_L}{Z_0}$$

and f_2, f_1 are boundary values of the frequency of the band required, $\mathrm{VSWR_{max}}$ is the maximum VSWR in the band $< f_1, f_2 >$, $INT\{x\}$ is the whole part of the expression x. The characteristic impedance of particular steps is calculated from the following relationships:

$$Z_{0i} = Z_0 z_i, \ \ for\, i = 1, 2, \ldots, n \tag{4.53}$$

where

$$z_i = V_i z_{i-1};\ z_0 = 1$$

and

$$V_{n+1-i} = V_i$$

The values of the coefficients V_i can be analytically exactly expressed only for at most four-step transformers.

$$Z_{n+1-i} = Z_i$$

n	maximum flat response	according to Chebyshev response
2	$V_1 = R^{\frac{1}{4}}$	$V_1^2 = \sqrt{C^2 + R} + C$
	$V_2 = R^{\frac{1}{2}}$	$V_2 = \frac{R}{V_1^2}$
3	$V_1^2 + 2\sqrt{R}V_1 - \frac{2\sqrt{R}}{V_1} - \frac{R}{V_1^2} = \frac{3\mu(R-1)}{4-3\mu^2}\delta$	
	$V_2 = \frac{\sqrt{R}}{V_1}$	(4.53a)
4	$V_1 = A_1 R^{\frac{1}{8}}$	$V_1 = \sqrt{R\left(B + \sqrt{B^2 + \frac{A^2}{R}}\right)}$
	$V_2 = R^{\frac{1}{4}}$	$V_2 = \frac{1}{A}$
	$V_3 = \frac{R^{\frac{1}{4}}}{A_1^2}$	$V_3 = \frac{A^2 R}{V_1^2}$,

where

$$\frac{1}{A_1^2} - A_1^2 = 2\frac{R^{\frac{1}{4}} - 1}{R^{\frac{1}{4}} + 1}$$

$$A^2 = \frac{1 - \frac{1}{R}}{2t_1 t_2} + \sqrt{\frac{\left(1 - \frac{1}{R}\right)^2}{4t_1^2 t_2^2} + \frac{1}{R}}$$

$$B = \frac{1}{2}\left(\frac{A}{A+1}\right)^2 \left[(t_1 + t_2)\left(A^2 - \frac{1}{A^2 R}\right) - 2A + \frac{2}{AR}\right]$$

$$C = \frac{(R-1)\,\mu^2}{2\,(2-\mu^2)};\ t_1 = \frac{2\sqrt{2}}{\left(\sqrt{2}+1\right)\mu^2} - 1;\ t_2 = \frac{2\sqrt{2}}{\left(\sqrt{2}-1\right)\mu^2}$$

$$\mu = \sin\left(\frac{\pi w}{4}\right);\ w = 2\frac{f_2 - f_1}{f_2 + f_1};\ V_i = \frac{Z_i}{Z_i - 1}$$

For $n = 1$ both solutions are identical:

$$z_1 = \sqrt{R} \tag{4.54}$$

For $n > 4$ different approximate solutions are used, which are tabulated (Matthaei and Young, 1964). Dimensions of particular segments of the line are determined from the ascertained value of the characteristic impedance Z_{0i} and their length equals 1/4 of a wavelength at a middle frequency. In the case of larger steps in the impedance values, a correction of the length of particular steps can be implemented in the microstrip line by taking into account the scattering capacitance of impedance steps (see 2.2.3, Microstrip discontinuities).

4.2.4 Low-pass filter

The low-pass filter is an important element in microwave systems and it is used for filtering undesirable harmonic frequency components. Most typically it has a transmission characteristic with a maximum flat or Chebyshev response in the transmission band (Fig. 4.11). All the zeros of the transmission functions are at infinity, the cascade implementation of the circuit is in the form of a ladder-type filter. The values of elements of the prototype low-pass filters g_i $(i = 0, 1, 2, \ldots, n+1)$ for the filter step $n \leq 15$ and for different rippling values of the transmission characteristics are for the Chebyshev

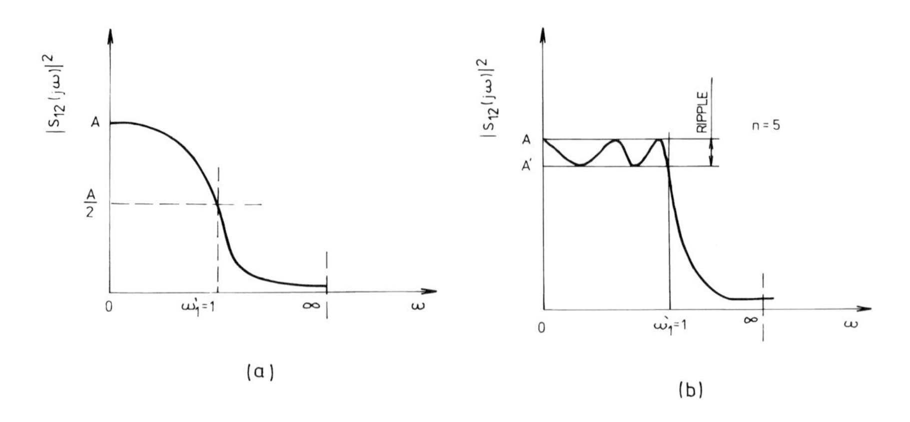

Fig. 4.11 (a) Maximally flat and (b) Chebyshev low-pass prototype filter transmission response.

filters and for the normalized limit angular frequency $\omega = 1$ and unit load presented for example in (Matthaei *et al.*, 1964) and they can also be calculated from simple recurrent relationships. For the termination of a prototype by a real resistance R, each impedance, admittance and impedance and admittance invertors are transformed, as follows:

$$L \to RL, \; C \to \frac{C}{R}, \; K \to RK, \; J \to \frac{J}{R} \tag{4.55}$$

The frequency transformation during the passage from one low-pass prototype to the other is in the form

$$p \to \alpha_t p \tag{4.56}$$

where p is the complex frequency and $\alpha_t = \omega_1'/\omega'$, ω' and ω_1' are limit frequencies of the original and transformed prototypes, respectively.

Fig. 4.12 shows a prototype low-pass filter starting by an inductance. Its microstrip version is formed by cascade coupling of capacitance and inductance segments of the line with impedance Z_C and Z_L. The method of the design in based on solving the problem of the prototype low-pass filter with lumped parameters which adheres to the given tolerance region. The values of the impendances Z_C and Z_L are chosen as necessary, typically Z_C as low as possible and Z_L as high as possible, depending on technology possibilities for the given substrate type. During this, the dimensions of line segments must, in addition, satisfy to the condition that they should be shorter than

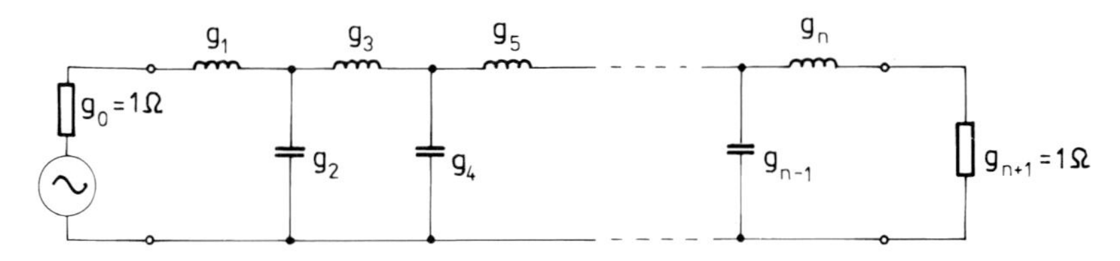

Fig. 4.12 Chebyshev low-pass prototype with element values $g_0 = 1$, $g_1 = \frac{2a_1}{\gamma}$, $g_k = \frac{4a_{k-1}a_k}{b_{k-1}g_{k-1}}$, $g_{n+1} = \begin{cases} 1 & \text{for } n \text{ odd} \\ \coth^2\left(\frac{\beta}{4}\right) & \text{for } n \text{ even} \end{cases}$ where $\beta = \ln\left(\coth\frac{L_A}{17.37}\right)$, $\gamma = \sinh\left(\frac{\beta}{2n}\right)$, $a_k = \sin\frac{(2k-1)\pi}{2n}$, $k = 1, 2, \ldots, n$, $b_k = \gamma^2 + \sin^2\frac{k\pi}{n}$, $k = 1, 2, \ldots, n$ and n is number of sections, L_A is passband ripple [dB].

1/4 of the wavelength in this line. The width of strips w_C and w_L corresponding to the impedances Z_C and Z_L, respectively chosen, is established on the basis of relationships for the analysis of the microstrip line (see part 2.1 transfer lines for the MIC) by the iteration with a simultaneous determination of the relevant wavelengths λ_C and λ_L. In the case of the low-pass filter according to Fig. 4.13, the length of inductive segments is of

$$l_k = \frac{\lambda_L}{2\pi}\arcsin\left[\frac{g_k R}{Z_l} - \frac{Z_C\pi}{Z_L\lambda_C}(l_{k-1} + l_{k+1})\right] \tag{4.57}$$

for $k = 1, 3, \ldots, n$ even; and that of the capacitive segments is of

$$l_k = Z_C\lambda_C\frac{Rg_k}{2\pi} - \frac{Z_C\lambda_C}{2Z_L\lambda_L}(l_{k-1} + l_{k+1}) \tag{4.58}$$

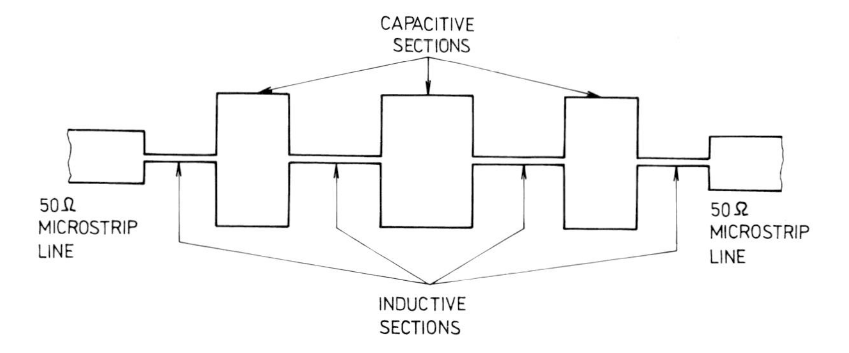

Fig. 4.13 Low-pass filter microstrip arrangment.

for $k = 2, 4, \ldots, n-1$, where l_k is the section length. The initial values for lengths of capacitive segments are calculated from relationship (4.41) with the help of the first term on the right-hand side of the expression. A further calculation by the iteration method is used up to the time when two subsequent values of the section length differ from each other by less than 1%. The length of particular sections should be corrected for the effect of the step change of the microstrip line width during the change of its impedance (see Fig. 2.42 Microstrip line discontinuities).

When it is necessary to change certain zeros of the transmission from infinity to a certain real frequency, then the elliptic or generalized Chebyshev prototype of the transmission function is employed. Both prototypes have comparable selectivities for the same degree of the filter, however, in the implementation based on a distributed circuit, in the case of the elliptic prototype the ratio of impedances in the structure can be as high as 1:10. For the generalized Chebyshev prototype this ratio is below 1:2 and thus, it is preferred in the MIC.

The generalized Chebyshev prototype low-pass filter is used for odd numbers of elements in such a way that the even number of zeros of the transmission is transferred from infinity to a finite frequency into the vicinity of the limit frequency of the low-pass filter. The typical response is shown in Fig. 4.14. The expression for $|S_{12}(j\omega)|^2$ is in the form

$$|S_{12}(j\omega)|^2 = \frac{A}{1 + \varepsilon^2 F_n^2(\omega)} \tag{4.59}$$

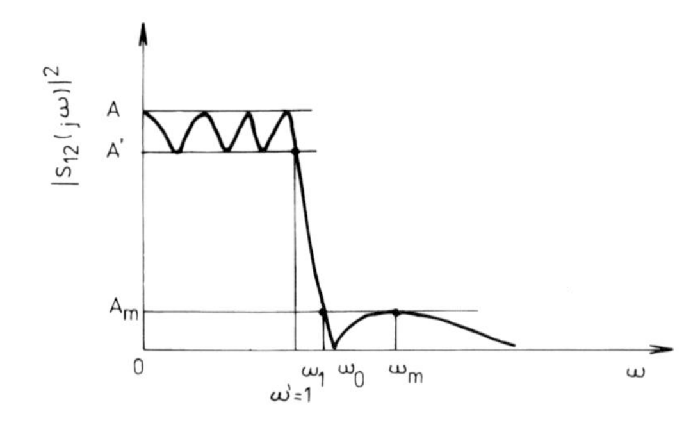

Fig. 4.14 Generalized Chebyshev prototype low-pass filter response ($\omega' = 1$ is band edge frequency, ω_1 is A_m short frequency, ω_0 is transmission zero frequency, ω_m is $A - m$ local maximum frequency and A_m is stop band local maximum signal amplitude).

where

$$\varepsilon = \sqrt{10^{\frac{RL}{10}} - 1} \tag{4.60}$$

and A is the amplitude of the input signal and RL [dB] is the minimum level of the back loss in the transmission band. For the function $F_n(\omega)$ it holds (Alseyab and Ashoor, 1986) that

$$F_n^2(\omega) = \cosh^2 \left\{ (n-q) \operatorname{arg\,cosh} \left[\omega \sqrt{\frac{\omega_0^2 - 1}{\omega_0^2 - \omega^2}} \right] + q \operatorname{arg\,cosh}(\omega) \right\} \tag{4.61}$$

where q is odd and it stands for the number of zeros of the transfer in infinity and ω_0 is the angular frequency, to which $(n-q)$ poles are transformed from infinity. The minimum attenuation in the band-stop filter occurs for the frequency

$$\omega_m^2 = \omega_0^2 + \frac{n-q}{q} \omega_0 \sqrt{\omega_0^2 - 1} \tag{4.62}$$

Some of the possible structures of prototype filters are shown in Fig. 4.15. The values of prototype elements of these structures for $n = 7 - 15$ and $q = 5$ are in Alseyab and Ashoor (1986), for $q = 3$ in Alseyab (1982). The method of implementing the filters is explained in Fig. 4.16a. The values of prototype elements are transformed into the distributed region by the Richardson transformation

$$p \rightarrow B \tanh(ap) \tag{4.63}$$

where the constants a and B are suitably chosen in order that it would be possible to form parallel resonance branches with the help of homogeneous open stubs with a characteristic admittance Y_0 (Fig. 4.16b). The characteristic admittance Y of the series resonance circuit

$$Y = \frac{pC}{1 + \left(\frac{p}{\omega_0}\right)^2} \tag{4.64}$$

where $\omega_0^2 = 1/LC$. By the use of the Richardson transformation and assuming that $B = \omega_0$, (4.65) is arranged to

$$Y = 0.5BC \tanh(2ap) \tag{4.65}$$

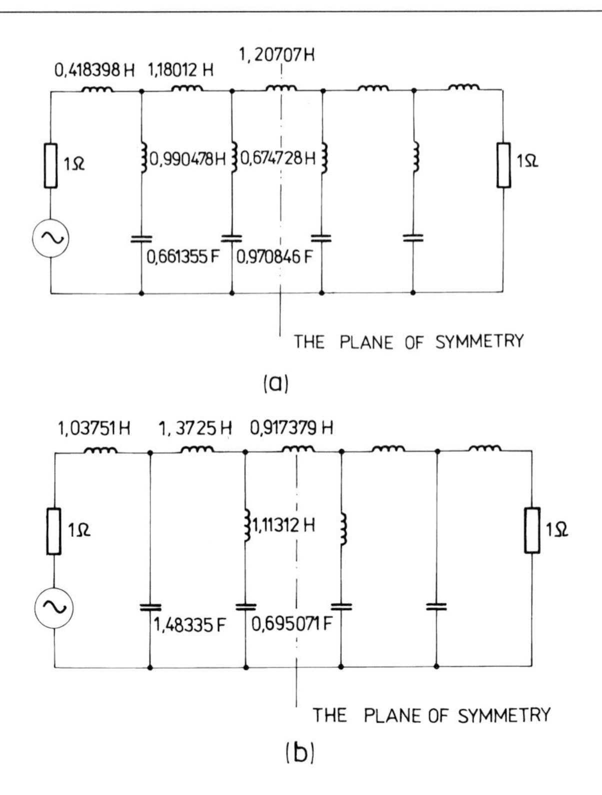

Fig. 4.15 Degree nine generalized Chebyshev prototype low-pass filter (a) prototype with one transmission zero at infinity ($\varepsilon = 0.1\,\text{dB}$, $L_m = 40\,\text{dB}$) (b) prototype with five transmission zeros at infinity ($RL \geq 20\,\text{dB}$, $L_m \geq 40\,\text{dB}$).

from which the characteristic admittance of the stub

$$Y_0 = BC \tag{4.66}$$

The constant a is determined by using the Richardson transformation for the limit frequency of the filter, so that

$$\omega_c = B \tan(2a\pi f_c) \tag{4.67}$$

$$a = \frac{1}{2\pi f_c} \arctan\left(\frac{\omega_c}{\omega_0}\right) \tag{4.68}$$

where ω_c is the limit angular frequency of the prototype ($\omega_c = 1\,\text{rad}\,\text{s}^{-1}$) and f_c is the limit frequency of the filter proposed.

The electric length of resonant stubs should be $\pi/2$ at a frequency

$$f_0 = \frac{1}{8a} \tag{4.69}$$

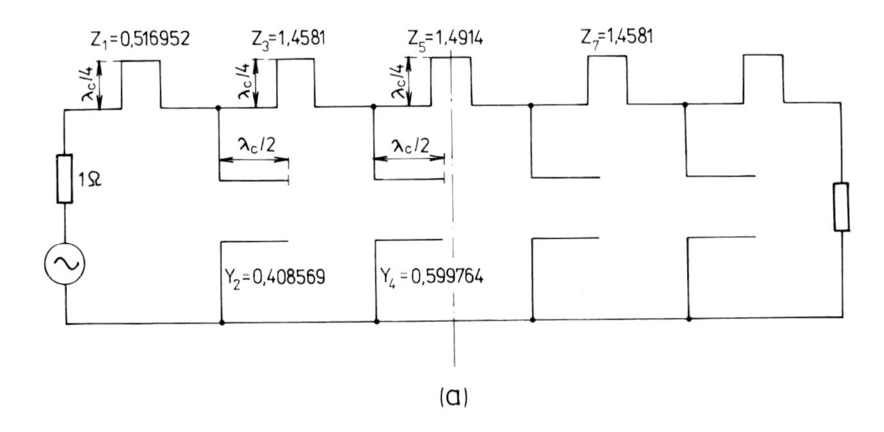

(a)

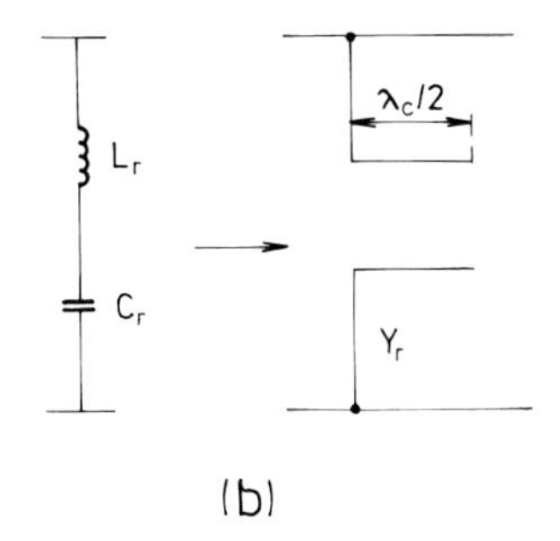

(b)

Fig. 4.16 Distributed quasi low-pass prototype ($\varepsilon = 0.1\,\text{dB}$, $L_s = 40\,\text{dB}$) (a) and the transformation of shunt series resonant circuits (b) Y_r is chracteristic stub admittance.

determined on the basis of (4.65) and (4.70).

The length l_r of resonators follows from the condition that it should be a quarter-wave for the frequency f_0, so that

$$l_r = \frac{0.25v}{f_0} \tag{4.70}$$

where v is the wave propagation velocity in the line.

The direct consequence of the Richardson transformation is that the transformed distributed circuit for parallel capacitors in the prototype is obtained as an open stub with a characteristic admittance

$$Y_{0p} = BC_p \tag{4.71}$$

which can be quarter-wave at a frequency $2f_0$. Then its length

$$l_p = \frac{0.125v}{f_0} \tag{4.72}$$

The series elements L_s are formed by short segments of high-impedance line of a length (Matthaei *et al*,. 1964)

$$l_s = \frac{vRL_s}{2\pi f_c Z_L} \qquad \text{for } l_s < \frac{v}{8f_0} \tag{4.73}$$

where Z_L is the characteristic impedance of the line.

4.2.5 RF chokes

The microwave choke is a filtration circuit element which serves for forming a defined input impedance, usually short-circuit or disconnected circuit. It is used for high-frequency short-circuits, inlets of the bias to semiconductor elements and at the output of high-frequency signals. It can be based on a cascade of segments of the line of high and low impendances with an electric length $\pi/2$ (Fig. 4.17).

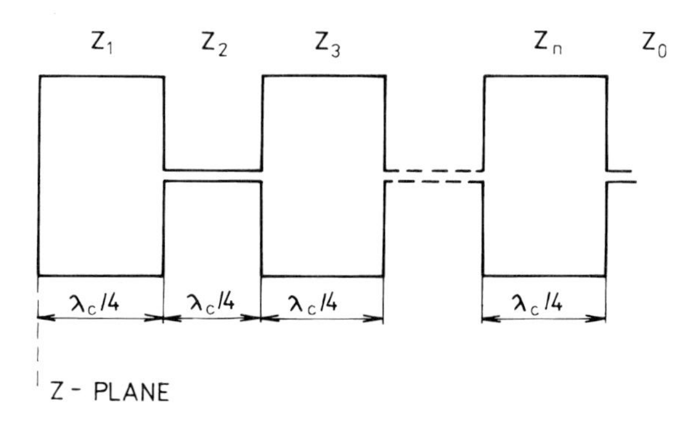

Fig. 4.17 Short circuiting at Z-plane RF choke arrangement (λ_c is wavelength in the capacitive or inductive line section).

For a larger band width, simple or multiple stubs are frequently used in the shape of circular sectors (Fig. 4.18).

The most simple type of the choke has impedance satisfying to the condition

$$Z_1 = Z_3 = Z_5 = \cdots = Z_c \qquad Z_2 = Z_4 = Z_6 = \cdots = Z_L$$

When the number of steps of the choke is higher than 2, then the input impedance is not very dependent on the section of the choke

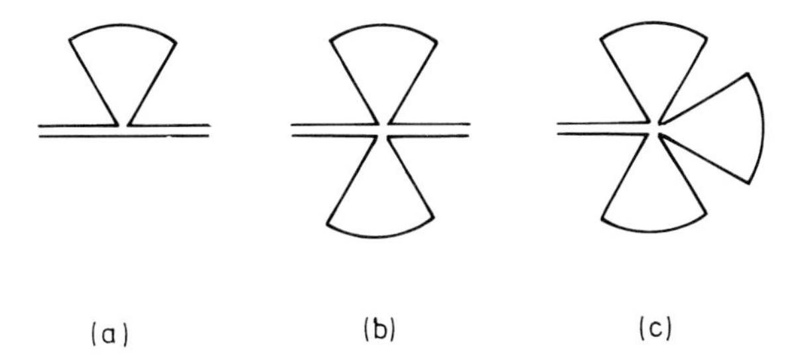

Fig. 4.18 RF chokes using simple or multiple radial stubs (a) simple radial stub (b) butterfly radial stub (c) clover radial stub).

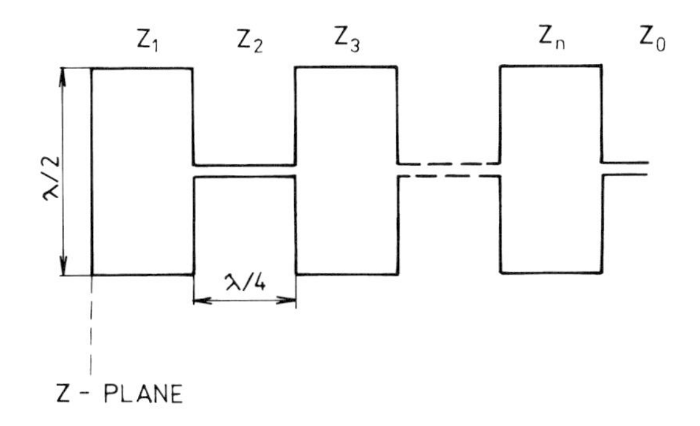

Fig. 4.19 Alternative RF choke arrangement.

termination. With respect to the fact that after the last section there is an impedance higher by about 50 Ohm it is suitable to terminate the choke by a low-impedance section. In certain cases, at higher frequencies it is impossible to implement low-impedance segments with a sufficiently low characteristic impedance, since the strip width w is restricted by the condition $w < \lambda/2$ where λ is the wavelength in the line at the operation frequency. Then a rather different configuration is chosen, as shown in Fig. 4.19. From the figure it is obvious, that this configuration is formed by parallelly coupled low-impedance open quarter-wave stubs representing a resonator transversely situated to the choke axis. The high-impedance segments of the line are connected to the site with the lowest impedance, in the resonance theoretically to the short circuit. Parallel connected quarter-wave stubs can be replaced by radial stubs having in their centre the impedance chosen, in this case the short circuit. The choke design in restricted by technology possibilities for the given substrate. For high-impedance segments of the line, the narrowest strip is chosen, which can be produced on the given substrate and its impedance Z_L is determined by the line analysis. When the choke is implemented, then it is necessary to satisfy to the condition (Bezoušek *et al.*, 1977)

$$Z_L > 1.8Z_0$$

where Z_0 is the impedance of the line which will be connected to the choke. The strip width w_c of the line with a low impedance Z_C is chosen

$$w_c \approx \frac{3}{8\lambda}$$

where $Z_L/Z_C > 2$ since otherwise this choke cannot be implemented. In the calculation of geometric dimensions of the choke it is also necessary to take into account the effect of discontinuities.

A design of a simple radial stub is described in Atwater (1983). Its geometry is shown in Fig. 4.20. The approximate analysis is

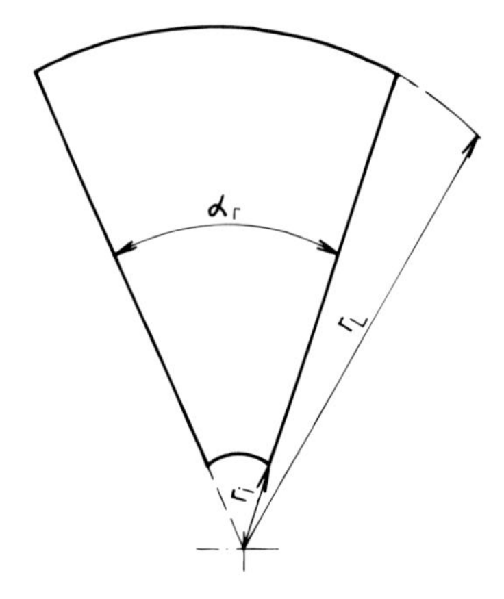

Fig. 4.20 Open radial stub geometry (r_L is outer radius, r_i is inner radius and α_r is radial stub angle).

based on a field distribution similar to that in a ring resonator. The propagation modes have in general a longitudinal component of the electric field E_z, and radial and azimuthal components of the magnetic field H_r and H_φ, respectively. The lowest mode which can be excited has only components of the field

$$E_z(r) = AH_0^{(1)}(kr) + BH_0^{(2)}(kr) \tag{4.74}$$

$$H_\varphi(r) = \frac{-k}{j\omega\mu_0}\left[AH_1^{(1)}(kr) + BH_1^{(2)}(kr)\right] \tag{4.75}$$

where $H_n^{(i)}$ is the Hankel function of the i-th kind, n-th order and $k = \sqrt{\varepsilon_r\varepsilon_0\mu_0}$ is the wave number. The terms $H_n^{(1)}$ represent waves directed radially inwardly, whereas terms $H_n^{(2)}$ waves directed radially outwardly. The constants A, B can be determined by establishing the input wave impedance

$$Z_i = \frac{E_z(r_i)}{H_\varphi(r_i)}$$

assuming that the loading wave impedance is known

$$Z_L = \frac{E_z(r_L)}{H_\varphi(r_L)}$$

The input wave impedance

$$Z_i = Z_{0i}\left[\frac{Z_L\cos(\Theta_i - \psi_L) + jZ_{0L}\sin(\Theta_i - \Theta_L)}{Z_{0L}\cos(\psi_i - \Theta_L) + jZ_L\sin(\psi_i - \psi_L)}\right] \tag{4.76}$$

where Z_0 is the wave impedance of the radial line given by the relationship

$$Z_0(kr) = \sqrt{\frac{\mu_0}{\varepsilon_r\varepsilon_0}\frac{J_0^2(kr) + N_0^2(kr)}{J_1^2(kr) + N_1^2(kr)}} \tag{4.77}$$

and angles

$$\Theta(kr) = \arctan\left(\frac{N_0(kr)}{J_0(kr)}\right) \tag{4.78}$$

$$\psi(kr) = \arctan\left(\frac{-J_1(kr)}{N_1(kr)}\right) \tag{4.79}$$

By the subscript i or L in parameters Z_0, Θ, φ in (4.76) the calculation of these parameters for radii r_i or r_L, respectively, is denoted. The total voltage between the conductors is of $-E_z h$, the total radial current in the microstrip conductor outwardly is $2\pi rH_\varphi$ and thus, the input terminal impedance

$$Z_T = \frac{-0.5hZ_i}{\pi r_i}$$

for a circular resonator. The microstrip radial stub is open (ideally $Z_L = \infty$) and $\alpha_r < 2\pi$. When having the input signal uniformly connected along the circuit with an internal diameter r_i, then

$$Z_T = j\frac{hZ_0(kr_i)\cos[\Theta(kr_i) - \psi(kr_L)]}{\alpha_r r_i \sin[\psi(kr_i) - \psi(kr_L)]} \tag{4.80}$$

Scattering fields, which are not included in the model, can be approximately described by replacing ε_r by the effective permittivity ε_{ef}.

It was checked that the width of the band is enlarged with increasing angle α_r. Unfavourable transient effects are eliminated by choosing $\alpha_r = \pi$. Larger width of the band can be achieved by using multiple stubs.

By arranging the wide band disconnected circuit at point A according to Fig. 4.21, coupling is made possible of the bias for parallelly connected active elements.

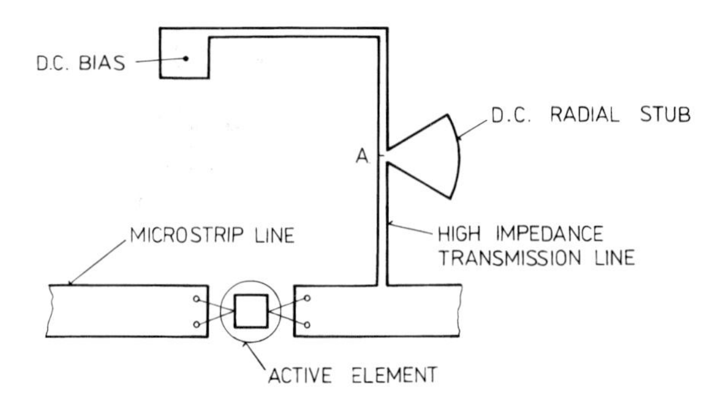

Fig. 4.21 Bias circuits for parallelly connected active elements using radial stubs.

4.2.6 Band-pass filters

In hybrid microwave integrated circuits of the microstrip form, band-pass filters are frequently used with parallel coupled sections. They are attractive first of all due to their small dimensions, high reproducibility and low price. In the design of filters for higher microwave frequency bands problems increase connected with precise determination of the line parameters and the effect of discontinuities increases which should be sufficiently precisely compensated for. When the computer design should be effective, then it is unavoidable to include into the computer program approximation models describing the behaviour of the circuit with a precision exceeding the errors resulting from technology allowances of the production and measurement.

The basis of the synthesis is a prototype low-pass filter, i.e. ladder-type circuit according to Fig. 4.22 A whose limit frequency and loading impedance are standardized to unity. For parallel coupled filters the Chebyshev or maximum flat prototype are frequently used. Values of elements of the actual band filter are obtained by a suitable frequency and impedance transformation of the prototype low-pass filter for the given step of the filter N and rippling of the transmission characteristic in the transmission band (Fig. 4.22 B,C) (Matthaei *et al.*, 1964). The inductive elements are transformed to series resonant circuits and the capacitive elements to parallel resonant circuits, resonating at a mean angular frequency ω_0. The parallel resonant circuit can be approximated in the case of a low load by a half-wave line with a characteristic impedance Z_0 (Fig. 4.22 D). With the help of ideal impedance invertors, all the resonance circuits are changed to circuits of the same type (Fig. 4.22 E) and with the help of the invertor K_{01} a simple impedance transformer is introduced, which transforms the resistance R to the value of the terminating impedance Z_0 (Fig. 4.22 F). All the parallel resonance circuits are weakly loaded by high-impedance invertors and they can be replaced by equivalent line segments (Fig. 4.22 G) with inserting before them a unit element with the impedance Z_0 (Fig. 4.22 H). When assuming the same velocities of the propagation of even and odd modes, the circuit consisting of the impedance invertor and two quarter-wave lines with the characteristic impedance Z_0 is equivalent to the line segment of a parallel coupled line. The impedance of the even and odd modes is given by the relationships shown in Fig. 4.22 I.

When it is necessary to obtain a band pass filter with a band width exceeding 20%, it is more suitable to use the following procedure for the calculation of the impedance of the even and odd modes of the propagation (Matthaei *et al.*, 1964)

$$Z_0^{-1} K_{k,k+1} = \frac{1}{\sqrt{g_k g_{k+1} \omega_1'}} \qquad \text{for } k = 0 \quad \text{and} \quad k = n \tag{4.81}$$

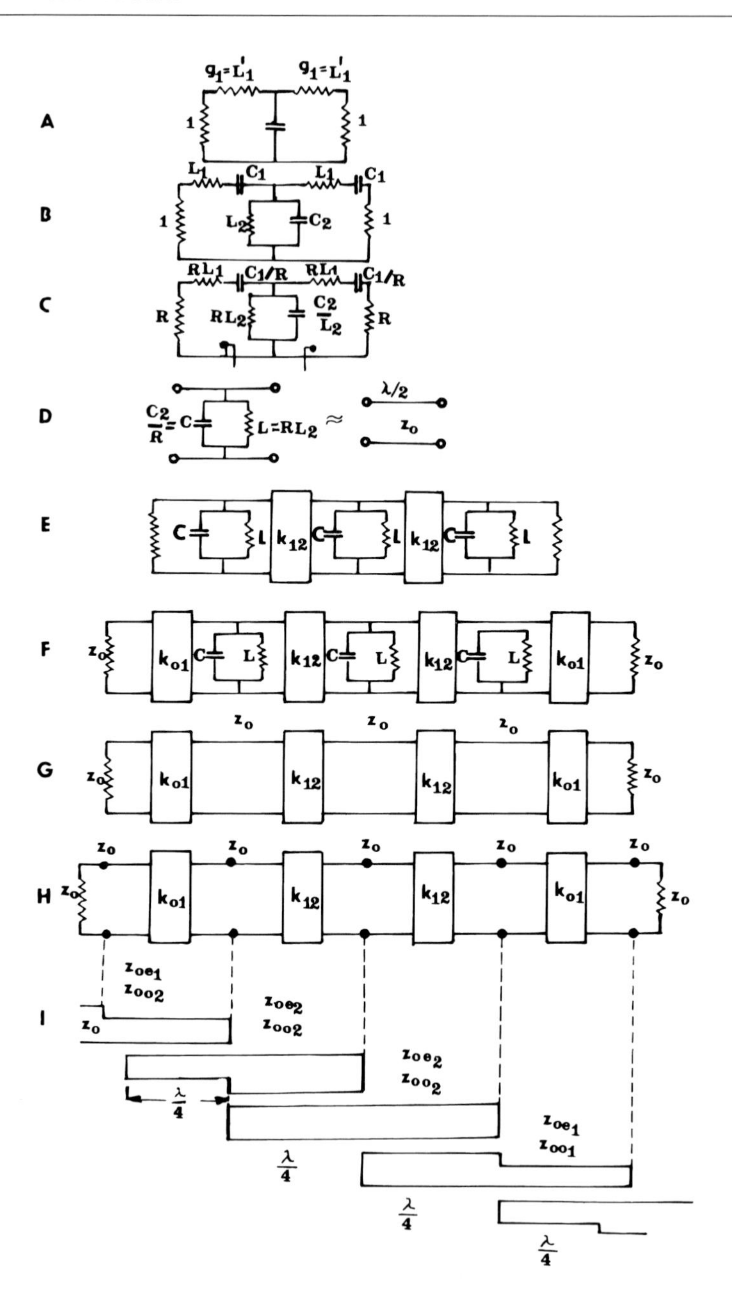

Fig. 4.22 Prototype low-pass filter.

and

$$
\begin{aligned}
Z_0^{-1} K_{k,k+1} &= \frac{1}{\omega_1' \sqrt{g_k g_{k+1}}} \\
N_{k,k+1} &= \sqrt{\left(\frac{K_{k,k+1}}{Z_0}\right)^2 + \frac{\tan^2 \Theta_1}{4}}
\end{aligned}
\tag{4.82}
$$

for $k = 1$ to $n - 1$ where

$$
\Theta_1 = 0.5\pi \left(1 - \frac{w}{2}\right) \tag{4.83}
$$

For impedances of the even and odd modes

$$
(Z_{0e})_{k,k+1} = hZ_0 \left(N_{k,k+1} + Z_0^{-1} K_{k,k+1}\right) \tag{4.84}
$$

and

$$
(Z_{0o})_{k,k+1} = hZ_0 \left(N_{k,k+1} - Z_0^{-1} K_{k,k+1}\right) \tag{4.85}
$$

respectively.

By choosing the parameter

$$
h = \frac{1}{0.5 \tan \Theta_1 + \left(\frac{K_{01}}{Z_0}\right)^2} \tag{4.86}
$$

the usually required identical width of strips in the input section is achieved.

It is frequently necessary to form a very tight coupling between certain resonators with the help of a multistrip coupling section (Ou, 1975). In this structure, k independent modes of the propagation are in general propagated (k is the number of strips in the coupling section). When alternately switching over the ends of resonators at both ends, as shown in Fig. 4.23, two basic modes of the propagation are

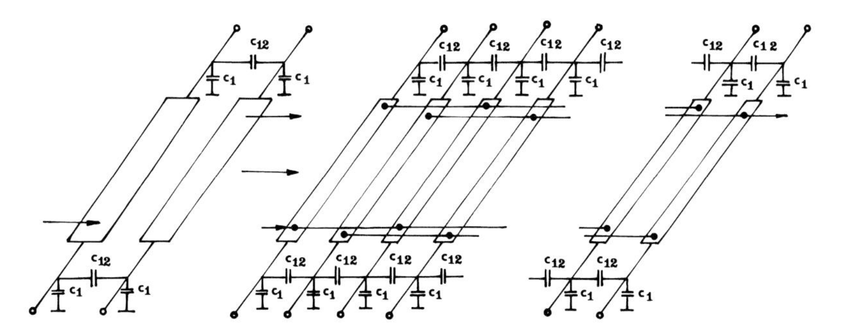

Fig. 4.23 Multistrip coupling section.

obtained, again. When neglecting the inductivity of overswitching elements and couplings for one resonator, the admittances of these modes can be described by relationships presented in Fig. 4.23.

The geometric dimensions of the filter structure are designed in accordance with equations in part 2.1.7 Coupled lines. In the calculation of the length of microwave resonators, it is also necessary to consider the effect of discontinuities i.e. of the scattering capacitance of the open ends of resonators and of the impedance step at the centre of resonators. The simplified representation (Fig. 4.24b) of an actual discontinuity (Fig. 4.24a) in a coupled filter has an equivalent diagram according to Fig. 4.24c. Then mean extension is most

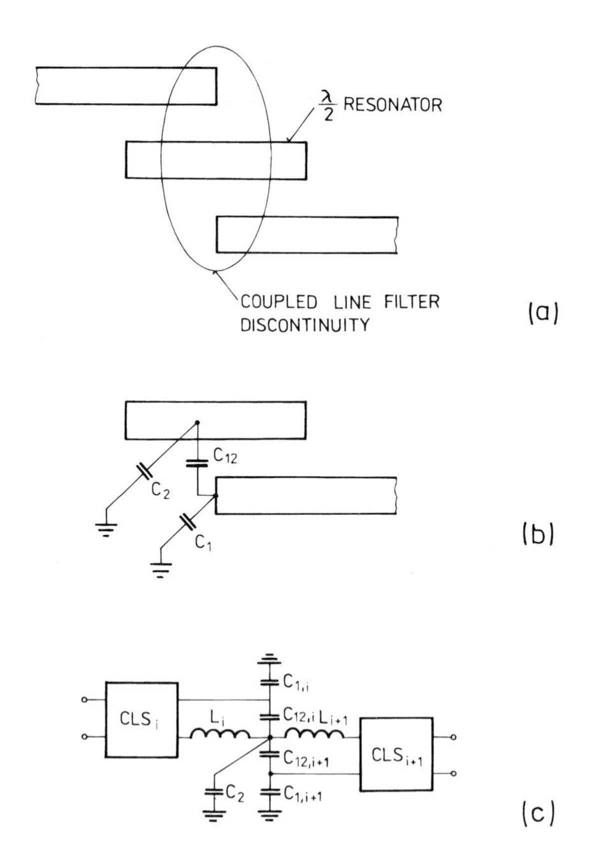

Fig. 4.24 Open end discontinuity in a coupled line filter (a) discontinuity geometry (b) simplified representation (c) equivalent circuit.(CLS is coupled line section, $C_{12,i}$ is strip to ground equivalent capacitance and $C_{12,i}$ is strip to strip equivalent capacitance.)

frequently determined for even and odd modes as a value for a simple strip (see part 2.2 Discontinuities). Relationships can also be used for parallel open ends with determining from the relative extension for the even and odd modes $l_{e,o}$ the value of the capacitance C_1

$$C_{e,o} = \frac{1}{\omega Z_{0e,o}(f)} \tan \frac{\omega \Delta l_{e,o} \sqrt{\varepsilon_{f_{e,o}}(f)}}{c} \tag{4.87}$$

where $\omega = 2\pi f$ and c is the light velocity in the vacuum. The value of the capacitance C_{12} can be neglected for wider slots, for narrower slots design graphs from Gupta (1979) can be used. The relative extension of the resonator for the odd mode

$$\Delta l_0 = \frac{C_0}{C_{\infty,o}} \tag{4.88}$$

where $C_0 = C_1 + 2C_{12}$ and $C_{\infty,0}$ is the capacitance of a unit length line segment for the unpaired mode. With respect to the fact that C_2, C_0, the effect is low and corrections can be introduced by extending the resonator length, with taking into account the effect of discontinuities is most`properly described by the relationship

$$l = \sqrt{l_e l_o} \tag{4.89}$$

where $l_{e,o} = (\lambda/4)_{e,o} - \Delta l_s$ and $\lambda_{e,o}$ is the wave length of the even and odd modes in the line.

4.2.7 Attenuators

In integrated circuits, in lines with the quasi-TEM mode, attenuators are used in the form of the resistance T or Π-element are used (Fig. 4.25). The values of resistances in both cases can be determined from the condition of the attenuator matching to the line of the characteristic impedance Z_0 and of the nominal attenuation L [dB] at a limit zero frequency:
for Π-element

$$R_1 = \frac{Z_0}{\sqrt{1-\beta}}$$

$$R_2 = \frac{Z_0}{\beta} \frac{1-\beta}{1+\frac{Z_0}{R_1}}$$

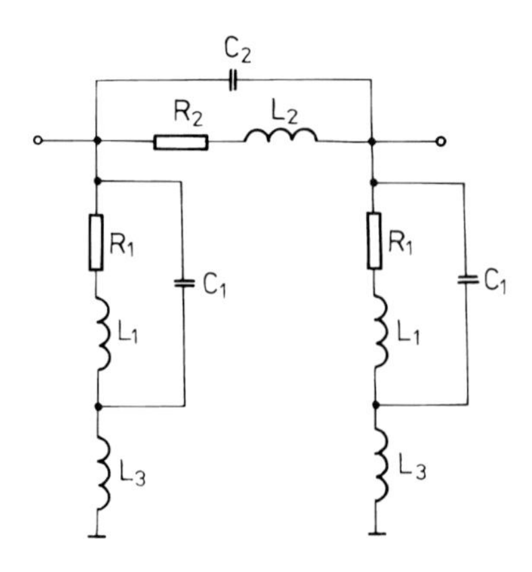

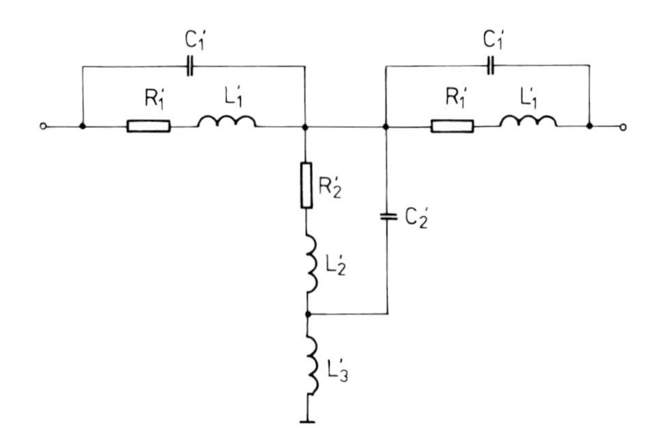

Fig. 4.25 Attenuators in the form of the resistance T or Π-element.

for T-element

$$R_1' = Z_0\sqrt{1-\beta}$$
$$R_2' = Z_0\frac{\beta}{1-\beta}\left(1-\frac{R_1'}{Z_0}\right)$$

where $\beta = 10^{-0.1L}$. (4.90)

The limit angular frequency ω_c of the attenuator

$$\omega_c \approx \begin{cases} \frac{1}{2}\left\{\max\left(\frac{L_1}{R_1}, \frac{L_2}{R_2}, \frac{L_3}{R_1}, C_1R_1, C_2R_2\right)\right\}^{-1} & \text{for } \Pi\text{-element} \\ \frac{1}{2}\left\{\max\left(\frac{L_1'}{R_1'}, \frac{L_2'}{R_2'}, \frac{L_3'}{R_2'}, C_1'R_1', C_2'R_2'\right)\right\}^{-1} & \text{for } T\text{-element} \end{cases} \tag{4.91}$$

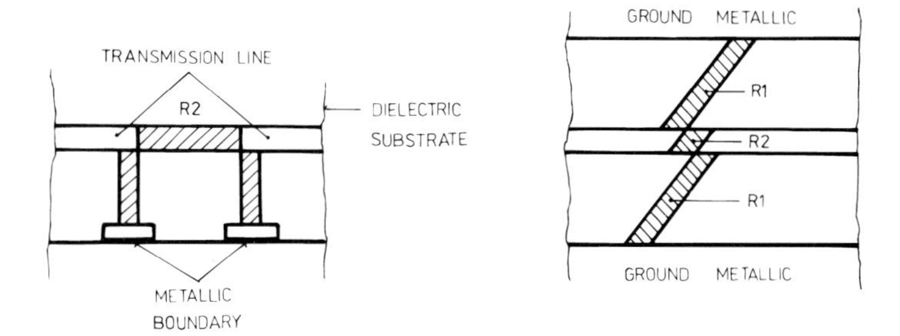

Fig. 4.26 Attenuator with thin-layer resistors with separated resistors.

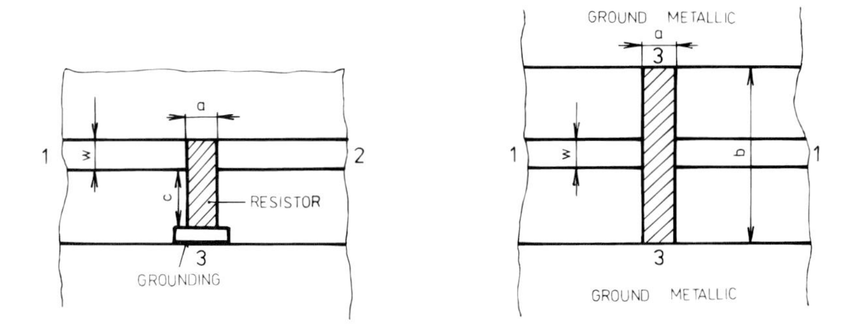

Fig. 4.27 Attenuator without the separation of particular resistors.

Above this frequency, the characteristics of the attenuator are already considerably degraded by parasitic reactances, linearly dependent on the circuit dimensions for a given geometry. For this reason, the integration of thin-layer resistances directly in the circuit is preferred, since the dimensions of inserted chip resistors with their contact areas enlarge the dimensions of the whole circuit, thus reducing the limit frequency. The attenuators made of inserted resistors are applicable to a frequency of about 2 GHz, depending on dimensions of resistors. The attenuator with thin-layer resistors can be arranged with separated resistors (Fig. 4.26) or without the separation of particular resistors (Fig. 4.27). In the first case, the procedure of the design is obvious. Skewing of resistive segments should compensate the finite dimensions of the resistors in the case of a coplanar line. In this way, the phase planes of all the resistors are at least approximately identical, which makes it possible to increase the limit frequency by about 30%. The attenuators of the second category make possible even a larger increase of the limit frequency. The

Table 4.1 Attenuator with compounded resistors (according to Fig. 4.28) $Z_0 = 50\,\Omega$

Attenuation	Contact resistivity		Size		Surface
L [dB]	1 – 2 [Ω]	1 – 3 (2 – 3) [Ω]	a/h	b/h	resistivity [Ω/mm^2]
1	5.74	437.37	0.159	2	50
2	11.46	221.02	0.246	1	49
3	17.1	150.45	0.341	1	48
4	22.62	116.17	0.504	1	48
6	33.20	83.60	0.755	1	48
8	43.06	68.82	1.045	1	48
10	52	61.1	1.323	1	48
16	72.6	52.58	2.25	1	48
20	81.8	51	2.88	1	48

dimension of the whole attenuator is reduced to that of only one resistor. The circuit geometry can be determined only by solving the Laplace's equation for the potential on the resistance surface. In addition, for technology reasons, the surface resistance of the resistive layer should be identical for all the attenuation values and simultaneously, it should be provided by the use of common techniques. These requirements are satisfied, for example, by the geometry according to Table 4.1. For low attenuation values ($L < 3\,\mathrm{dB}$) the problem is usually encountered of the feasibility of obtaining the resistive area, whereas for large values $L > 10\,\mathrm{dB}$, the limit frequency of the attenuator decreases (which can be compensated for by dividing this attenuator to a cascade of attenuators with lower attenuations). For an attenuator with the geometry corresponding to Fig. 4.27 and for a substrate with $\varepsilon_r \approx 10$ it is approximately possible to consider

$$f_c \approx \frac{8}{a} \qquad [\mathrm{GHz;mm}] \tag{4.92}$$

Increases of the limit frequency by reducing the dimensions are naturally limited by the possibility of implementing the resistive area and by restricting the RF power transmitted. For a checked cooling ability ($1\,\mathrm{W.mm^{-2}}$) of a Al_2O_3 substrate the power range of microstrip attenuators is approximately given by

$$P_{\max} < a\,(w + b) \qquad [\mathrm{W;mm;mm}] \tag{4.93}$$

For a coplanar line on Al_2O_3, at a dielectric thickness of 0.64 mm

$$P_{\max} < 0.25ac \qquad [\text{W;mm;mm}] \tag{4.94}$$

4.2.8 Power dividers

The power dividers include all the circuits dividing the power from one line without losses and reflections into N lines in a ratio desired. The particular input ways, however, need not necessarily be independently matched. Depending on this, these are either bilaterally or unilaterally matched dividers. The bilaterally matched dividers have, in addition to this, an important characteristic — the output ways are isolated from each other. With the help of the circuit scattering matrix, these characteristics can be expressed as follows:

$$\begin{aligned} &S_{N+1,N+1} = 0 \text{ a unilaterally matched divider for } N \text{ ways} \\ &S_{i,i} = 0 \text{ for } i = 1, 2, \ldots, N+1 \\ &S_{i,k} = 0 \text{ for } i, k = 1, 2, \ldots, N \text{ for a bilaterally matched divider.} \end{aligned} \tag{4.95}$$

From the standpoint of the circuit topology, power dividers with central branching and those with stepwise branching are considered. In the first case all the branches have one mutual point, in the second case, the whole divider consists of individual two-way dividers. The dividers with the central branching are more compact, however, in general their design is not simple. The second type of the divider has larger dimensions and it is easy to design. Both types of dividers are, however, based on dividers with the unilaterally matched central branching. This divider is defined by a frequency band, desired SWR of the common input and N coefficients of power dividing n_1 to n_a adhering to the condition of lossless circuit

$$\sum_{i=1}^{N} n_i = 1$$

According to Fig. 4.28 it consists of two different parts. In part I, there are the same voltages in the excitation from a common branch in each phase in all the branches (coherent part), in part II, this condition is not adhered to and particular branches are essentially independent. The output transformation of admittances of particular

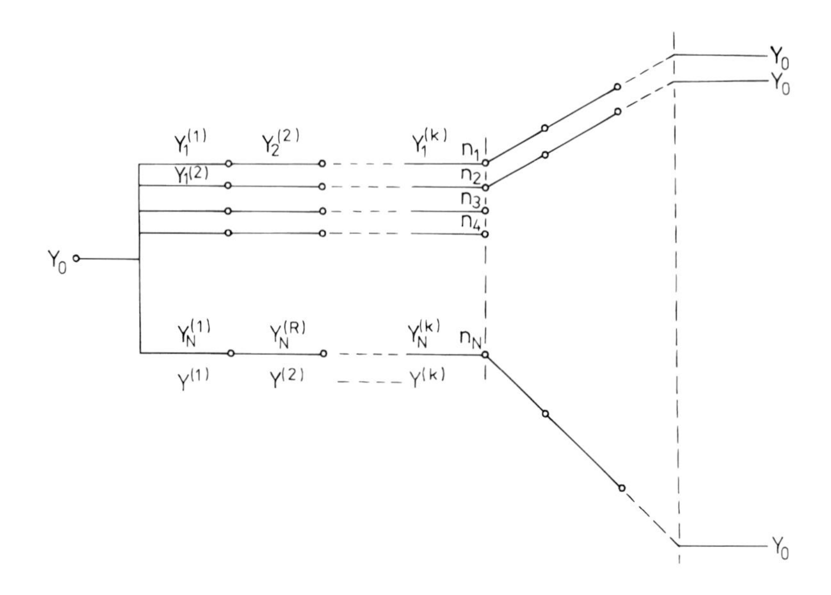

Fig. 4.28 Power divider.

branches to characteristic admittances of outputs Y_0 is accomplished here. Since in part I the voltages in all the branches are the same in an identical phase plane, the admittances of particular branches could be in the ratio of coefficients n_i

$$\frac{Y_i}{Y_k} = \frac{n_i}{n_k}, \qquad i, k = 1, 2, \dots, N \tag{4.96}$$

This also holds at a boundary between parts I and II. In order that a minimum transformation ratio in part II could be achieved for each distribution of coefficients n_i, the admittances at this site are chosen

$$Y_i = \frac{n_i Y_0}{\sqrt{n_{\max} n_{\min}}} \tag{4.97}$$

where $n_{\max} = \max(n_i)$ and $n_{\min} = \min(n_i)$. Further it holds that

$$\frac{Y_{\max}}{Y_0} = \frac{Y_0}{Y_{\min}}$$

$$\sum_{i=1}^{N} Y_i = \frac{Y_0}{\sqrt{n_{\max} n_{\min}}}$$

and in the case of uniform dividing

$$n_{\max} = n_{\min} = \frac{1}{N}$$
$$Y_i = Y_0$$
$$\sum_{i=1}^{N} Y_i = NY$$

Thus, the first part is formed by an impedance transformer between the input admittance Y_0 and output admittance $Y_0/\sqrt{n_{\max}n_{\min}}$, which is solved as a step transformer according to part 4.2.3. The impedance transformers. Particular steps (j) are always formed by parallel branches with admittances $Y_{(i)}^{(j)}$ (Fig. 4.28)

$$Y_i^{(j)} = n_i Y^{(j)} \tag{4.98}$$

Particular branches of part II are also solved as step transformers from admittance Y_i to Y_0.

Bilaterally matched dividers with central dividing have been yet based on unilaterally matched dividers. The topology described is complemented by resistances between neighbouring branches at sites of ends of transformation segments of part I. The values of these resistances are in general determined by a computerized optimization method. The values of these resistances can be analytically calculated only for the case of one-step two-way power divider:

$$R = Z_0\sqrt{\frac{n_1}{n_2} + \frac{n_2}{n_1}} \tag{4.99}$$

For multibranch symmetric dividers ($n_{\max} = n_{\min}$), this problem was numerically solved by many authors, for example.

In the case of unilaterally matched dividers, the main problem of their implementation is in the compensation of a discontinuity of central branching, which presents considerable difficulties in the case of a higher number of unidentical branches. In a two-way divider, the connection configuration in the T-shape is usually chosen with using means for compensating for this discontinuity according to Section 2.2.3. Individual procedures are employed in the case of multistep dividers, since the general procedure has not been yet

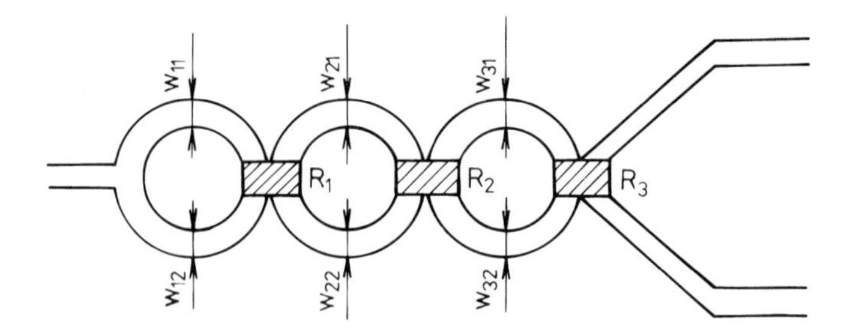

Fig. 4.29 Two-way power dividers at lower frequencies.

promising. In the case of bilaterally matched power dividers, there is also a problem of relationships between neighbouring branches since at the sites of resistors, the neighbouring branches should be situated very near in order that the length of these resistors could be negligible in comparison with the wavelength. In two-way power dividers at lower frequencies, this problem is solved by a convex shape of particular branches (Fig. 4.29). In multibranch dividers, including two-way dividers at higher frequencies, it is necessary to conduct the branches in parallel. Dimensions of particular lines are designed by methods of the synthesis of multistrip lines respecting their mutual coupling. This synthesis is, however, known only for a limited group of types of lines. In all the cases the mode is of a governing importance, which has identical cases in all the lines. With respect to the fact the phase velocities of particular modes are different, the differences should be compensated in suitable manners, since otherwise the characteristics of the divider would be degenerated (decreased insulation of branches). For a two-way divider in Fig. 4.30,

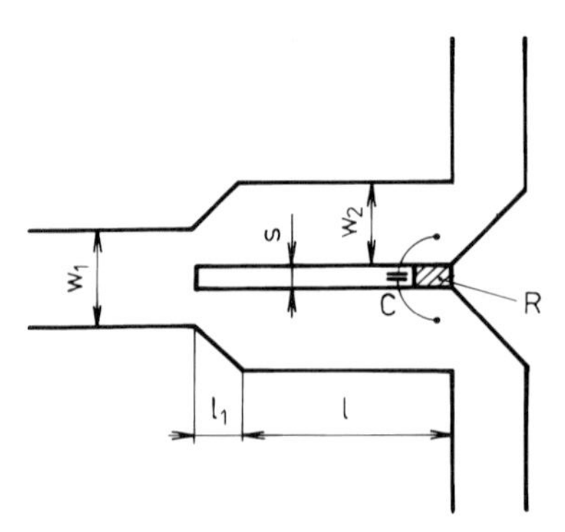

Fig. 4.30 Two-way divider with lengthening of the slot by the length l_1.

the suitable form of this compensation is a lengthening of the slot by l_1. The length l is determined by a quarter-wave of the even mode at a middle frequency. Then the length $l + l_1$ is a quarter-wave of the odd mode. This compensation form is suitable for higher frequencies, where $l_1 < w_2 + \frac{s}{2}$. At lower frequencies, a capacitor with a capacitance C is added in parallel to the resistor R (see relationship 4.107 Section 4.2.9). The dimensions of the resistor and thus also the width of the slot are determined on the basis of an analysis similar to that in the case of the terminating resistance in part 4.2.1 reflectionless termination.

The power dividers with central dividing have five additional ways and the dividing ratio $n_{\max}/n_{\min} \leq 4$. For higher numbers of ways

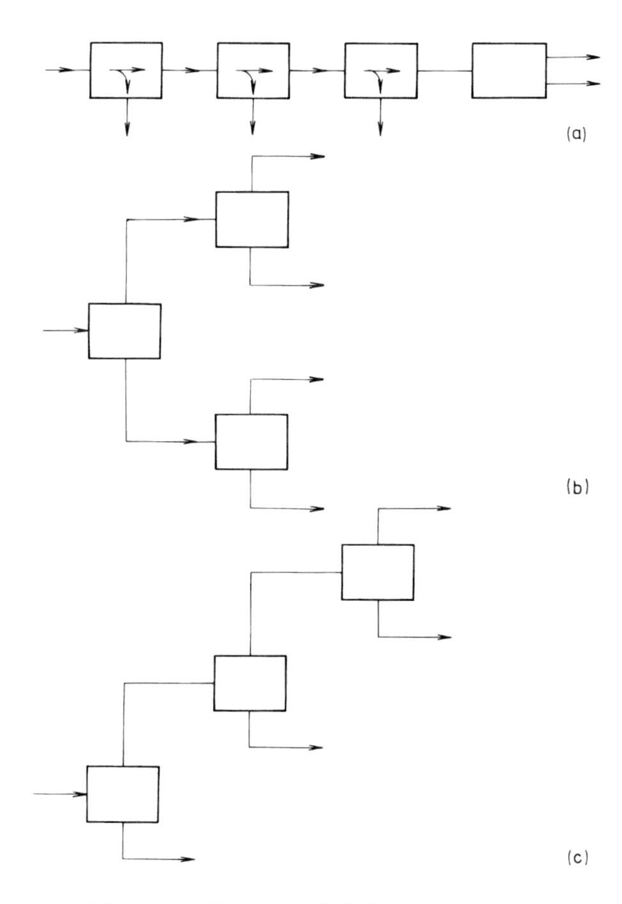

Fig. 4.31 Example of dividers connecting.

with uniform dividing it is more advantageous to use dividers with stepwise dividing, stepwise branching with couplers (Fig. 4.31) or in the case that the number of ways equals a power of two, a symmetric binary divider. In the case that the ratio of the output powers high, it is advantageous to use a cascade of non-symmetric binary dividers (Fig. 4.31c). In this case, branching from the common way starts at $n_{\max}$ and continues through decreasing n_i to $n_{\min}$. In this way, the dividing ratio of particular dividers is minimized.

4.2.9 Directional coupling element and hybrid elements

The directional coupling element (directional coupler) is a reciprocal lossless four-port which is in the ideal case characterized by scattering coefficients (Fig. 4.32)

$$S_{ii} = 0, \qquad i = 1, 2, 3, 4 \tag{4.100a}$$

$$S_{14} = S_{23} = 0 \tag{4.100b}$$

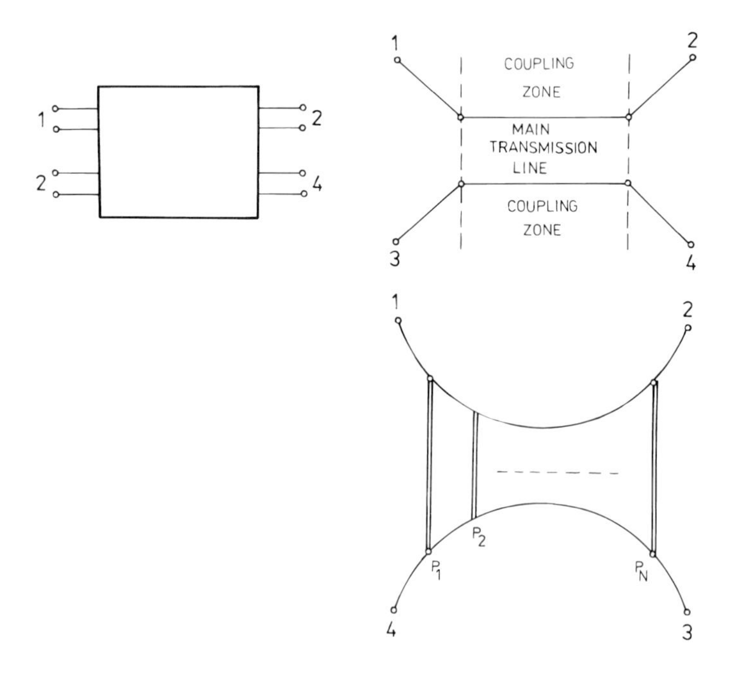

Fig. 4.32 Directional coupling element.

Coupling attenuation L and voltage coefficient of coupling k is further defined

$$L = -10\log|S_{13}|^2 = -10\log|S_{24}|^2 \qquad [dB]$$
$$k = |S_{13}| = |S_{24}| \tag{4.101}$$

In a real case, equation (1.100b) is not adhered to and the isolation attenuation I and the directivity S are defined by

$$I = -10\log|S_{14}|^2 = -10\log|S_{23}|^2 \qquad [dB]$$
$$S = I - L \tag{4.102}$$

The synthesis of the directional coupler is based on dividing a symmetric four-port to two-ports for even and odd modes. The resulting effect is achieved by combining both modes (Fig. 4.33). When

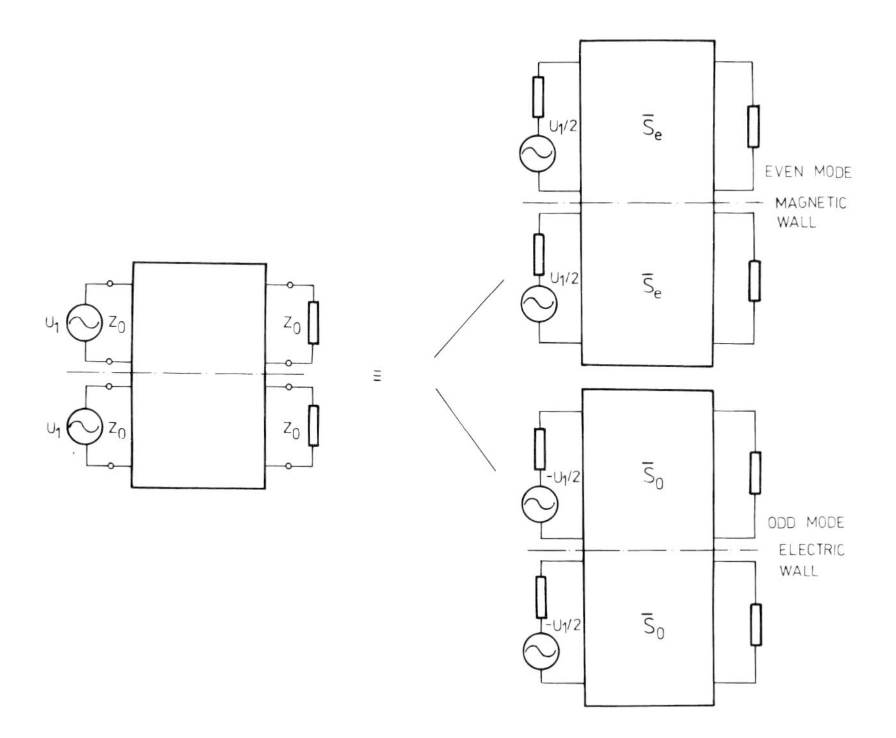

Fig. 4.33 The synthesis of the directional coupler.

denoting elements of matrices of the even and odd two-port S_{eik} and S_{oik}, $i, k = 1, 2$, then

$$\begin{aligned} 2S_{11} &= S_{e11} + S_{o11} \\ 2S_{13} &= S_{e11} - S_{o11} \\ 2S_{14} &= S_{e21} - S_{o21} \end{aligned} \tag{4.103}$$

During this

$$\begin{aligned} S_{11} = S_{14} = 0 &\rightarrow S_{e11} = -S_{o11} \\ S_{e21} &= S_{o21} \\ |S_{13}| = k &\Rightarrow \left| \frac{S_{e11} - S_{o11}}{2} \right| = k \Rightarrow |S_{e11}| = k \end{aligned}$$

These conditions are adhered to for example by a two-port with parameters

$$S_{e11} = k, \qquad S_{o11} = -k \tag{4.104}$$

representing an impedance transformer with a transformation ratio of $(1+k)/(1-k)$ (even two-port) and $(1-k)/(1+k)$ (odd two-port).

Coupler with back wave, Lange's hybrid. In the case of this coupler (Fig. 4.32), the coupling part is formed by coupled lines, which are continuously coupled along the whole length of the coupling region. The impedance transformers are implemented as a cascade of quarter-wave segments of homogeneous lines. During this, the number of steps and magnitudes of impedance values of the even and odd modes of particular steps are determined from the following parameters required: coupling attenuation L (voltage coupling $k = 10^{-0.05L}$), coupling ripple ΔL and frequency band with the use of equations in Section 4.2.3 and on the basis of the following transpositions:

$$\begin{aligned} &\text{even mode — transformation ratio } R = \frac{1+k}{1-k} \\ &\text{odd mode — transformation ratio } R = \frac{1-k}{1+k} \end{aligned} \tag{4.105}$$

For a one-step coupler simple relationships are obtained in this way

$$Z_{0e} = Z_0 \sqrt{\frac{1+k}{1-k}}; \qquad Z_{0o} = Z_0 \sqrt{\frac{1-k}{1+k}} \tag{4.106}$$

This procedure includes an assumption that particular steps are quarter-wave ones at an intermediate frequency. It is not always simple to satisfy this condition, since in coupled lines of MIC, the values of the velocities of the even and odd modes are different. The form of the compensation for this difference depends on what types of coupled lines were chosen for the coupler. Coupled microstrip lines are usually chosen. They are suitable for couplings between 6 and 15 dB. The compensation for differences in velocity values is provided by additional capacitors

$$C = \frac{\cot\left(\frac{\pi}{4}\frac{v_{fe}}{v_{fo}}\right) - 1}{\omega_0 Z_{0o}} \tag{4.107}$$

at both ends of particular coupling segments.

For stronger couplings, it is possible to use multistrip structures. These structures are characterized by a higher number of eigenmodes, which should be suppressed by special junctions. A typical exämple of this coupler type is the Lange's hybrid, which is shown in two variants in Fig. 4.34. This is a single step coupler with a 3 dB

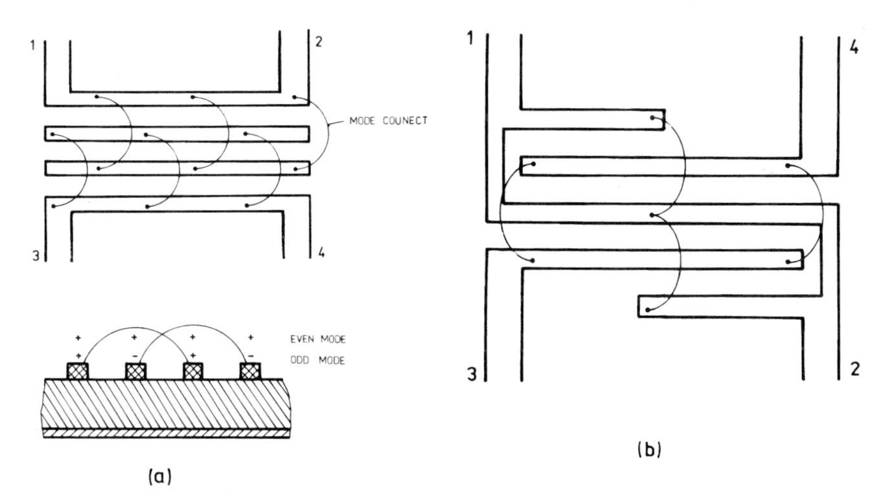

Fig. 4.34 Lange's hybrid (a) four-strip coupling (b) special arrangement of branches 3 and 4.

coupling with four-strip coupling sections. In the case b), branches 3 and 4 are modified by a special arrangement, which facilitates the design of certain more complex circuits (for example of symmetric mixers and amplifiers). Different phase velocities of modes can be compensated for, again, by capacitive loading to the coupling segment, however, this is not always simple with respect to the small width of strips.

Coupler with coupling branches. This type of the coupler has two lines (Fig. 4.32c) coupled at isolated points by connecting lines P_1 to P_N. In comparison with the back wave coupler, branches 3 and 4 are modified. For equivalent diagrams of transformers for even and odd modes see Fig. 4.34. The difference between them is that parallel stubs of the even mode transformer are disconnected at a distance of $l_i/2$, whereas the stubs of the odd mode are short-circuited at this distance. The theory of these ladder networks has been extensively elaborated. The optimum length of all the segments (series as well as parallel) is 1/4 of the wave length. The impedance of particular segments is determined depending on requirements for the width of the band $\langle f_1, f_2 \rangle$, coupling L and its ripple ΔL in the given band according to tables (for example Matthaei *et al.*, 1964). A typical use of this type of the coupler is a single-step hybrid element, i.e. a coupler with coupling $L = 3$ dB. In contrast to the Lange's hybrid it can be more easily implemented, however the bandwidth is at most 20%. At a higher frequency, its parameters are degraded by discontinuities at junctions of particular lines (decreasing insulation). The impedance of particular segments is determined in this simple case as

$$\begin{aligned} Z_S &= \frac{Z_0}{\sqrt{2}} \\ Z_{P1,2} &= Z_0 \end{aligned} \tag{4.108}$$

4.2.10 Differential phase shifter

The differential phase shifter is a lossless, symmetric, matched two-port, with a transmission S_{21} having its phase larger by a constant angle φ_d than the phase of the transmission of a matched line with a certain length l_r. This length is termed the reference one. Thus, for elements of the scattering matrix of an ideal differential phase shifter it holds that

$$S_{11} = S_{22} = 0$$
$$S_{21} = S_{12} = e^{\frac{j\omega l_r}{v_f} + \varphi_d} \tag{4.109}$$

where l_r is the reference length, φ_d differential shift and v_f phase velocity.

The differential phase shifter is used for achieving a constant phase difference (independent of the frequency) between two ways. Ideal characteristics as expressed in relationships (4.109) can be only approached in a limited frequency band, for example by a circuit consisting of segments of coupled lines according to Fig. 4.35. All the

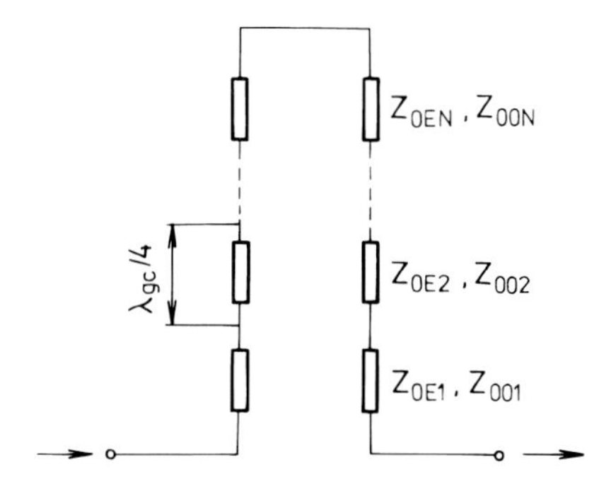

Fig. 4.35 Differential phase shifter.

segments are quarter-wave ones at an intermediate frequency. With the number of steps the bandwidth increases and rippling of differential shift φ_d decreases, however, with increasing the reference length l_r. In the determination of the impedance of the even and odd modes of particular segments the method of distribution of a symmetric two-port to two one-ports is used for even and odd excitation modes according to Fig. 4.36. For the reflection coefficients Γ_e, Γ_o of these one-ports

$$0.5\,(\Gamma_e + \Gamma_o) = S_{11} = 0 \Rightarrow \Gamma_o = -\Gamma_e$$
$$0.5\,(\Gamma_e - \Gamma_o) = S_{21} \Rightarrow \Gamma_e = S_{21} = e^{\frac{j\omega l_r}{v_f} + \varphi_d} \tag{4.110}$$

The first relationship is adhered to when choosing

$$Z_{0ei} Z_{0oi} = Z_0^2$$

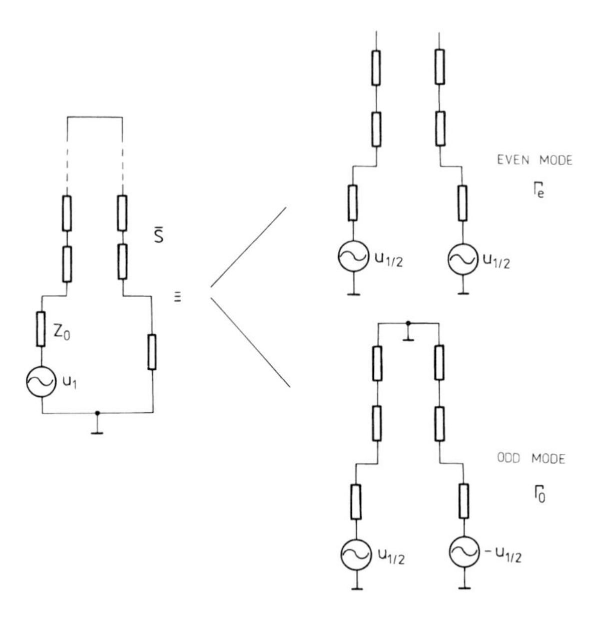

Fig. 4.36 Determination of the impedance for even and odd modes.

for all the steps $i = 1, 2, \ldots, N$. Thus, it is still necessary to determine all the even impendances in such a way that the second relationship in (4.110) could be satisfied to as much as possible. This can be in general achieved only by a numerical optimization. For certain values of the phase shift φ_d and ripple $\Delta\varphi_d$ in normalized bands this was performed in Shelton and Mosko (1966).

In practice, coupled microstrip lines can be prevalently considered, where the transverse sizes are determined from known Z_{0ei} and Z_{0oi}. The length of section equals the quarter-wave of the even mode at an intermediate frequency. Different velocities of the even and odd modes are compensated for by adding capacitors in parallel with the slot at both ends of particular segments. The capacitances of the capacitors are determined from (4.107), again. The capacitors from neighbouring segments at the site of their contact can be calculated.

4.2.11 Baluns and magic-tee

By the balun we mean a two-port forming a lossless and reflectionless transition from a non-symmetric line to a symmetric line. These circuits are frequently used particularly in symmetric mixers and in fur-

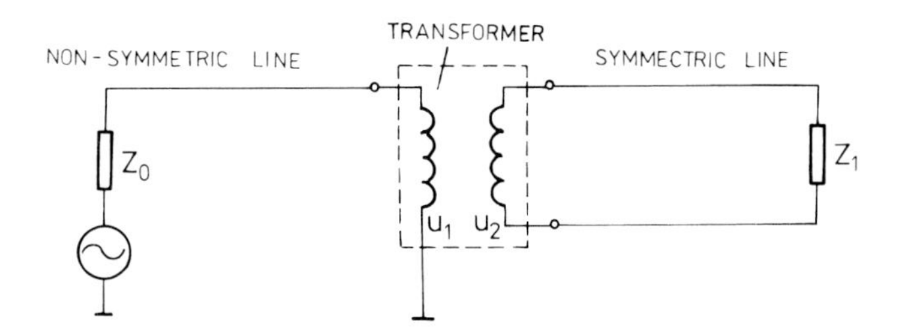

Fig. 4.37 Lumped symmetrizing transformer.

ther symmetric circuits. In the lower frequency range of microwaves (up to 2 GHz) in hybrid circuits and in intermediate frequency range (from 2 to 8 GHz) in monolithic circuits it is possible to use lumped symmetrizing transformers (Fig. 4.37). The transformers for hybrid circuits are coiled on ferrite cores made of special high-frequency material. The transformers of monolithic circuits are air transformers with spiral coiling. The number of turns and the inductances must satisfy relationships presented in Fig. 4.37.

In hybrid circuits for the middle frequency region a symmetrizing circuit is frequently used with a cophasic power divider and differential phase shifter 180° according to Fig. 4.38. The simplest phase

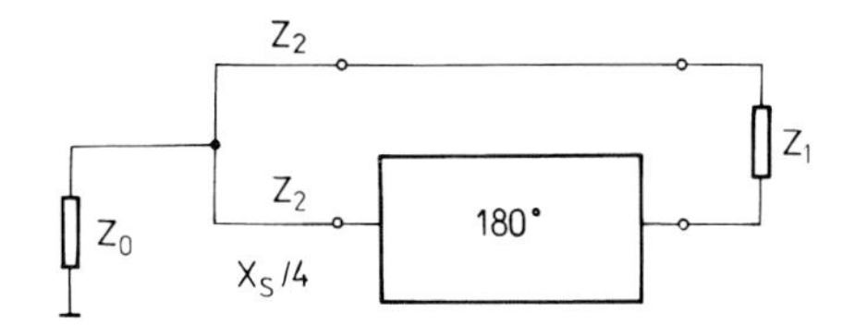

Fig. 4.38 Differential phase shifter 180°.

shifter can be a line of length $\lambda_g/2$. A larger bandwidth is achieved by using shifters according to part 4.2.10 Differential phase shifter.

In the upper frequency region, it is possible to use a direct transition by a stepwise transformation of one type of the line to the other. A transition can serve as an example from a microstrip to a symmetric two-strip line according to Fig. 4.39. The transition region should be essentially larger than the wavelength.

The frequently used type of the microwave symmetrizing circuit is the magic T. In an ideal case, this is a lossless four-port with the

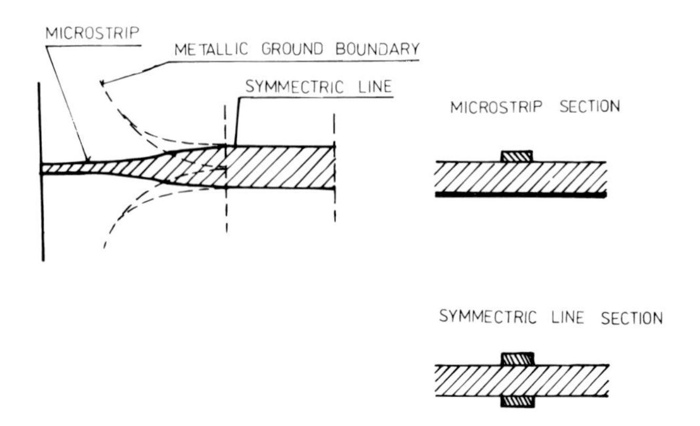

Fig. 4.39 Transition from microstrip to symmetric two-strip line.

scattering coefficients

$$\begin{aligned}
S_{ii} &= 0 \qquad \text{for } i = 1, 2, 3, 4 \\
S_{12} &= S_{21} = 0 \\
S_{31} &= -S_{41} \\
S_{32} &= S_{42} \\
|S_{31}| = |S_{41}| &= |S_{32}| = |S_{42}| = \frac{1}{\sqrt{2}}
\end{aligned} \tag{4.111}$$

In the MIC, shaping in the form of a circular T is usually used according to Fig. 4.40. The bandwidth of this circuit is limited by using quarter-wave and three quarter-wave segments of the line and thus, it typically does not exceed 20%. The output waves in branches 3 and 4 have the same amplitude and reverse phase during the excitation from branch 1. They can be used for supplying a symmetric line with a characteristic impedance $2Z_o$.

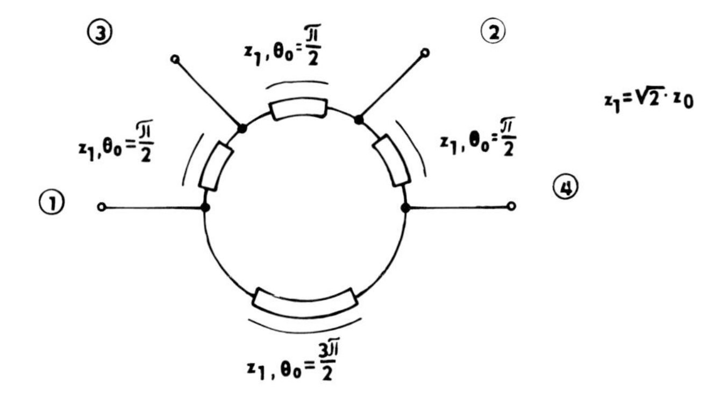

Fig. 4.40 MIC in form of circular T.

4.3 BASIC ACTIVE NON-LINEAR CIRCUITS

P. Bezoušek, F. Hrníčko and M. Pavel

4.3.1 Detectors

This part considers different aspects of designing a diode detector of the RF signal. It considers only detectors using the Schottky-barrier diode as a basic non-linear element for the detection. During this, for the detection of weak signals, the ZBS diode or the point-contact Schottky diode are preferably used (see part 3.1). The principle of the detection is based on using the non-linearity of the diode voltampere characteristic

$$I_j = I_s \left(e^{\frac{eU_j}{nkT}} - 1 \right) \tag{4.112}$$

where I_j is the current passing through the diode, U_j is the voltage on the diode transition, I_s is the diode saturated current and n is the factor ideality. For a harmonic voltage at the transition

$$U_j = U_0 + U_1 \cos \omega t \tag{4.113}$$

where U_0 is the direct component of the voltage at the transition, U_1 is the amplitude of the first harmonic of the voltage at the transition,

ω is the angular frequency and then the current passing through the diode can be expressed by the expansion

$$I = I_s \left[e^{\frac{eU_0}{nk}} \left\{ B_0 \left(\frac{eU_1}{nkT} \right) + 2 \sum_{n_i=1}^{\infty} B_n \left(\frac{eU_1}{n_i kT} \right) \cos n\omega t \right\} - 1 \right] \tag{4.114}$$

where $B_{n_i}(x)$ is the modified Bessel function of the n_i-th order of the argument x. The first two components of the current

$$\begin{aligned} I_0 &= I_s \left[e^{\frac{eU_0}{nkT}} B_0 \left(\frac{eU_1}{nkT} \right) - 1 \right] \\ I_1 &= 2I_s e^{\frac{eU_0}{nkT}} B_1 \left(\frac{eU_1}{nkT} \right) \end{aligned} \tag{4.115}$$

where I_0 is the direct component and I_1 is the amplitude of the first harmonic component of the current in the diode.

In the case of a very weak signal $U_1 \ll nkT/e$, the relationship for the direct current components in (4.115) is simplified, so that

$$I_0 \approx R_j^{-1} \left(U_0 - U_{00} \right) \tag{4.116}$$

where $U_{00} = eU_1^2/2nkT$ is the open-circuit voltage, $R_j = nkT/(eIs)$ is the differential resistance of the diode at a zero bias. At a small signal, the diode transition behaves as a resistance R_j in series with the voltage source U_{00} when disregarding the transition capacitance. The relationships (4.116) describe the quadratic detector, since the voltage detected U_{00} exerts a quadratic dependence on the amplitude of the RF signal U_1. At a higher signal, the open-circuit voltage detected will be given by

$$U_{00} = -\frac{nkT}{e} \ln \left[B_0 \left(\frac{eU_1}{nkT} \right) \right] \tag{4.117}$$

Equation (4.117) can be approximately expressed for $U_1 \gg nkT/e$ as

$$U_{00} \approx \frac{nkT}{e} - U_1 \tag{4.118}$$

which is an equation of the linear detector, since the voltage detected is a linear function of the RF signal amplitude. The expressions presented are derived under an assumption that the quantities n and I_s are independent of the transition voltage, which is not completely adhered to in the case of the Schottky-barrier diode. In this case, it is necessary to replace the expressions (4.116) for R_j and U_{00} in the low-signal approximation by the following relationships, where the derivatives are obtained based on the voltampere characteristic of the given diode:

$$R_j = \left.\frac{\partial I_j}{\partial U_j}\right|_{U_j=0}$$
$$U_{00} = U_1^2 \left(\frac{\partial I_j}{\partial U_j}\right)^{-1} \left.\left(\frac{\partial I_j^2}{\partial U_j^2}\right)\right|_{U_d=0} \tag{4.119}$$

where U_j is a voltage at the diode transition, I_j is the conductive current through the transition.

The important parameter of the detecting diode is the voltage sensitivity, which is defined as a ratio of the voltage detected $|U_{00}|$ (in the region of the quadratic detection) and RF power P_j consumed on the diode

$$\gamma = \frac{|U_{00}|}{P_j} \tag{4.120a}$$

When assuming a regular voltampere characteristic (4.116), we can express the voltage sensitivity of the diode as

$$\gamma \approx \frac{R_j e}{nkT} = I_s^{-1} \tag{4.120b}$$

The ability of the diode to detect very weak signals is expressed with the help of the tangential sensitivity. This is a power, which would form after the detection of a detected voltage higher by a factor of 2.5 than the output noise voltage. The original definition of this quantity is empirical in nature. This quantity expresses the detector threshold sensitivity. It can be calculated from an equivalent schematic diagram of the Schottky diode in Fig. 4.41 assuming the

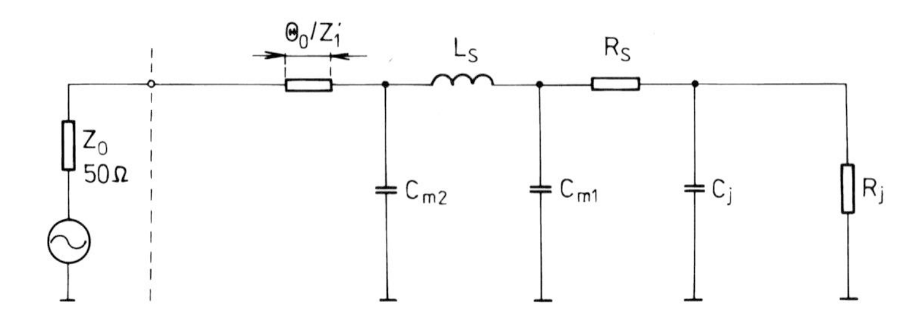

Fig. 4.41 Equivalent schematic diagram of the Schottky-diode.

presence of a small signal at a zero bias or with the help of an effective temperature T_e of the Schottky diode defined in part 3.1. According to this definition, the noise voltage at the detector output will be in the band $\langle f_1, f_2 \rangle$ (determined by the measuring equipment filter)

$$u_n = \sqrt{4k\left(R_j + R_s\right) \int_{f_1}^{f_2} T_e\left(f\right) \mathrm{d}f} \tag{4.121}$$

The output RF power P_1, necessary for the excitation of the output signal detected is, according to the definition of the tangential sensitivity,

$$2.5\frac{u_n}{\gamma} = 2.5\frac{\sqrt{4k\left(R_j + R_s\right) \int_{f_1}^{f_2} T_e \, \mathrm{d}f}}{\gamma} \tag{4.122}$$

The tangential sensitivity is typically expressed in the logarithmic form as

$$TS = 10 \log\left(\frac{P_1}{1\,mW}\right) \quad [\mathrm{dB\,m}] \tag{4.123}$$

The detector as a whole is characterized by its voltage sensitivity γ_{det}, bandwidth for the sensitivity decrease by 4 dB or possibly by a product of these quantities δ. The quantities given are defined by

$$\begin{aligned} \gamma_{det} &= 4\gamma \frac{R_j Z_0}{\left(R_j + Z_0\right)^2} = \frac{4e}{nkT} \frac{R_j^2 Z_0}{\left(R_j + Z_0\right)^2} \\ B &\approx \frac{R_j + Z_0}{2\pi R_j Z_0 \left(C_j + C_m\right)} \\ \delta &= \gamma_{det} B \approx \frac{2}{\pi} \frac{e}{nkT} \frac{R_j}{\left(R_j + Z_0\right)\left(C_j + C_m\right)} \end{aligned} \tag{4.124}$$

where C_m, C_j, R_j, Z_0 are quantities obvious from Fig. 4.41.

From the standpoint of maximizing the quantity δ, it is necessary to have as low capacitances C_j, C_m as possible and $R_j \gg Z_0$. The bonding capacitance C_m has a lower limit value of 0.03 to 0.05 pF when using the chip diode inserted into the MIC. The capacitance of the transition C_j can be decreased by reducing the diode diameter, which simultaneously increases the R_j, which is advantageous from the standpoint of the coefficient δ. This upper limit of the resistance and thus also lower limit of the diode diameter is necessary due to parameters of the evaluating circuit. The output resistance Z_1 of amplifiers of the detecting voltage depends on their bandwidth and for bands of about 10^6 Hz it is usually of $Z_1 = 1\,\mathrm{k}\Omega$. It is, however, necessary to adhere to the condition $R_j \leq Z_i$.

By choosing the impedance of the RF signal source Z_0, it is possible to change within a wide range the bandwidth of the detector B and its sensitivity without essential changes of the coefficient δ, which is used in designing wide-band detectors.

All the relationships presented, following from parameters of the detector were derived by neglecting the series resistance of the diode and further parasitic elements and thus, they can serve only as an approximate guideline when designing the detector, particularly when selecting diodes and when choosing the circuit connections.

As an example, two limit variants of the detector are considered: a) The narrow-band detector with a maximum sensitivity in a microstrip line $50\,\Omega$ at a frequency $f_0 = 10\,\mathrm{GHz}$. The diode is chosen in a chip form with parameters $R_j = 1\,\mathrm{k}\Omega$, $C_j = 0.1\,\mathrm{pF}$. The detector input circuit will be formed by a simple impedance transformer according to Fig. 4.41. The electric length Θ and characteristic impedance of the transformer can be easily determined with the help of the Smith chart or numerically in such a way that the circuit at a frequency f_0 is matched to the diode impedance Z_d including all the parasitic elements:

$$\begin{aligned} \Theta_1 &= 75.6^\circ \\ \Theta_0 &= 70.2^\circ \end{aligned} \tag{4.125}$$

$$\begin{aligned} Z_d\,(f_0) &= 17.8 - j74 \quad [\Omega] \\ l &= \frac{\Theta_0 v_f}{\omega_0} \end{aligned} \tag{4.126}$$

where l is the length of the transformer, v_f is the phase velocity of the propagation along the transformer line.
b) The wide-band detector with a minimum detection ripple in a band of 0 to 20 GHz in a microstrip line $Z_0 = 50\,\Omega$. The diode in the chip form is chosen, again, with parameters $R_j = 1\,\text{k}\Omega$, $C_j = 0.1\,\text{pF}$. The parasitic elements are estimated in the same way as in the preceding case: $C_{ml} = 0.05\,\text{pF}$, $L_s = 0.5\,\text{nH}$. In this case, the attainable width of the band for a decrease by 3 dB on the impedance of $50\,\Omega$ is approximately given by:

$$B = \frac{1050}{2\pi 50\,(C_j + C_{m1})\,1000} = 22.4 \quad [\text{GHz}] \qquad (4.127)$$

which corresponds to the limit of the applicability of the resistive matching. With respect to the requirement for a maximum planarity, it is, however, necessary simultaneously to use reactance elements. A simple circuit satisfying to these requirements is shown in Fig. 4.42. The determination of the values of elements of this circuit is however,

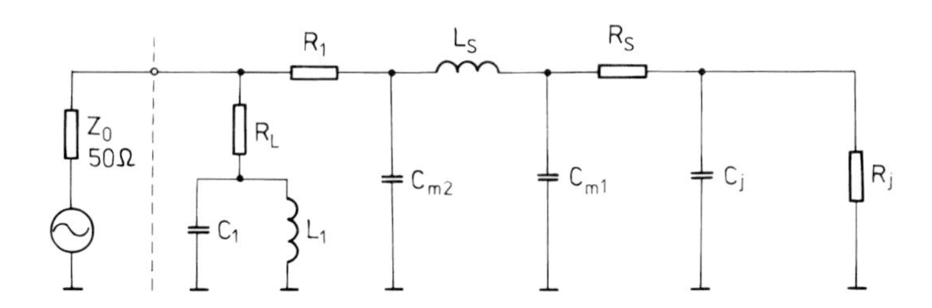

Fig. 4.42 Wide-band detector.

not simple. In this case, it is necessary to use the method of the computerized optimatization (for example Bezoušek and Tubl, 1989). Its results are shown in a schematic diagram in Fig. 4.42.

4.3.2 Mixers

The mixers represent a widely used and diverse group of microwave circuits. This part considers resistive mixers, i.e. those mixers where the basic non-linear element exerts a resistive non-linearity in contrast to reactance mixers, whose importance in microwave frequency bands currently decreases. Depending on the choice of the basic non-linear element, the resistive mixers can be divided into diode and transistor ones. Depending on the degree of the circuit symmetry, symmetric, non-symmetric, double balanced and twice double balanced mixers are considered. Depending on the processing of an image signal by the mixer or of a lateral band mixers with a) suppressing, b) utilizing and c) separation of the image signal can be differentiated from each other.

For an outline and characteristics of particular types of mixers see Table 4.2.

Processes inside of the basic non-linear element are described below. In the case of both diode or transistor, the basic characteristic of this element is a non-linear relationship between the current and voltage in this element. This relationship can be frequently expressed by

$$I = F\left(U_1, U_2\right) = \sum_{k=0}^{\infty} \sum_{l=0}^{k} a_{kl} U_1^k U_2^{k-1} \tag{4.128}$$

where a_{kl} are real constants. So, for example for a mixer with the Schottky-barrier diode it can be considered as a current passing through the transition I_{jR} ((1) Section 3.1) and $U_1 + U_2$ as a total voltage at the transition U_j. For a mixer with single base MESFE transistors, the current I can be interpreted as a current of the channel I_d (see part 3.5) and the sum of voltage values $U_1 + U_2$ can be considered as a gate voltage U_g. In the case of double-base transistors MESFE, U_1, and U_2 can represent voltages on particular gates.

In the analysis of the mixer, an assumption is typically adopted that U_1 and U_2 are values of a simple harmonic voltage at different frequencies

$$\begin{aligned} U_1 &= U_0 + U_{10} \cos\left(\omega_1 t\right) \\ U_2 &= U_0 + U_{20} \cos\left(\omega_2 t\right) \end{aligned} \tag{4.129}$$

As a result of the non-linearity described by the relationship (4.128), the current I will have components for all the values of the combination frequency $m\omega_1 + n\omega_2$, where m, n are integers and their amplitudes will be given by the relationship

$$I_{m,n} = \sum_{k=0}^{\infty} \sum_{l=0}^{\infty} A_{k,l,m,n} U_{10}^{(|m|+2k)} U_{20}^{(|n|+2l)} \tag{4.130}$$

Table 4.2 Partition of mixers and their characteristics

Type of mixer	Diode mixers	Transistor with one base transistor	Mixers with two base transistor
Simple	frequency bands f_S, f_{LO}, f_{IF} disjunct		band f_{IF} disjunct
	$L \simeq 4 \div 5\,\text{dB}$	$L \simeq 0 \div 4\,\text{dB}$	
	low dynamics $P_{1\,\text{dB}} < 1\,\text{mW}$	middle dynamics $P_{1\,\text{dB}} \simeq 5\,\text{mW}$	middle dynamics $P_{1\,\text{dB}} \simeq 4\,\text{mW}$ unique
	unique using in MIC unsuitable for MMIC		using in MIC and MMIC
Symmetric	band f_{IF} disjunct with F_S or f_{LO}		full covered of bands f_S, f_{LO} and f_{IF}
	band width 20%	band width 40% $f_{LO} > 100\,\text{MHz}$	
	$L \simeq 5\,\text{dB}$	$F \simeq 4 \div 6\,\text{dB}$, $L \simeq 0 \div 4\,dB$	
	middle dynamic $P_{1\,\text{dB}} \simeq 2\,\text{W}$	good dynamic $P_{1\,\text{dB}} \simeq 10\,\text{mW}$	
	using in MIC	small using in MIC	suitable for using in MIC and MMIC
	not suitable for using in MIC		
Double balanced	band f_{IF} disjunct with F_S and f_{LO}	$f_{IF} > 100\,\text{MHz}$ full cover	
	band width up to 90%	band width to 90%	
	$L \simeq 8\,\text{dB}$	$F = 6\,\text{dB}$, $L \simeq 0 \div 6\,\text{dB}$	
	good dynamic $P_{1\,\text{dB}} \simeq 6\,\text{mW}$	excellent dynamic $P_{1\,\text{dB}} \simeq 20\,\text{mW}$	
	using in MIC and MMIC	small using in MIC suitable for using in MMIC	
Twice double balanced	cover of band f_{IF}, f_S, f_{LO} band width up to 90%		
	$F = 9 \div 10\,\text{dB}$ $L \simeq 9\,\text{dB}$	$F = 8 \div 9\,\text{dB}$, $L \simeq 2 \div 8\,\text{dB}$	
	good dynamic to 10 mW suitable for MIC	excellent dynamic to 30 mW not suitable for MIC	
	small using in MMIC		

where $A_{k,l}$ are coefficients and m, n are integers $(-\infty, \infty)$. From this it can be seen that even combination frequencies so called combination frequencies with even m or n are an even function of a relevant amplitude U_1 or U_2 and the odd ones are an odd function. This is a very important feature, which is employed when designing different symmetric connections. When reversing the mixing diode polarity at a certain site of the circuit, the functions defined by (4.128) and (4.130) will be reversed

$$\begin{aligned} F(U_1, U_2) &\rightarrow -F(-U_1, -U_2) \\ I_{mn} &\rightarrow (-1)^{m+n+1} I_{mn} \end{aligned} \tag{4.131}$$

Thus, the values of the current at a combination frequency with a sudden sum $m+n$ change their sign when changing the diode polarity, whereas the values of the frequency with the odd sum do not change the polarity. When further denoting S_1 and S_2 polarities of signals at a frequency ω_1 and ω_2, S_d the polarity of the diode and $_{m,n}$ the polarity of the combination frequency $(S_1, S_2, S_d, S_{m,n} = \pm 1)$ then the general relationship will be valid

$$S_{m,n} = S_1^m S_2^n S_d^{(m+n+1)} \tag{4.132}$$

In practice, only three or four values are typically used out of the whole set of values of the combination frequency. According to the generally accepted terminology, from the standpoint of the amplitude the dominant component is termed the local oscillator f_o. When using the mixer in a receiver then the input signal is usually referred to as a high-frequency signal or only signal f_s and the difference signal is termed the intermediate frequency f_{IF}. A further combination frequency is the image signal or only the image f_z. The schematic diagram of the situation is shown in a spectral diagram in Fig. 4.43. When using the mixer for a reverse process, i.e. for transmission or modulation, and when the input frequency is the intermediate frequency, then both output frequencies, i.e. the signal and image are usually termed the upper and lower sidebands, respectively. It is easy to find from relationship (4.130) that amplitudes of both lateral bands are identical in a particular non-linear

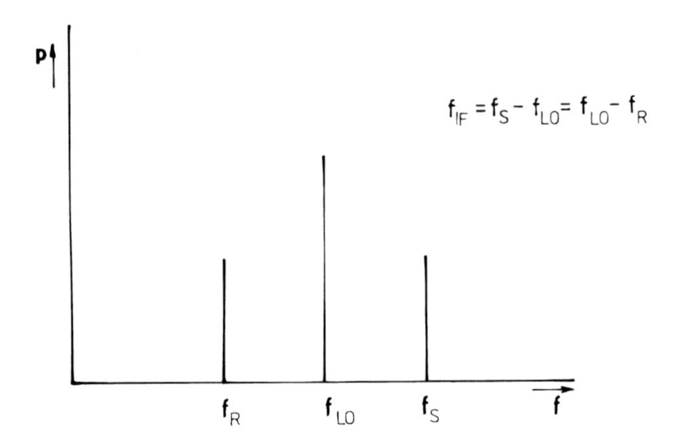

Fig. 4.43 Spectral diagram.

element ($|I_{1,1}| = |I_{1,-1}|$). Similarly in a receiving mixer, the transmission amplitude from signal as well as image into the intermediate frequency equals ($|I_{-1,1}| = |I_{1,-1}|$). Thus, the image frequency is the most important competing frequency of the RF signal. For its elimination it is insufficient to use only different symmetric connections of non-linear elements, however, it is necessary to use more complex systems based on differences in the frequency or on the phase of these values of the combination frequency.

A further important knowledge can be derived from assuming a very weak signal. When the amplitude of one of the signals is essentially lower than that of the second signal, for example $|U_2| \ll |U_1|$, then it is possible to linearize the relationships (4.130), so that for the values of the combination frequency, which are of interest, it is possible to write linear relationships

$$I_{-1,1} = \left(\sum_{k=0}^{\infty} A_{k0,-1,1} U_{10}^{(1+2k)} \right) U_{20} \tag{4.133}$$

with a similar approach to $I_{1,1}$ and $I_{1,-1}$.

Thus, from the standpoint of the transmission to the combination frequency, the non-linear element is manifested as linear in spite of the fact that the condition for the existence of this transmission is the element of non-linearity. An important parameter expressing characteristics of the mixer in the linear transfer regime are conversion losses L (or gain in transistor mixers) characterizing the transfer of the power from the input to the output. The conversion loss is a ratio of the power attainable at the mixer output to the attainable power of the RF signal source coupled to the mixer input. This loss depends on the one hand on the non-linearity of the non-linear element itself, on the generator impedance and, on the other hand, on the external circuit impedance and on further values of the combination frequency (particularly of the image frequency). This results in, that the conversion losses of the mixer depend on a particular implementation of the mixer circuit as well as on the external circuit in which it is incorporated. In order that it is possible to express in a certain manner the characteristics of the mixer, conversion losses are presented for the mixer usually for optimum matching of the input and for the termination of the residual values of the combination frequency by the characteristic impedance.

The noise characteristics of mixers are expressed with the help of the noise figure F of the mixer (definition of the noise figure of a two-port see in Section 4.3.4). The following quite general relationship holds between the effective temperature T_e of the mixing diode, conversion losses and noise figure of the diode mixer:

$$F = L\,(T_e + F_{IF} - 1) \tag{4.134}$$

where F_{IF} is the noise figure of the IF amplifier.

The mixing losses L and effective temperature T_e depend on the nature of the non-linear element, amplitude of the local oscillator $|U_1|$, operation point and structure of the mixer. In spite of this, for orientational purposes, instead of the effective temperature T_e of the diode mixer it is possible to substitute in the first approximation for example the effective temperature of the Schottky diode derived in Section 3.1. It ranges between 0.8 and 1 in practice.

Thus, for the use of diode mixers in practice it is possible to write as a good approximation

$$F\,[dB] \approx L\,[dB] + F_{IF}\,[dB] \tag{4.135}$$

In transistor mixers, the noise figure is in no direct relationship to conversion losses (gain) and thus, it is impossible to use the simplified equation (4.135).

A further part considers the description of the design and characteristics of main mixer types used in the MIC and MMIC techniques.

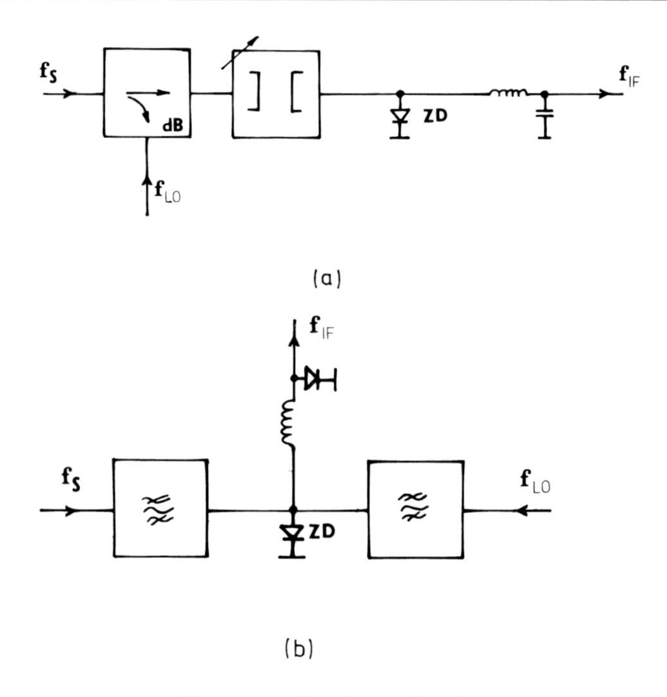

Fig. 4.44 Simple diode mixer.

The simple diode mixer (Fig. 4.44) contains only one Schottky-barrier diode and combination of ways of the local oscillator and RF signal is commonly achieved by a directional coupler (Fig. 4.44a). The combination with the help of a frequency diplexer is less common (Fig. 4.44b). The former solution has the disadvantage of a considerable loss of the local oscillator power (typically 10 dB) the latter solution requires in addition to the complex nature of the circuit that the frequency bands of the RF signal and of the local oscillator should not overlap (high intermediate frequency). In general the simple mixer is rarely used due to the disadvantage mentioned.

The symmetric mixer with two mixing diodes (Fig. 4.45) has an advantage in comparison with the simple one that it has naturally separated ways of the signal and local mixer without any loss of the power. There are many variants of solving the symmetric mixer under MIC conditions. Only two basic variants are mentioned here. The first variant (Fig. 4.45a) is a mixer, where the combination of both signals is accomplished by an either hybrid or magic T-element (Section 4.2) in the microstrip form. In the use of the hybrid T-element it

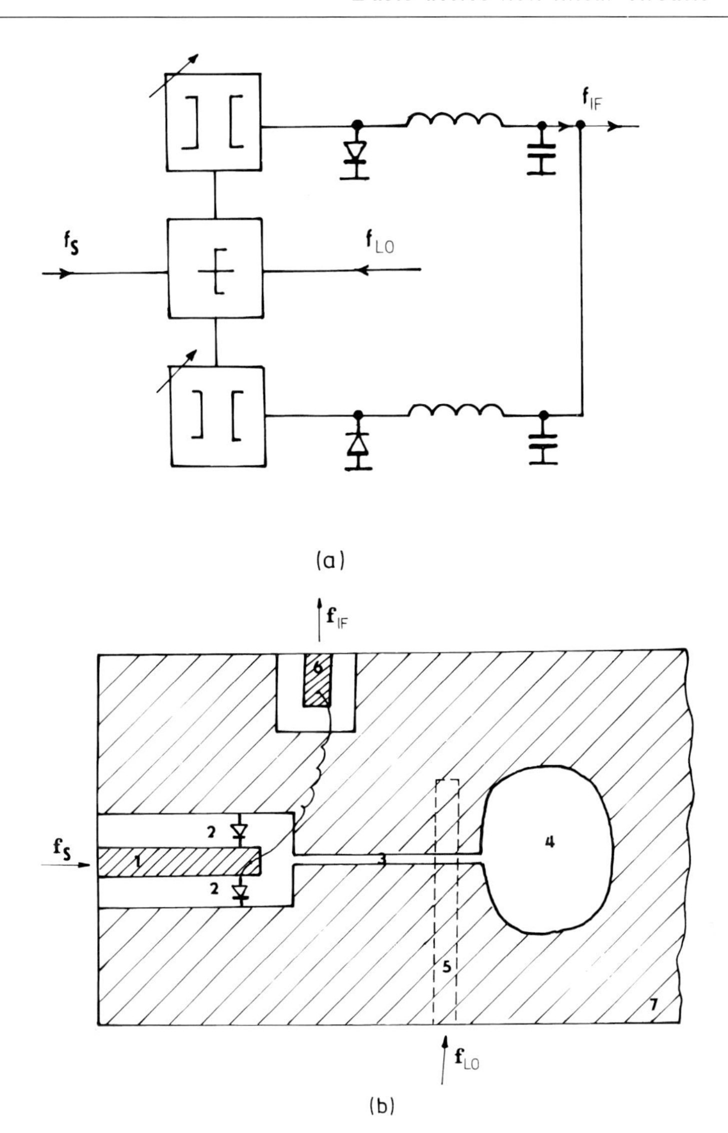

Fig. 4.45 (a) Symmetric mixer with two mixing diodes. (b) Other variant of the two-diode mixer.

is advantageous to insert the differential shift $\pi/2$ into the branches, which results in a consequent separation of both ways even in the case of the imperfect matching of diodes. If this shift were not used, the waves reflected from particular diodes would always be summed in the output branch opposite to that which they were excited from.

The use of the magic T principally solves this problem, however, this circuit has an essentially narrower band. The solution mentioned is very simple from the standpoint of the circuit, the whole circuit is situated on one side of a dielectric substrate. Its disadvantages include rather larger dimensions and smaller width of the band, by about 20%.

The other variant of the two-diode mixer is shown in Fig. 4.45b. This is a combining circuit solved in a combination of coplanar and slot lines. This circuit has a wider band, however its design is technologically more complex, since it assumes bilateral MIC. In this part it is shown how it is possible, with the use of general characteristics of a non-linear element expressed by relationship (4.114), to derive the output as well as suppression of the mf signal and of the other combination signals in symmetric connections. With respect to the symmetry of the connection, the amplitudes of all the combination components will be the same on both diodes. The polarities of the input signal and local oscillator are shown by arrows with markings S and LO, those of the diode and intermediate frequency by symbols D and IF. In the case of the intermediate frequency ($m = 1$, $n = -1$) it can be seen that the mf signals from particular diodes are summed in the central conductor of the input line. On the other hand, the direct components and the second harmonic ($m; n = 0.2$) of the output and LO from both diodes are mutually eliminated at this point.

The double balanced mixer with a number of four diodes (Fig.4.46) is characterized by its wide band. It is possible to achieve a band-

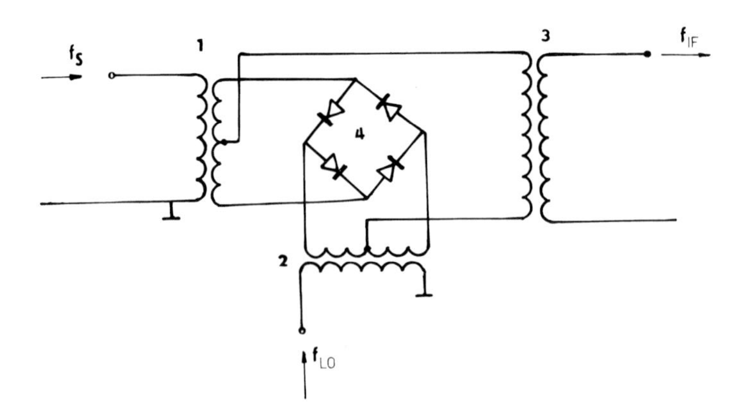

Fig. 4.46 Double balanced mixer.

width as large as four octaves. The disadvantage is the requirement for a higher power of the local oscillator. The accomplishment of symmetrization transformers depends on the frequency (Section 4.2.11). In the problem of non-linear elements, monolithic groups of four LBS diodes are most frequently used to reduce the spread of diode parameters in the group and to minimize requirements for the local oscilator power.

The twice double balanced mixers with two groups of four diodes (Fig. 4.47) have as their main advantages a wide band and good insulation of all the inputs with a possibility of a high intermediate frequency, which makes possible overlapping of frequency band of the RF signal, local oscillator and intermediate frequency. The disadvantage is a complex construction and considerable requirements for the power of the local oscillator. The symmetrizing transformers can be solved by several methods, again, (according to 4.2.11). One of them is shown in Fig. 4.47b. This is an analogy of a continuous change from the microstrip to symmetric line in a mixer with one group of four diodes.

Transistor mixers are particularly used in the technique of monolithic MIC, particularly due to the fact that in this technology, it is simpler to produce a transistor as a high quality Schottky's barrier diode. The disadvantage of transistor mixers with MESFE transistors is a high flicker noise, which makes impossible their use from frequencies to 100 – 200 MHz including the intermediate frequency. In one-substrate transistors, only different symmetric connections are commonly used, since in the monolithic performance, it is more simple to produce several transistors in a circuit as a separating circuit in the distributed form (by a coupler or filter). Out of many connections used, three types are mentioned below.

The double balanced mixer with one-substrate transistors (Fig. 4.48) has the advantage of non-linearity of the conversion characteristic. The transistors T_1 and T_2 serve for a preliminary amplification and separation of the RF signal. Symmetric supply is achieved

a) by a symmetric solution of exciting steps
b) by symmetrizing transformers of spiral type
c) by active symmetrizing stages.

The simple mixer with a two-substrate transistor (Fig. 4.49) is particularly used in the hybrid MIC technique or its simple nature. It utilizes the non-linearity of the conversion characteristic and RF

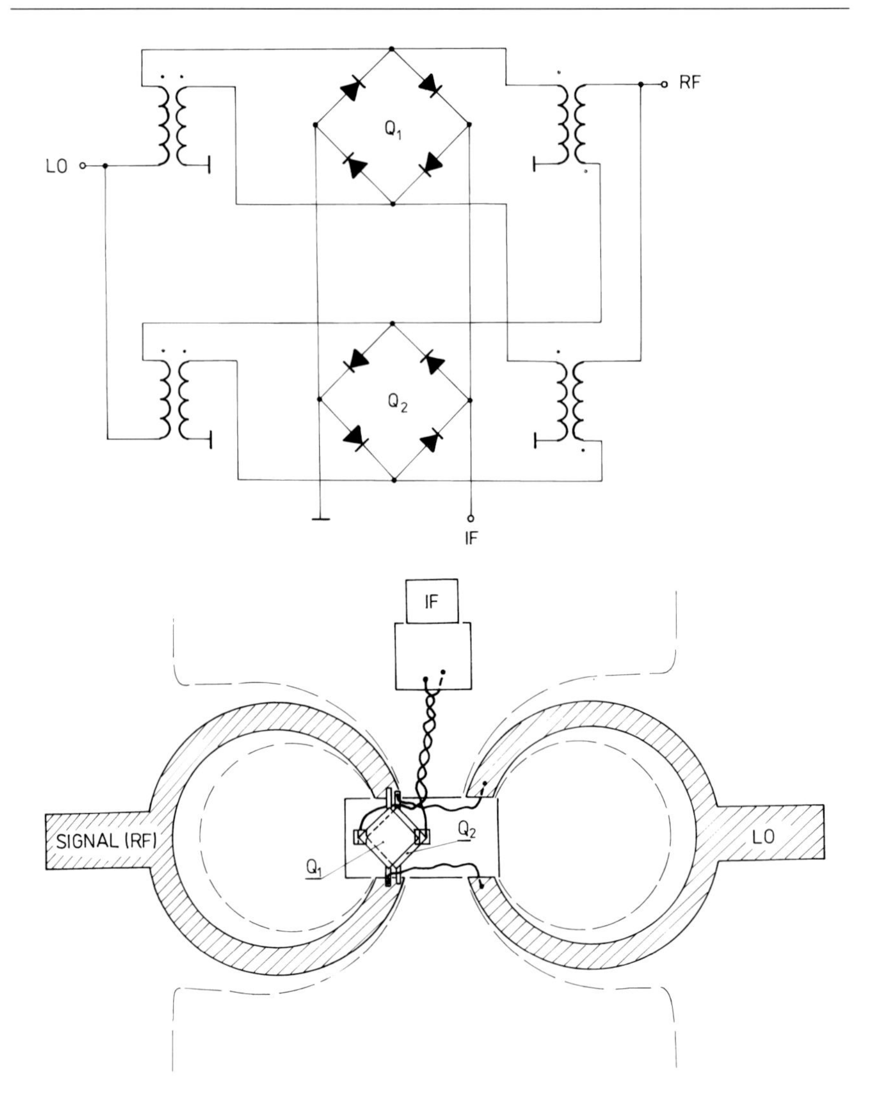

Fig. 4.47 Double balanced mixer and its equivalent circuit.

separation of both substrates. The main disadvantage is the penetration of the signal frequency and of the local oscillator to the output, which should be filtered.

The balanced mixer with two-substrate transistors (Fig. 4.50) is characterized by suppressing the penetration of the signal and of the local oscillator to the output. The separation of output of these signals is provided by using two-substrate transistors.

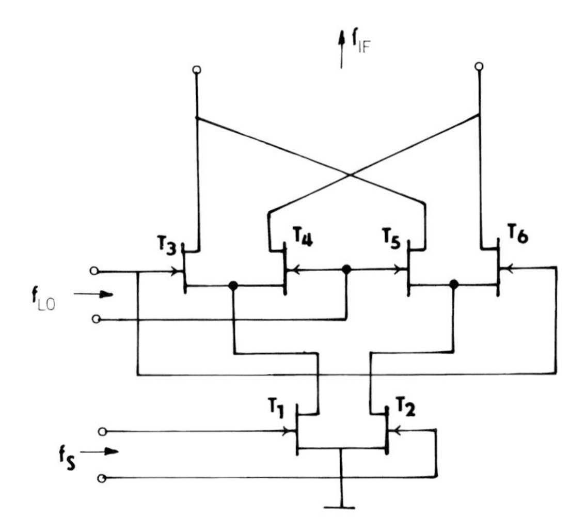

Fig. 4.48 Double balanced mixer with one-substrate transistors.

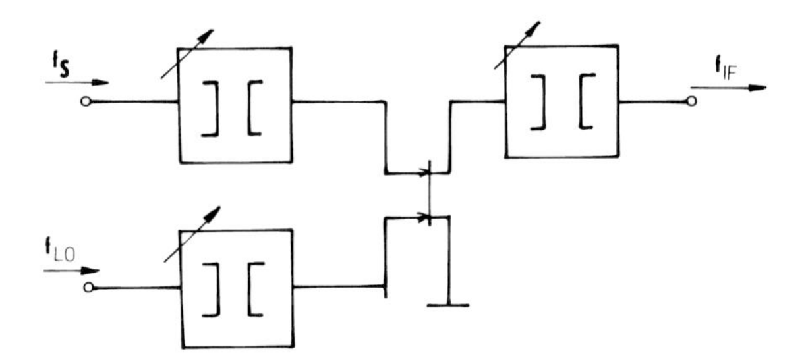

Fig. 4.49 The simple mixer with a two-substrate transistor.

All the mixers mentioned have principally identical characteristics of the RF signal frequency and mirror one. This results in signals coming into the mixer at the RF frequency signal as well as those of the mirror signal are converted to the intermediate frequency with the same conversion losses without the possibility of differentiating their origin. And, on the other hand, the RF signal received in principle generates an image signal $f_R = 2f_{LO} - f_S$ is absorbed by the external circuit, which increases conversion losses of the mixer. These disadvantages can be avoided by higher systems of mixers, which can be made of any type of the mixers discussed. The mixers suppressing the image frequency are implemented with the help of a filter inserted before the mixer, which passes through the RF signal and retains the image signal. A fairly simple solution is to insert a ferrite insulator

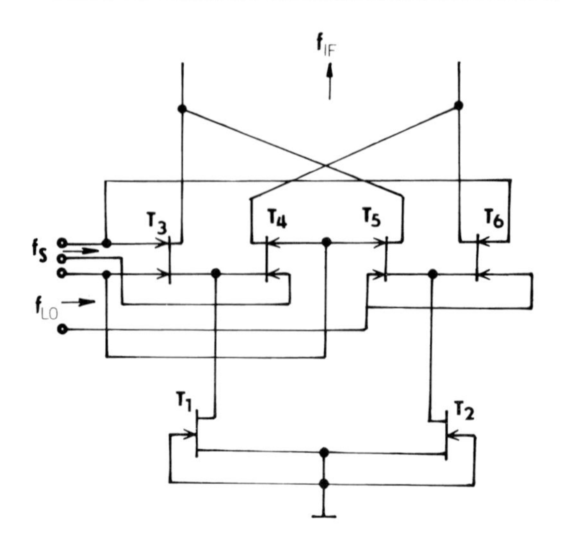

Fig. 4.50 The balanced mixer with two-substrate transistors.

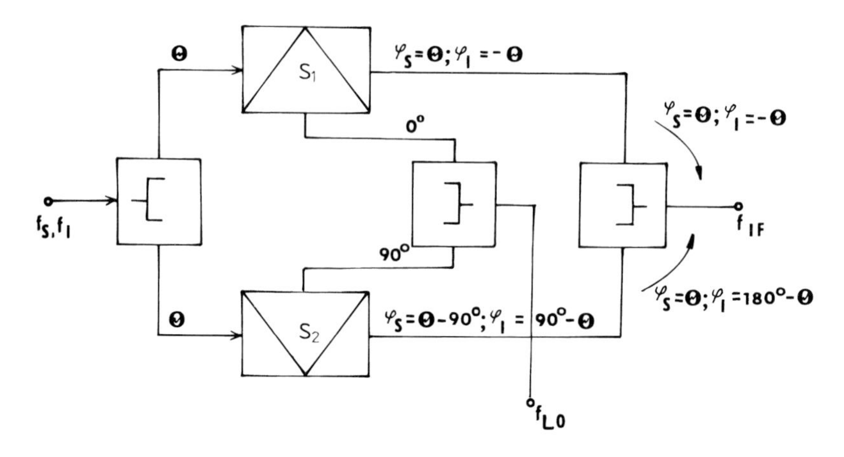

Fig. 4.51 Two mixers S_1, S_2 with phase shifts in branches by 90°.

between this filter and mixer with only one restriction that the band of the signal and image should not overlap. This solution is suitable for hybrid MIC at a high intermediate frequency. The other solution, shown for example in Fig. 4.51 is based on using two mixers S_1 and S_2, which are interconnected by cophasic power dividers with phase shifts in branches by 90°. The separation principle is obvious from the relationships presented. In this type of mixer, the mixing losses as well as noise number are different from corresponding quantities of a simple mixer only by losses of additional circuits (thus, a weak degradation of these characteristics is encountered).

The mixers using the image frequency are types designed in such a way that the image generated in non-linear elements of the mixer by the process $f_R = 2f_{LO} - f_S$ encounters no losses in the resistive load of the mixer and it is back used for the generation of the mf frequency $f_{IF} = f_{LO} - f_R$. This can be achieved either with the help of a filter or by a system of mixers with cophasic and quadrative power dividers. In the first case, it is, however, necessary to remove the insulator between the filter and mixer and to situate the filter at such a phase distance that the reflected image signal is returned into the mixer at an optimum phase. This can be most properly achieved experimentally, since the phase ratios are not simple in a real mixer. In the case of a system of two mixers, the cophasic power divider must not be matched from the output side and it phase distance from both mixers should adhere to certain conditions. In this way, in practice it is possible to achieve a reduction of conversion losses and improvement of the noise figure by about 1 to 2 dB with simultaneous suppressing of the parasitic reception at the image frequency.

The mixer with the separation of the image frequency (Fig. 4.52) is

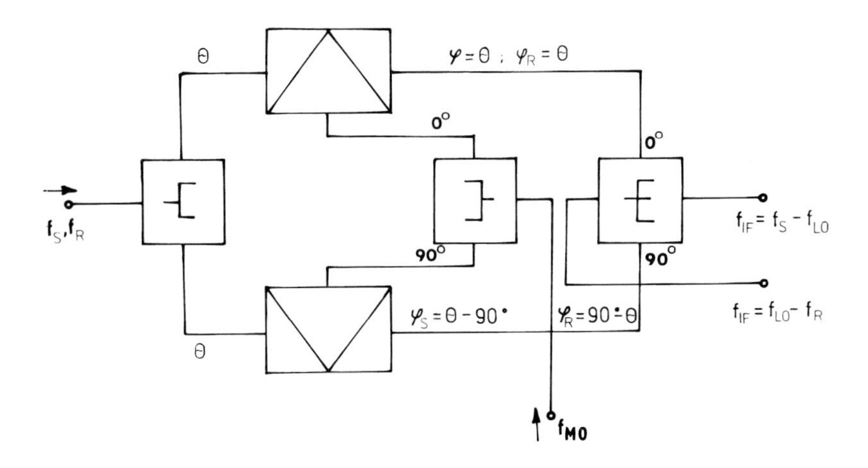

Fig. 4.52 The mixer with the separation of the image frequency.

different from that in Fig. 4.51 only by that the output quadrature divider is replaced by a hybrid T-element. The philosophy of the whole connection is obvious from the mentioned relationships for phases of the RF signal and image. In this way it is possible to achieve as large a separation as 20 dB in practice.

When designing a mixer, it is first necessary to select the mixing element and the type of the mixer according to Table 4.2 with taking into account the basic requirements. The choice, is, however, frequently limited by the availability of the mixing element. In a simple or symmetric mixer, matching ways of the MIC for the power chosen are then either experimentally or theoretically proposed. In the theoretical proposal, the method of the equilibrium of harmonic components is typically used. By the linearization of the relationship (4.133), it is possible to derive the input impedance for the signal frequency with subsequent optimization of matching from the input by a method used for linear circuits. In mixers with higher forms of symmetry, the possibilities of matching are rather restricted and thus, it is necessary to choose the non-linear element carefully. The producers of groups of four diodes usually specify the suitability of their products for certain frequency bands and line impedance. In transistors, the theoretical analysis is first necessary (by the form of the equilibrium of harmonic components, again). In spite of the fact the calculations can be considerably simplified by the use of the symmetry principle, they still remain very complicated due to the presence of many non-codivisible values of the frequency combinations.

4.3.3 Oscillators

The most extensively occurring type of oscillators in the MIC and MMIC are transistor oscillators, which are considered in this section. General conditions of the oscillation, of the oscillation origination and of their stability will be first described. After that the basic parameters of oscillators will be defined and the classification of oscillators according to their connection will be presented. The design of oscillators and examples of certain typical solutions will conclude this section.

Microwave oscillators are typically theoretically treated and described in the frequency with respect to the complicated nature of equivalent diagrams of microwave circuits. Solutions in the time domain are, however, also known, which are rather suitable for the investigation of principal dependences, since they present requirements for considerable simplifications of the circuit part. The present explanation will be limited to the frequency region.

For the purpose of the oscillator analysis, the transistor can be represented for example by a T two-port with an admittance matrix $\mathbf{Y}_T$, which has elements which are in general a function of the basic oscillator frequency, amplitude and phase between all the harmonic components.

Different connections of oscillators with one transistor can be in general expressed by a circuit according to Fig. 4.53, where P and

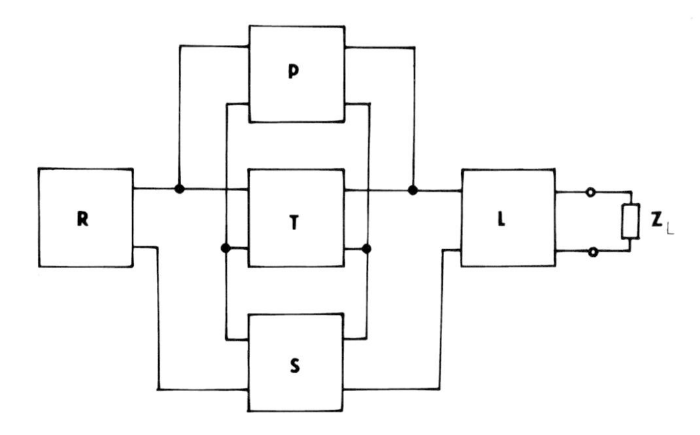

Fig. 4.53 Oscillator with one transistor.

S are feedback two-ports (parallel and series coupling), R is an one-port, which requires resonance load, L is the input transformation two-port and Z_L is the external useful load. In the solution of this circuit, it is theoretically necessary to solve the circuit equations at a basic frequency and of all its harmonics.

The necessary condition for the oscillations requires that the equation for the RF currents and voltages in the circuit in Fig. 4.53 should have a non-zero solution (in the absence of the RF source). This leads to a system of equations

$$Z_{osc}^{(k)} + Z_z^{(k)} = 0$$

$$Z_{osc}^{(k)} = Z_{L12}^{(k)} - \frac{Z_{L21}^{(k)}}{Z_{L12}^{(k)}} \left(Z_B^{(k)} + Z_{L11}^{(k)} \right) \tag{4.136}$$

where

$$Z_B^{(k)} = Z_{A12}^{(k)} - \frac{Z_{A21}^{(k)}}{Z_{A12}^{(k)}} \left(Z_R^{(k)} + Z_{A11}^{(k)} \right)$$

$$Z_A^{(k)} = \left(Y_T^{(k)} + Y_P^{(k)} \right)^{-1} + Z_S^{(k)} \qquad \text{for } k = 1, 2, \ldots \infty$$

$Y_{T_{(k)}}^{(k)}$ is the transistor admittance matrix at the k-th harmonic, Y_P is the admittance matrix of the parallel coupling at the k-th harmonic, $Z_{S_{(k)}}^{(k)}$ is the impedance matrix of the series coupling at the k-th harmonic, Z_{Lmn} are elements of the impedance matrix of the transformation circuit at the k-th harmonic, $Z_{Am,n}^{(k)}$ are elements of an auxiliary impedance matrix $Z_A^{(k)}$, $Z_R^{(k)}$ is the impedance of the resonance circuit at the k-th harmonic and $Z_{osc}^{(k)}$ is the impedance of the oscillator at the k- th harmonic.

All the quantities are functions of the basic angular frequency of the oscillator ω, the elements of the admittance matrix of the transistor are, in addition, a function of amplitudes and phases of basic harmonic components of the voltage. Thus, the relationship considered is a system of equations for the frequency, amplitudes and phases of harmonic components. For providing that the solution of the system of the equations represents oscillations, it is at least necessary to adhere to the condition of stable oscillations, which expresses the ability of the systems to return to the original condition after vanishing of an infinite small disturbance (Liapunov's stability). This leads to the following condition when considering a restriction to the first harmonic component of the oscillator current I_1:

$$\frac{\partial R}{\partial I_1}\frac{\partial X}{\partial \omega} + \frac{\partial R}{\partial \omega}\frac{\partial X}{\partial I_1} < 0 \tag{4.137}$$

where

$$R = \Re\left\{ Z_{osc}^{(1)} + Z_L^{(1)} \right\}$$

$$X = \Im\left\{ Z_{osc}^{(1)} + Z_L^{(1)} \right\}$$

The relationships presented (4.136) and (4.137) are usually analysed with the use of a method of the equilibrium of harmonic components. Sometimes it is possible to acquire valuable data by the method of the analysis at the fundamental frequency (FFA — Fundamental Frequency Analysis) i.e. with neglecting all the higher harmonic components with $k > 1$. Equation (4.136) with condition (4.137) either need not have any solution or they can have several solutions. The mentioned systems, however, do not delimit the method by which it is possible to bring the circuit into the oscillating condition. The circuits are most important in practice, which are able to achieve a stable, unambiguously defined oscillation condition after being switched on. For this purpose, equation (4.136) and condition (4.137) should have one solution only and, in addition, it is also necessary to satisfy to the condition of the circuit instability at a low signal. This can be simply expressed by the relationship

$$\Gamma_{osc}\Gamma_L > 1 \tag{4.138}$$

where

$$\Gamma_{osc} = \frac{Z_{osc}^{(1)} - Z_0}{Z_{osc}^{(1)} + Z_0}$$
$$\Gamma_L = \frac{Z_L^{(1)} - Z_0}{Z_L^{(1)} + Z_0}$$

reflection coefficients of the oscillator and load at the fundamental frequency for a low signal in the line with a characteristic impedance $Z_0 \neq Z_L$.

Under the given conditions, the circuit in Fig. 4.53 usually spontaneously starts to oscillate due to noises. The exceptions are cases, where this circuit tends to pass to relaxations or other complex oscillating conditions. This can be avoided by choosing a suitable connection.

In addition to the frequency f_0 and power P_0, the basic oscillator parameters are the external quality coefficient Q_{ext}, width of the mechanical and electric overtuning Δf_m and Δf_e and factors affected by the load and supply voltage. Particular quantities are defined by the following relationships:

$$Q_{ext} = \frac{\omega_0 \partial \Im \left\{ Z_{osc}^{(1)} \right\}}{\Re \left\{ Z_{osc}^{(1)} \right\} \partial \omega} \tag{4.139}$$

Δf_m = $(\max f_o - \min f_o)$ at changes of circuit dimensions
Δf_e = $(\max f_o - \min f_o)$ at a change of the bias of the electric and overtuning element

$$\text{pushing factor} = \frac{\partial f_0}{\partial\,(\text{SWR} - 1)}$$

$$\text{pulling factor} = \frac{\partial f_0}{\partial U_0}$$

where SWR is standing wave ratio in the input line (only for a real load), U_0 is the supply voltage of the oscillator. In certain applications, the stability of the oscillator frequency can be of a great importance. The long-term stability is expressed by a frequency change depending on temperature or time in millionth parts of the frequency per unit time or temperature $\Delta f_{T,t}$ [ppm/C or ppm/day, year, etc.]. For expressing the short-term stability, it is necessary to know statistical characteristics of the phase fluctuation or of the oscillator frequency, or the course of the oscillator RF spectrum. For simplicity's sake, it is frequently necessary to use only one piece of information, effective deviation of the frequency Δf_{rms}, for example at a distance of 1 kHz from the carrier frequency and in a band width of 1 or 100 Hz.

With respect to a considerable variety of requirements, transistors attainable and circuit technologies, in practice there are many possibilities of connecting the oscillators. Table 4.3 summarizes at least the most important characteristics of basic connections, classified from the standpoint of the situation of the resonant circuit and method of implementing the feedback.

In oscillator design, it is first necessary to choose the proper oscillation element. When choosing the transistor, then the basic condition is that the limit transistor frequency $f_{\max}$ should be higher than the oscillator frequency f_0. For wide-band overtuned oscillators it must hold that: $f_{\max} \geq f_{0\,\max}$, where is the highest oscillator frequency. It is, however, simultaneously unsuitable to use transistors with too high $f_{\max}$ with respect to f_0. Each microwave circuit has always many resonant frequencies and thus, equation (4.118) can easily have several solutions which, however, leads to an ambiguity of the origin of the oscillation conditions and to other forms of instability. From the power standpoint, transistors with the saturated power $P_{sat} = 10 + 50\,\text{mW}$ with $P_{sat} > 1.5\,\text{P}_0$ are the most suitable ones. When the saturated power is according to this inequality

Table 4.3 Summary of basic connections of oscillators

Backward coupling	Serial suitable for common base connection	Parallel suitable for common emitter connection
Location of resonator		
Reject resonator (R)	output power up to = 0.9 P_{sat} of transistor tuning band width to 20% suppression of parasitic frequencies on the output max 20 dB simple structure and design, suitable for tuning by varactor	
	frequency to 0.8 $f_{\max}$	frequency to 0.9 $f_{\max}$
		suitable for DR
In backward coupling (P or S)	output power up to 0.7 P_{sat} of transistor tuning band width to 90% suppression of parasitic frequencies on the output max 30 dB complicated structure and design, suitable for tuning by varactor frequency to 0.7 $f_{\max}$	
	suitable for YIG	suitable for DR
In output transformer (L)	output power from 0.5 to 0.6 P_{sat} of transistor tuning band width to 70% suppression of parasitic frequencies on the output 40–50 dB complicated structure and design, suitable for tuning by varactor frequency to 0.8 $f_{\max}$	
	suitable for DR	suitable for YIG

Where DR is a dielectric resonator and YIG is ferimagnetic resonator tuning by magnetic field

larger than the mentioned range, then it is more suitable to insert after the oscillator a power amplifier, since transistors with a higher saturated power usually have special capsules making the formation of the oscillator more difficult. In addition to this, in electrically tunable oscillators it is more suitable (due to the non-linearity of the tuning element — either varactor or resonator YIG) to keep the oscillator power at a low level. On the other hand, the decrease of the oscillator power results in deteriorating amplitude and phase stability (the separation signal-noise is reduced). In practice, it is sometimes possible to consider a choice between the bipolar transistor and field — controlled transistor. In principle, the transistors of these two types are equivalent in their use in oscillators, only the parameters $f_{\max}$, P_{sat}, the capsula, etc. are of importance. The only exception are oscillators with a low phase noise. Since the transistors on silicon have a lower flicker noise (by 10 to 20 dB) than transistors on GaAs or on other mixed semiconductors, for these applications, the Si bipolar transistors are preferred.

For all the types of oscillators transistors in the form of a chip or encapsulated transistors without a considerably defined shared electrode are suitable. The resonant circuit is chosen according to requirements for Q_{ext} and for the possibility of evertuning the oscillator. It is particularly necessary to provide that

$$Q_0 \geq 2Q_{ext} \tag{4.140}$$

where Q_0 is the unloaded coefficient of the resonator quality.

When requiring the electric overtuning up to 20%, it is necessary to use a varactor as a part of the resonant circuit (when using a varactor with a superabrupt transition, it is possible to achieve as much as 60%). When requiring the electric overtuning of one octave and above, it is necessary to use the YIG resonator. The most frequently used resonance circuits are outlined in Table 4.4.

Depending on the transistor and resonant circuit chosen and according to further requirements for characteristics of oscillators, some of the basic connections according to Table 4.3 are chosen.

For the oscillator design itself, it is also possible successfully to use a simplified method, which is based on a knowledge of low-signal parameters of the transistor. The proposed linear circuit according to Fig. 4.53 is analysed and the optimization method is used to search for the solution most properly satisfying to the conditions

$$\begin{aligned} \Im\left\{\Gamma_{osc}\Gamma_L\right\} &= 0 \\ \left|\Gamma_{osc}\Gamma_L\right| &= K = const\,(\omega) > 1 \end{aligned} \tag{4.141}$$

where the quantity K should be understood as a constant during changes of parameters of the tuning element, which is simultaneously maximized by an optimizing program. This procedure leads to an oscillator with approximately constant and maximum output power.

Table 4.4 Summary of basic resonant circuits using in design of hybrid integrated oscillators

Type of circuit	freq.	Tuning		Q_O
		mechanical	electrical	
Lumped resonant circuit	to 4 GHz	20%	–	100/4 GHz
Lumped resonant circuit tuned by varactor	to 4 GHz	20%	$10 \div 70\%$	50/4 GHz 400/5 GHz
Microstrip resonator	$2 \div 20$ GHz	20%	–	$150 \div 200/10$ GHz
Microstrip resonator tuned by varactor	$2 \div 20$ GHz	20%	$10 \div 40\%$	200/5 GHz 100/10 GHz
Dielectric resonator	$2 \div 27$ GHz	5%	–	$Q_O f > 40000$ GHz
Dielectric resonator tuned by varactor	$2 \div 27$ GHz	5%		$Q_O f > 10000 \div$ 20 000 GHz
Resonator YIG	$1 \div 18$ GHz	–	90%	$Q_O f \approx 100$ GHz

As an example, several typical connections of transistor oscillators in the MIC or MMIC performance are presented. Fig. 4.54 shows a classic connection falling into the group of oscillators with a series feedback and rejection resonator formed by a microstrip stub P_R,

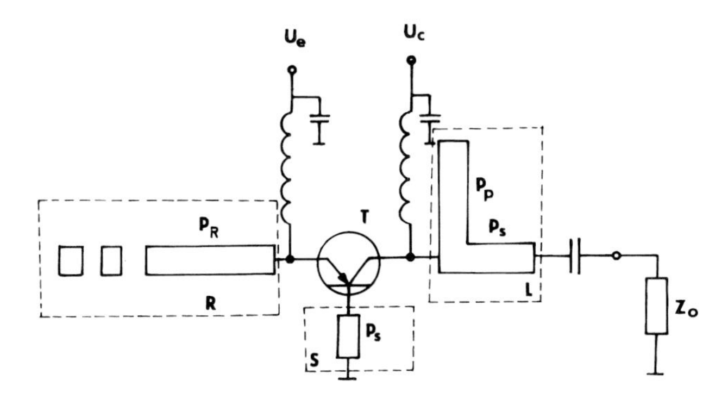

Fig. 4.54 Oscillator with series feed-back and rejection resonator formed by a microstrip stub P_R.

whose length and thus also the oscillator frequency can be mechanically varied within small ranges. The feedback two-port is reduced to a segment of short-circuited microstrip line P_s of an inductive nature $(1 < \lambda_g/4)$.

In the case of the oscillator in Fig. 4.55, the resonator P_R as well

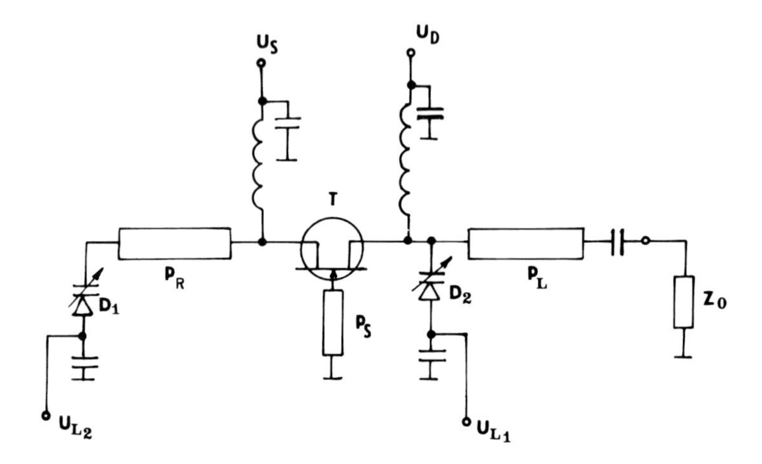

Fig. 4.55 Other configuration of oscillator.

as output transformer P_L are simultaneously tuned by varactors D_1 and D_2, respectively. This makes it possible to achieve a wider band of overcoming with an easier achievement of a relatively constant output power. For this purpose it is necessary, however, to achieve an agreement between the tuning voltages U_{L1} and U_{L2}, which can sometimes be hardly obtained in practice, particularly in the case of requirements for a high tuning rate. Thus, one of the varactors is frequently omitted.

An example of the oscillator tuned by the YIG resonator is the circuit shown in Fig. 4.56. Here, the tuning element simultaneously forms a feedback two-port, which considerably facilitates the light-band tuning and contributes to the uniformity of the output power.

Fig. 4.57 shows an oscillator stabilized by a dielectric resonator simultaneously forming a parallel feedback. This is a frequently used connection, which is characterized by a high frequency stability and high output power.

In addition to oscillators with one transistor, oscillators with two transistors (exceptionally also with several transistors) are also commonly used. The symmetric oscillator (Fig. 4.58) is of the almost

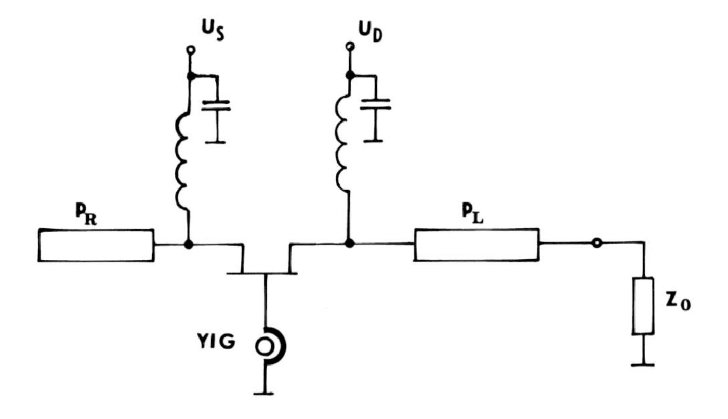

Fig. 4.56 Oscillator tuned by a YIG resonator.

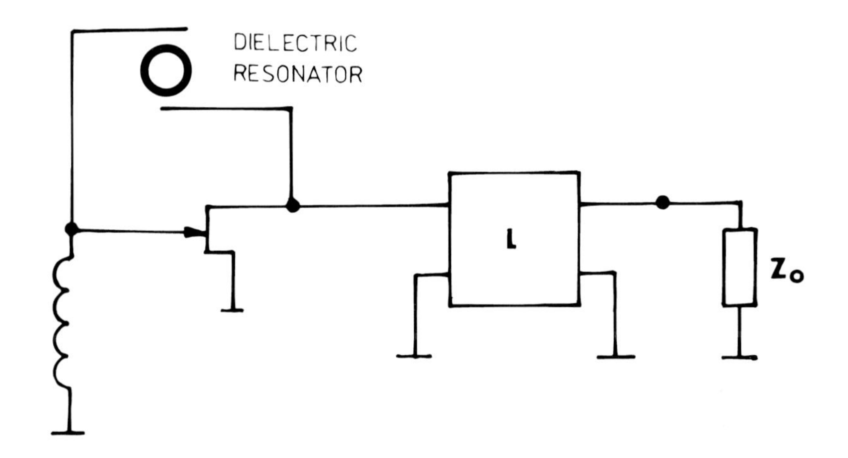

Fig. 4.57 Oscillator stabilized by a dielectric resonator.

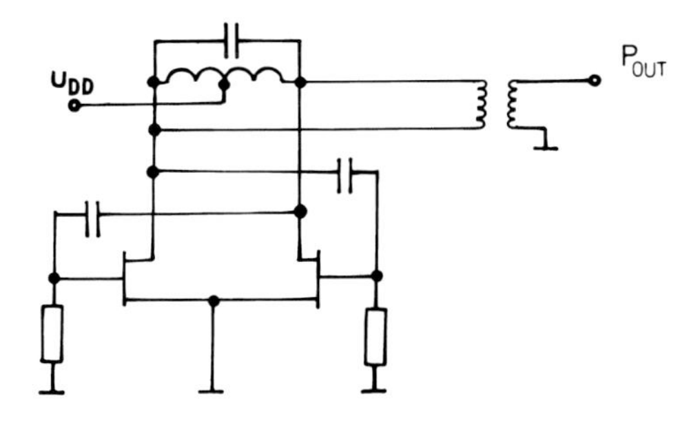

Fig. 4.58 Symmetric oscillator.

importance. This is actually an oscillator with a parallel feedback and with a resonant circuit situated in the output transformer, where the parallel feedback is implemented by a further transistor. This makes it possible to solve a very wide band element, particularly suitable for the monolithic integration. The symmetry of the connection facilitates the inlet of supplying voltages in wideband applications.

4.3.4 Amplifiers

Similarly as mixers or oscillators, amplifiers also form a very diversified group of microwave circuits. Various approaches to their categorization also correspond to this fact.

Depending on the method of use they are divided into low-noise, intermediate-power and power ones. The low-noise amplifiers are basic elements of microwave receivers, determining their sensitivity and thus also their quality. The intermediate-power amplifiers are used for different applications in processing microwave signals. The power amplifiers find their main use at end steps of microwave sources and transmitters.

All these three classes of amplifiers can be further divided, depending on the semiconductor elements used, to transistor amplifiers, parametric amplifiers and amplifiers using a negative differential resistance of certain diodes (avalanche, Gunn and tunnel diodes). Out of these classes the transistor amplifiers are by far most frequently used. The parametric amplifiers are used in low-noise applications, either as cryogenic for achieving extremely low-noise numbers or in the upper frequency region of millimetre waves, where quality transistors are not yet available. At very high values of the frequency, avalanche diodes in the IMPATT regime and Gunn diodes have been recently used for the above mentioned reasons. In the cases of using active elements other than transistors, the waveguide and coaxial techniques are, however, mostly preferred rather than the MIC technology. This results from attempts to reduce losses. The MMIC are also not used in these amplifiers due to technology problems. Thus, this chapter considers only transistor amplifiers. They can be divided from the standpoint of the method of connecting the transistor into amplifiers without the external feedback, feedback amplifiers and travelling wave amplifiers. There are still further approaches to the categorization, as e.g. depending on the number of steps, type of transistor, bandwidth, type of the line used, etc. These approaches are, however, of importance from rather technology than designing standpoint.

The basic problems of proposing and designing the amplifiers can be explained on an example of low-noise amplifiers first of all of the most frequently used type of a simple amplifiers without the external feedback. In further types, either low-noise or power ones, certain special features in the proposal or possibly design will be shown. In this respect the intermediate-power amplifiers exert no special features and thus, they will not be discussed in a special chapter. In the whole explanation below the use of MESFE transistors is assumed in connection with their earthed emitter. All the design procedures, however, remain valid for the other connections and/or for the other types of transistors, too.

4.3.4.1 Low-noise amplifiers

Basic problems of low-noise amplifiers

The function of these amplifiers used as an input part of microwave receivers is a sufficient amplification of the input signal with a minimum deterioration of the separation of the signal from the noise level. The amplifier linearity is usually a complementing requirement, providing an undistorted transmission over a wide scale of input powers.

The first problem, which is of a primary importance in the design, is the choice of a suitable transistor and method of its connection. In addition to requirements for the parameters achieved, the availability of transistors, technology possibilities and economy standpoint usually serve as the criteria in the selection. The result is always a compromise and thus, there is no unambiguous instruction for the procedure. Only the main principle can be mentioned.

The most commonly used low-noise transistors are MESFE type transistors. In lower frequency bands, bipolar transistors are also used, having as their advantage, in addition to their low cost, a possibility of easier matching to the commonly used 50Ω loads. For the amplifier quality, particularly in the region of millimetre waves and at higher bands of centimetre waves, HEM type transistors are used due to their essentially better noise as well as transmission characteristics. In certain cases, amplifiers also occur, using the field-controlled transistors with a double gate. These transistors makes it possible to achieve a higher gain at one stage and particularly the control of the gain and possibility of its modulation by the external voltage, however, they have rather worse noise characteristics. Thus, they are typically included into higher amplifying stages.

The decisive parameter of the low-noise amplifiers is their noise figure. In multistage amplifying chains it is given by the known relationship

$$F_z = F_1 + \frac{F_2 - 1}{G_1} + \frac{F_3 - 1}{G_1 G_2} + \cdots + \frac{F_n - 1}{G_1 ... G_{n-1}} \tag{4.142}$$

where F_z is the chain noise figure, F_i is the noise figure of the i-th stage and G_i is its achievable gain at a given input impedance. When the i-th step is formed by one-stage amplifier without external feedback, then it is possible to show with the Friis formula, again, that the noise figure F_i is given by the equation

$$F_i = L_{1i} F_{Ti} + \frac{L_{2i} - 1}{G_{Ti} L_{2i}} \tag{4.143}$$

where G_{To} is the achievable transistor again at the i-th stage F_{To} is its noise figure and L_{1i} or L_{2i} stand for the intrinsic losses of the input or output matching circuits, respectively. When taking into account the fact that the typical intrinsic losses of connectors and transformers in the MIC performance range between 0.1 and 0.3 dB, it is possible to conclude that in narrow-band amplifiers, the transistor used should have its noise figure better by 0.3 to 1 dB in comparison with the requirement for the amplifier. In wide-band amplifiers, where it is necessary to take into account a deterioration of parameters due to a deteriorated matching to the optimum impedance or possibly also due to losses in separating elements, it is possible to expect a further deterioration of the amplifier noise figure by 0.5 to 3 dB. The requirement for the bandwidth is affected by the choice of the transistor and type of the amplifier also from the standpoint of suitable impedance ratios. Thus, in wide-band amplifiers, the chip-type transistors are almost exclusively used, where the characteristics are not degraded by parasitic capacitances of the capsule. This, however, also brings higher technology requirements in the production.

Synthesis of simple one-stage amplifier without external feedback

The transistor choice is also tightly related to the choice of the amplifier design type. The one-stage simple amplifier without external feedback is the most simple one. It consists of an input and output matching transformer, supplying circuit and transistor.

The supplying circuits are beyond the scope of this work. A number of types are presented for example in Arden (1976). Here, it is necessary only to remind readers, that they should be connected in such a way, as not to affect the properties of the amplifier. For this purpose, sites with microwave short circuiting are for example used for its purpose in matching circuits.

For the synthesis of matching circuits, different methods have been elaborated from the most simple graphic methods (Wienert, 1968) through methods considering design procedure similar to procedure in the synthesis of filters (Carbin, 1878) to the optimizing method, which take an advantage of optimizing programs providing a direct optimization of microwave characteristics of the amplifiers depending on the geometry of microwave circuits.

The mathematically most elegant methods are those falling into the second group. They make it possible to provide a majority of solutions in their analytical form with the use of mathematically well managed procedures. They are, however, rather complicated and in the conclusive stage of the calculation representing in the MIC or MMIC performance, it is usually necessary to employ a number of simplifications degrading the precision of the results. For this reason they are currently seldomly used and numerical optimizing methods are preferred.

The graphic solutions are used in their simplest form most typically neglecting the internal feedback of the transistor, for obtaining input configurations for the optimizing methods. Neglecting the internal feedback of the transistor makes it possible to independently solve the input and output circuits of the amplifiers. For $s_{12} = 0$ the transistor actually represents a perfect insulator.

In the case of lossless matching transformers the transmission gain of the amplifier equals the unilateral gain of the transistor (Wienert, 1968)

$$G_z = \frac{|s_{21}|^2 (1 - |\Gamma_s|) \left(1 - |\Gamma_l|^2\right)}{|1 - \Gamma_s s_{11}|^2 \, |1 - s_{22}\Gamma_l|^2} \tag{4.144}$$

where Γ_s or Γ_l are the coefficients of the reflection of the input or output matching elements, respectively, s_{ij} are elements of the transistor scattering matrix.

The expression (4.144) can be expanded to give a product of three components representing separated effects of the input and output matching circuits and the transistor transmission itself. Then in this form

$$G_z = G_1 G_2 \left|s_{21}\right|^2 \tag{4.145}$$

where

$$G_1 = \frac{1 - \left|\Gamma_s\right|^2}{\left|1 - s_{11}\Gamma_s\right|^2} \tag{4.146a}$$

$$G_2 = \frac{1 - \left|\Gamma_l\right|^2}{\left|1 - s_{22}\Gamma_l\right|^2} \tag{4.146b}$$

For the noise figure of the amplifier in the lossless matching the known relationship

$$F_z = F_{opt} + \frac{R_n}{12.5} \frac{\left|\Gamma_s - \Gamma_{opt}\right|^2}{\left|1 + \Gamma_{opt}\right|^2 \left(1 - \left|\Gamma_s\right|^2\right)} \tag{4.147}$$

is valid, where F_{opt}, Γ_{opt} and R_n are noise parameters of the transistor.

The first step in the graphic design of the input circuit of the amplifier is plotting of circles of the constant noise figure F_z and constant gain G_i into the Smith chart for the required frequency values. The centres S_F and radii ρ_F of circles of the constant noise figure are given by the equation

$$\begin{aligned} S_F &= \frac{\Gamma_{opt}}{1 + \alpha} \\ \rho_F &= \frac{1}{1 + \alpha} \sqrt{\alpha^2 + \alpha \left(1 + \left|\Gamma_{opt}\right|^2\right)} \\ \alpha &= \frac{12.5 \left(F_z - F_{opt}\right) \left|1 + \Gamma_{opt}\right|^2}{R_n} \end{aligned} \tag{4.148}$$

These equations can be obtained by converting equation (4.147) to a canonic equation of a circle in plane Γ_s. By a similar arrangement of (4.146a), relationships for centres S_G and radii ρ_G of the circle of the constant gain G_1 are obtained in the form

$$S_G = G_1 \frac{\left(1 - |s_{11}|^2\right) s_{11}^*}{1 - |s_{11}|^2 \left(1 - G_1 \left(1 - |s_{11}|^2\right)\right)}$$

$$\rho_G = \frac{\sqrt{1 - G_1 \left(1 - |s_{11}|^2\right)} \left(1 - |s_{11}|^2\right)}{1 - |s_{11}|^2 \left(1 - G_1 \left(1 - |s_{11}|^2\right)\right)} \tag{4.149}$$

The circuit will be further found, which will transform the impedance of the signal source in the frequency band required to the region of the impedance as close as possible to the impedance with the minimum noise figure and simultaneously with a satisfactory gain. After the synthesis of this input circuit, the frequency dependence of the gain G_1 is read off from the Smith chart. Now, for known $|s_{21}|^2$, it is possible to determine requirements for the frequency dependence of the gain G_2 in the given frequency band. Similarly as for the input circuit, the synthesis is provided of the circuit matching the load impedance to the transistor output in such a way that the circuit adheres to requirements for G_2, the centres and radii being determined by relationships analogous to (4.149).

For narrow-band amplifier, it is sufficient to consider only the mean frequency in the synthesis discussed. In wide-band amplifiers, the design should be provided for the central frequency and for both marginal frequencies of the band.

The amplifier configuration obtained in this way is then used as an input information of optimizing programmes for the synthesis of amplifiers. In these programs, geometric dimensions of matching circuits serve as the optimizing variables. Based on them, reflection coefficients Γ_s, Γ_l are calculated and based on their knowledge, the amplifier quantities of interest are obtained — the gain, noise figure and SWR. These are then optimized by different optimizing methods for the frequency course desired.

Stability of amplifiers

Certain problems in the synthesis of amplifiers are encountered when using conditionally stable transistors. It is known in general that the necessary and sufficient condition for the possibility of oscillations is an either negative or zero real component of the circuit impedance. It is equivalent to the conditions

$$\left| s_{11} + \frac{s_{21}s_{12}\Gamma_l}{1 - \Gamma_l s_{22}} \right| \geq 1 \qquad (4.150a)$$

$$\left| s_{22} + \frac{s_{21}s_{12}\Gamma_l}{1 - \Gamma_s s_{11}} \right| \geq 1 \qquad (4.150b)$$

By the analysis of these equations, it is possible to find regions of reflection coefficients Γ_s, Γ_l of the transistor load, for which a stable amplifier is obtained. With respect to a formal agreement of relationships (4.150a) and (4.150b), it is sufficient to consider only one of them, for example (4.150b). The results for the second reflection coefficients are obtained by a formal change of inputs and outputs.

The case of the equality in (4.150b) represents a circle with a centre S_{Γ_s} and radius ρ_{Γ_s}. By a transformation of the equations to their canonic form, circle parameters are obtained:

$$\begin{aligned} S_{\Gamma_s} &= \frac{s_{11}^* - \Delta^* s_{22}}{|s_{11}|^2 - |\Delta|^2} \\ \rho_{\Gamma_s} &= \left| \frac{s_{12}s_{21}}{|s_{11}|^2 - |\Delta|^2} \right| \qquad (4.151) \\ \Delta &= s_{11}s_{22} - s_{12}s_{21} \end{aligned}$$

The circle obtained is named the circle of stability and it divides the plane of the reflection coefficients Γ_s to the region of the stability and that of the potential instability of a two-port represented by a transistor. When Γ_s occurs in the potential instability region, the circuit oscillation can start depending on the choice of Γ_l.

Figure 4.59 shows possible mutual positions of the stability circle and unit circle $|\Gamma_s| = 1$ with marking the regions of the two-port stability assuming that $|s_{22}| < 1$ and $|s_{11}| < 1$. It is possible to show that the pertinence of the transistor to a certain type shown in Fig. 4.59 cannot be changed by its cascade connection with lossless two-ports. This means that by choosing the matching lossless circuits it is impossible to achieve the necessary stability of the amplifier equipped with a conditionally stable transistor. In this way it is possible only to effect the position of the instability regions.

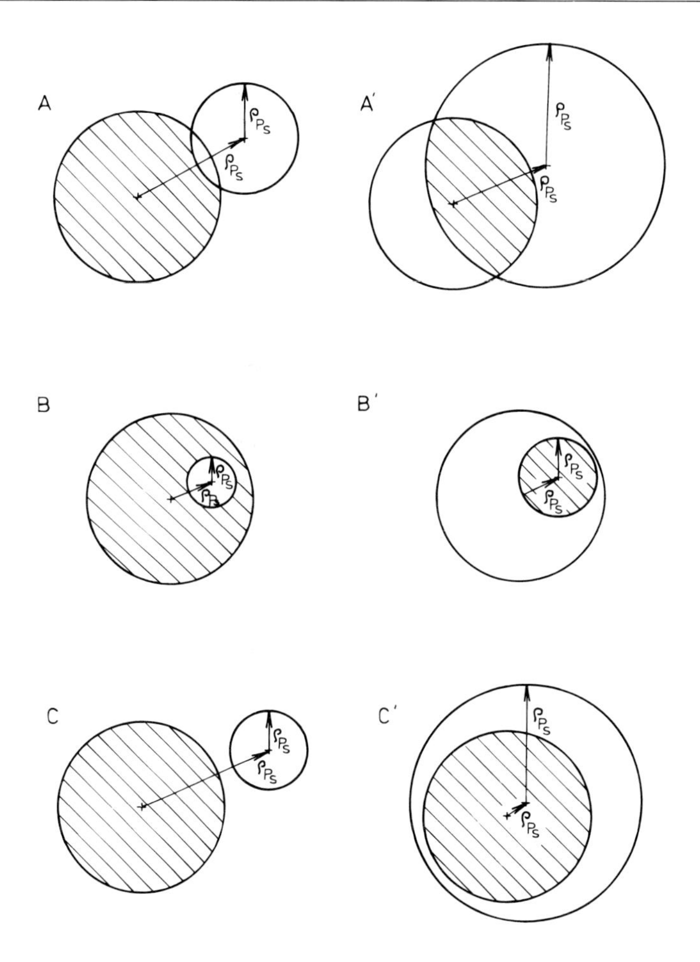

Fig. 4.59 Possible mutual positions of the stability circle.

The definite stability of the amplifier can be provided in the case of using a conditionally stable transistor by using lossless elements in matching transformers. This, however, deteriorates the noise characteristics and the gain. The other way is the use of an external feedback.

Amplifiers with external feedback

In addition to a possibility of improving the amplifier stability, the feedback design yields means for extending the operation band and possibly to reducing the SWR of amplifying stages. For an example of a feedback amplifier see the diagram in Fig. 4.60. The elements

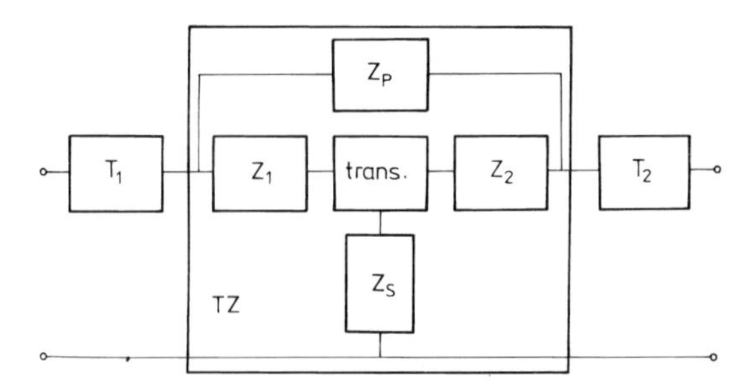

Fig. 4.60 Feedback amplifier.

Z_1, Z_2 correspond to line segments necessary for connecting the transistor into the circuit. Z_p and Z_s are parallel and series feedbacks, respectively. The circuits T_1 and T_2 provide the impedance matching at the input and output of the amplifier, respectively.

As can be seen from the figure, it is possible to consider the feedback amplifier as an amplifier without the external feedback, however, equipped with a new element formed by the transistor itself and circuits Z_1, Z_2, Z_s and Z_p instead of the original transistor.

The last stage of the amplifier synthesis, the synthesis of the input and output circuits is thus the same as in the case of the simple amplifier. It is, however, preceded by a design of the element TZ. This is implemented by optimization methods, again. The noise and transmission characteristics of the two-port TZ, similarly as the characterization of its stability can be simply expressed based on a knowledge of parameters of particular components with the help of elementary matrix operations (Hartman, 1973). The frequency dependences of these parameters are then optimized depending on the geometry or possibly on further physical characteristics of elements Z_1, Z_2, Z_s and Z_p for obtaining the course desired.

The optimization process of the synthesis of the two-port TZ and elements T_1 and T_2 can also be combined into one procedure, however, on account of reducing the comprehensive nature of and by increasing the requirements for the optimizing program.

As shown for example in Euberg (1974), by a suitable choice of feedback elements it is also possible to demonstrate a reduction of the minimum noise figure of the element TZ in comparison with the transistor used. This fact is, however, not properly applicable for improving noise characteristics of the circuit synthesized, since it is accompanied by reducing the gain. This can be instructively seen in following the measure of the noise. It is invariant with respect to changes of lossless elements Z_1, Z_2, Z_s and Z_p (Vendelin, 1975). When using the lossless elements it will be even increased use to introducing new noise sources. Feedback amplifiers with loss elements are thus used in cases which are not connected with high requirements for noise characteristics.

Travelling wave amplifiers

The travelling wave amplifiers are a quite special type of amplifier. They are used for an ultrawide-band amplification. They also make possible a design of wide-band amplifiers when using transistor with a very low gain.

A schematic diagram of the travelling wave amplifier is in Fig. 4.61. The amplifier shown is made of identical sections, which is the most

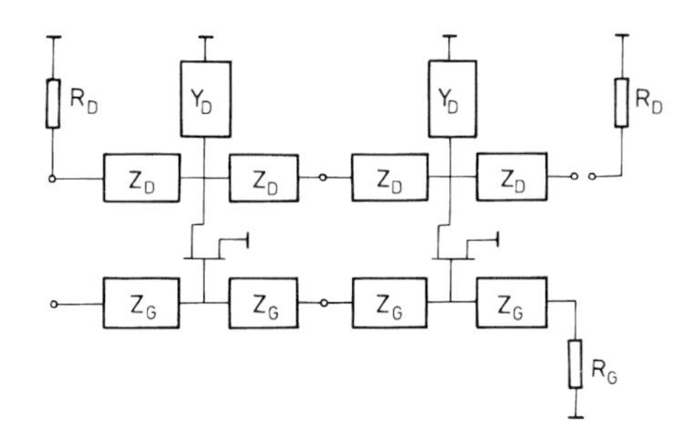

Fig. 4.61 Schematic diagram of travelling wave amplifier.

frequent case. The identity of the section is, however, no general condition.

The travelling wave amplifier can be considered as a system of voltage-controlled generators represented by transistors situated between the input and output lines. It is possible to tell with a certain simplification that these generators are excited by a voltage wave, travelling from the source in the input line. In the output line, they generate a current wave, which is suitably phase-summed and continues into the load. The parallel elements Y_D play a role of elements compensating the effects of impedances of transistors on the transmission line in this process.

The design and synthesis of a travelling wave amplifier is a rather complicated matter. The first step of the procedure is the circuit analysis with respect to its transmission and noise characteristics.

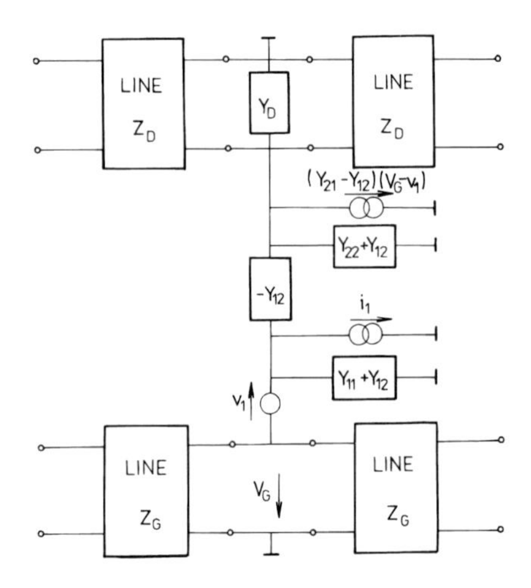

Fig. 4.62 The k-th basic section of the amplifier.

For the k-th basic section of the amplifier represented by a four-port shown in Fig. 4.62

$$\begin{bmatrix} V_{D_k} \\ I_D \\ V_{G_{k-1}} \\ I_{G_{k-1}} \end{bmatrix} = [\mathbf{A}_k] \begin{bmatrix} V_{D_k} \\ I_{D_k} \\ V_{G_k} \\ I_{G_k} \end{bmatrix} \tag{4.152}$$

where $\mathbf{A}_k = \mathbf{A}_1\mathbf{A}_2\mathbf{A}_1$. The matrix $\mathbf{A}_2$ represents the transistor with a compensation element Y_D. The matrix $\mathbf{A}_1$ describes characteristics of input and output circuits formed by the lines Z_D and Z_G. For electric lengths of these lines Θ_D and Θ_G, the matrix $\mathbf{A}_1$ can be expressed in the form

$$\mathbf{A}_1 = \begin{bmatrix} \cos\Theta_D & jZ_D\sin\Theta_D & 0 & 0 \\ j\frac{\sin\Theta_D}{Z_D} & \cos\Theta_D & 0 & 0 \\ 0 & 0 & \cos\Theta_G & jZ_G\sin\Theta_G \\ 0 & 0 & j\frac{\sin\Theta_G}{Z_G} & \cos\Theta_G \end{bmatrix} \tag{4.153}$$

During this, the matrix $\mathbf{A}_2$ is in the form

$$\mathbf{A}_2 = \begin{bmatrix} 1 & 0 & 0 & 0 \\ Y_D + Y_{22} & 1 & Y_{21} & 0 \\ 0 & 0 & 1 & 0 \\ Y_{12} & 0 & Y_{11} & 1 \end{bmatrix} \tag{4.154}$$

It is obvious that for a multitransistor travelling wave amplifier it is possible to write

$$\begin{bmatrix} (V_D)_0 \\ -\frac{(V_D)_0}{R_D} \\ (V_G)_0 \\ (V_G)_0 \end{bmatrix} = \left(\prod_{k=1}^{N} [A_k] \right) \begin{bmatrix} (V_D)_N \\ -(I_D)_N \\ (V_G)_N \\ \frac{(V_G)_N}{R_G} \end{bmatrix} \qquad (4.155)$$

By inserting relevant values into this equation and by a number of manipulations, it is possible to obtain relationships for S-coefficients of the amplifier. This process is similarly described for example in Niclas (1983).

A similar method can be employed for obtaining relationships for noise parameters on the basis of a procedure for noise currents and voltages (Niclas, 1983a).

For available noise and transmission parameters of a distributed amplifier, optimization procedures can be used to optimize these parameters depending on elements Y, Z_D, Θ_D, Z_G, Θ_G, R_D, R_G on components forming these elements. Results obtained by considerably simplified transistor models are usually used as input configurations in these computer procedures.

Travelling wave amplifiers can in principle be implemented in the hybrid as well as monolithic form. They are, however, more suitable for the MMIC. As a matter of fact, the amplifiers of this type cannot be simply divided into particular independently designed and adjustable sections. For this reason, the additional tuning after their synthesis is very difficult. Thus, a high precision of the performance and of the design is required, which can be more easily achieved in MMIC than in MIC.

Many stage amplifiers

In a majority of cases, it is necessary to achieve gains higher than those, which can be achieved in an one stage amplifier. For this purpose, particular stages are arranged to form cascades. The simple combination of the independently designed and adjusted stages is, however, possible only in the case that these stages have negligible SWR. When this is not the case, then it is necessary to consider the effects of particular stages on each other.

In particular amplifiers, either with or without feedback, this problem is most properly solved by providing the design for the cascade as a whole, which takes into account the above-mentioned interactions. During this it is unnecessary to pass from the output of the preceding transistor to the input of the subsequent one through the 50 Ω impedance. This improves the characteristics of the amplifier. The whole procedure is made by essentially identical methods as in the case of one-stage amplifiers.

With the increasing number of stages, the effect also increases of the spread of parameters of transistor and parasitic effects, which are neglected in the design. This considerably complicates the adjustment of the amplifier itself by tuning elements serving for the compensation for these imprecisions. In the design of many stage amplifying chains, several amplifiers with a lower number of stages are thus proposed, which are then combined to form a cascade with the use of separating insulators.

The other ways, commonly used mainly in wide-band amplifiers is the use of the balance design. The principle of the function is in a parallel connection of two simple one-stage amplifiers with the help of separating elements with characteristics described below to form one stage of the balance amplifier having low SWR at its input.

For a schematic diagram of the balance amplifier see Fig. 4.63. The separating element divides the power wave a_1 entering port 1

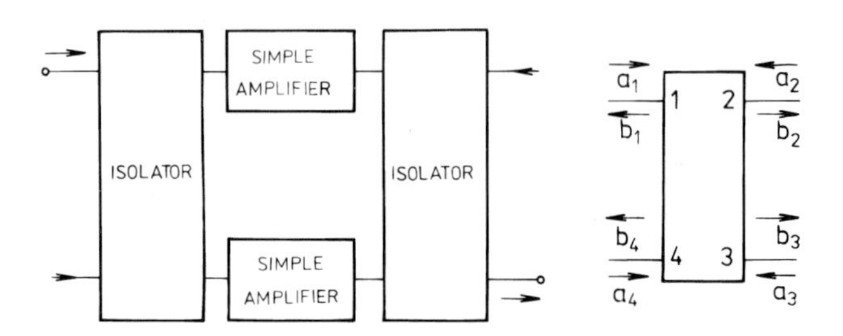

Fig. 4.63 Balance amplifier.

into two waves b_2, b_3 of the same amplitudes, however, phase-shifted by 90°, outgoing in ports 2 and 3. During this, in ports 1 and 4 there are zero output signals b_1 and b_4, respectively. These characteristics have a fourfold symmetry with respect to the input port.

When coupling to ports 2 and 3 circuits with identical reflection coefficient, then the reflected waves interfere due to the above mentioned characteristics in such a way that $b_1 = 0$ and $b_4 = a_2 + a_3 = 2a_2$. In the amplifier shown in Fig. 4.63, when using identical simple amplifiers the reflected power is absorbed in terminating resistors. It is reflected back into the circuit only due to an imperfect agreement between the used simple amplifiers and imperfectnesses of separating elements.

There are a number of types of microwave circuits, which have characteristics required in the case of separating elements. One of the most easily implementable is the hybrid element named 'branch-line'. In spite of its relative simplicity, it has not long found its applications in balance amplifiers due to its large dimensions. It starts to be used in designing amplifiers for higher frequency bands, where the larger size can be accepted. There is, however, still a drawback given by a small bandwidth, which can be enlarged only by a many stage design of this element. This however, results in a further increase of the dimensions and particularly in losses.

In the first balance amplifiers, different types of branch couplers were also used. With respect to the arrangement of their outlets, it was, however, necessary to use a laminated structure and thus, they are essentially no more used today.

For narrow-band applications it is also possible to use a 3-dB power divider. Branch couplers with an interdigital structure are, however, most frequently used, also referred to the 'Lange's hybrids'. Their advantages are particularly in small dimensions, wide-band features and possibility of their implementation by technology procedures common in more complicated MIC circuits.

In addition to the low SWR, the balance design of the amplifier also possesses further advantages. The effect of non-matching of external circuits is reduced in contrast to the simple design as if this circuits were separated by 3dB attenuation elements. These characteristics make possible independent adjustment of balance amplifying stages and their combination into cascades without a risk of mutual interactions. In addition, the power dividing into two branches also leads to improving the dynamic characteristics of the amplifier (by 3 dB) and its power resistance.

Dynamic characteristics of amplifiers

In addition to the noise number, gain and output power, in many applications of amplifiers, the measure of the distortion of the signal transmitted is also of importance. The ideal undistorted transmission is provided only in the case of strictly linear amplifiers. However, the amplifiers exert this feature only in the region of low output powers. With increasing power, the transistors of the last stages enter into the non-linearity region, which results in a decrease of the gain. The output power, for which this drop is just by 1 dB, is one of the most commonly presented dynamic parameters of amplifiers. On the basis of its knowledge, it is, however, yet impossible to tell anything about the magnitude of parasitic signals occurring due to the non-linearity. The complete information about these characteristics is included in an analytical expression of the dependence of the amplitude of the output signal E_0 on the amplitude of the input signal E. By expanding this functional dependence into a power series the relationship

$$E_0 = A_0 + A_1 E + A_2 E^2 + \cdots + A_n E^n + \cdots \tag{4.156}$$

is obtained.

When assuming a sine signal to enter the amplifier input

$$E = E_1 \cos \omega_1 t \tag{4.157}$$

it is possible to express the output signal E_0 as a sum of signals with frequency values corresponding to harmonic values of the input signal frequency. The relationship between the magnitude of the signal outgoing in the form of harmonics and magnitude of the basic output signal is characterized by the quantity THD (total harmonic distortion) defined by the relationship

$$THD = 100 \frac{\sqrt{\sum_{i=2}^{\infty} (A_i)^2}}{A_1} \tag{4.158}$$

However, in most cases rather the magnitude of particular parasitic signals than the total contribution of the signal concentrated in the harmonics is of interest. The importance of these data is particularly considerable in the case when several signals of different frequencies enter the amplifier input. In this case, at the output we obtain the whole spectrum of signals originating by mixing on the amplifier non-linearity in the same way as for one signal.

For the case of the same value of signals $E_1 \cos \omega_1 t$, $E_1 \cos \omega_2 t$ with very similar frequency values at the output similarly as in the

mixer we obtain:
a) basic signals

$$\begin{aligned} E_{V_1} &= A_1 E_1 \cos \omega_1 t \\ E_{V_2} &= A_2 E_1 \cos \omega_2 t \end{aligned} \tag{4.159a}$$

b) intermodulation products of the second order

$$\begin{aligned} E_{V_3} &= \frac{1}{2} A_2 E_1^2 \cos 2\omega_1 t \\ E_{V_4} &= \frac{1}{2} A_2 E_1^2 \cos 2\omega_2 t \\ E_{V_5} &= A_2 E_1^2 \cos (\omega_1 + \omega_2) t \\ E_{V_6} &= A_2 E_1^2 \cos (\omega_1 - \omega_2) t \end{aligned} \tag{4.159b}$$

c) intermodulation products of the third order

$$\begin{aligned} E_{V_7} &= \frac{1}{4} A_3 E_1^3 \cos 3\omega_1 t \\ E_{V_8} &= \frac{1}{4} A_3 E_1^3 \cos 3\omega_2 t \\ E_{V_9} &= \frac{3}{4} A_3 E_1^2 \cos (2\omega_2 - \omega_1) t \\ E_{V_{10}} &= \frac{3}{4} A_3 E_1^3 \cos (2\omega_2 + \omega_1) t \\ E_{V_{11}} &= \frac{3}{4} A_3 E_1^3 \cos (2\omega_1 - \omega_2) t \\ E_{V_{12}} &= \frac{3}{4} A_3 E_1^3 \cos (2\omega_1 + \omega_2) t \end{aligned} \tag{4.159c}$$

together with intermodulation products of higher orders.

Thus, the magnitude of the amplitude of particular parasitic components of the output signal has such a dependence on the input power that the contribution of intermodulation products of higher orders increases with this one. With respect to the fact that the coefficients A_i respectively rapidly decrease with increasing i, for the case of weak signals, particularly intermodulation products up to the third order are of essential importance. The largest portion of the parasitic output power is concentrated in intermodulation products of the second order. These are, however, manifested mainly in wideband amplifiers. In narrow-band amplifiers, signals E_{V_9} and $E_{V_{11}}$ are

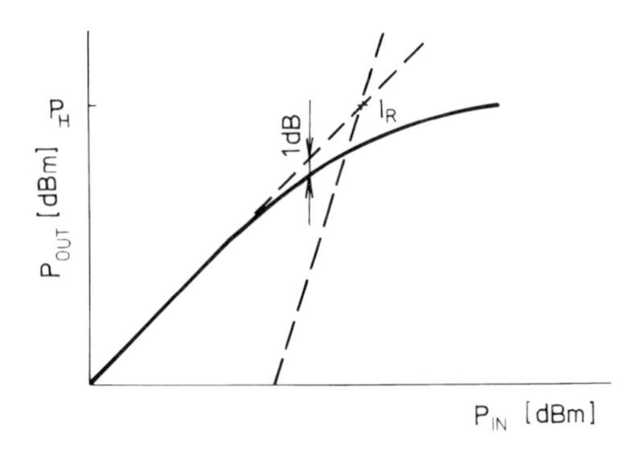

Fig. 3.64 Power dependence.

most pronounced, which are intermodulation products of the third order.

Particular intermodulation products can be characterized by points of intersections I_R of straight lines, which are the interpolations of low-power segments of dependences of output powers of the basic signal and of the relevant intermodulation products (Fig. 4.64). This point is termed the block point. As a parameter of the amplifier, the value of the output power is most frequently presented for the points of block E_{V_9} or $E_{V_{11}}$, which are named the third-order point of block.

The dynamic characteristics of the amplifier are first of all determined by the transistors used and they are affected by choosing their operation points. In the field-controlled transistors, they are essentially independent of the impedance of input circuits, however, they are dependent on output circuits. For a detailed analysis see for example (Minisian, 1980).

4.3.4.2 Power amplifiers

The main difference between amplifiers operating with small signal and power amplifiers is that in the power amplifiers it is necessary to take into account the transistor non-linearity. This can be instructively seen from the standpoint of thermodynamic principles. When considering the amplifier as a converter, transforming the energy from a direct source to the high-frequency region, the relationship

$$P_i + P_{dc} = P_0 + P_d \tag{4.160}$$

follows from the energy conservation law, where P_i is the input RF power, P_{dc} is the power supplied from a direct current source, P_0 is the output RF frequency power and P_d is the power consumed in the amplifier. When expressing the output power with the help of the gain, then

$$P_d = P_{dc} - (G - 1)\, P_i \tag{4.161}$$

From this relationship it follows that with increasing P_i and simultaneously a constant gain G and power P_{dc}, the power P_d should decrease, with obtaining a negative power at a certain moment. When the power P_{dc} is actually limited by characteristics of the transistor, then it is obvious that the gain should be a decreasing function of P_i from a certain moment. With respect to the fact that the relationships for the gain were derived without assuming low signals and that they are independent of the input power, the logic consequence of the above consideration is that the S-parameters of the transistor are changed depending on the input high-frequency power.

For a quality design of a power amplifier, it is first necessary to acquire data about the transistor behaviour at high powers. Low-signal S-parameters are sufficient only in the case that the amplifier operates in the region of the transistor linearity.

At first sight, the simplest way of obtaining power parameters could be the measurement of S-parameters in a $50\,\Omega$ measuring system at a high power of sources. Many methods applicable for this approach are known (Johnson, 1979; Mitsui *et al.*, 1977). The data obtained in this way are, however, insufficient. With respect to the fact that the characteristics of the transistor depend on the total input power, which is dependent on impedances coupled to the transistor, its features should be characterized at reflection coefficients Γ_s and Γ_1, which are coupled to the transistor during the application. The most frequently measured quality is G/P_o or S-parameters corresponding to G/P_0. These measurements are described for example in Lusack *et al* (1974).

In the case of the knowledge of power parameters, for the synthesis of matching circuits similar mathematical procedures are used as for low-signal amplifiers.

In a number of cases, requirements for the output power exceed possibilities of one transistor. Then it is necessary to parallel couple simple amplifiers. One of the most common types of parallel couplings is the combination of amplifiers into balance stages. Then, their balance coupling is possible, again. By repeating this procedure, it is possible to obtain the stage with 2^n parallel coupled amplifiers. Instead of the N-fold doubling, it is also possible to use some of a number of types of multiway dividers. The multiway dividers are particularly advantageous in parallel coupling of many transistors, since they make possible coupling with relatively low losses. The last step necessary in the design of the power amplifier is completing of the many stage amplifier. This is natural heat transport with the help of materials with good thermal conductivity forced air circulation or cooling with liquid are used. Particular calculation and design of cooling are, however, rather complicated and exceed the scope of this book.

REFERENCES 4.1

Aprille T.J., Trick T.N. (1972a) A computer algorithm to determine the steady-state response of nonlinear oscillators, *IEEE Trans. Circuit Theory, vol. CT-19, 7*, pp. 354–60.

Aprille T.J., Trick T.N. (1972b) Steady-state analysis of nonlinear circuits with periodic inputs, *Proc. IEEE, vol. 60, 1*, pp. 108–15.

Baucells J., Mediavilla A., Tazon A. (1988) Nonlinear analysis, in GaAs MESFET Circuit Design, *R. Soares, ed., Artech House, Boston.*

Camacho–Peñalosa C. (1983) Numerical steady-state analysis of nonlinear microwave circuits with periodic excitation, *IEEE Trans. Microwave Theory Tech., vol. MTT-31, 9*, pp. 724-30.

Chu T.S., Itoh T. (1986) Generalized scattering matrix method for analysis of cascaded and offset microstrip step discontinuities, *IEEE Trans. Microwave Theory Tech., vol. MTT-34, 2*, pp. 280–4.

Chua L.O., Lin P. (1975) *Computer-Aided Analysis of Electronic Circuits: Algorithms and Computational Techniques*, Prentice-Hall, Englewood Cliffs, NJ.

Collin R.E. (1966) *Foundations of Microwave Engineering*, McGraw-Hill, New York, pp. 170–82.

Colon F.R., Trick T.N. (1973) Fast periodic steady-state analysis for large-signal electronic circuits, *IEEE J. Solid-State Circuits, vol. SC-8, 8*, pp. 260–9.

Desoer C.A., Kuh E.S. (1969) *Basic Circuit Theory*, McGraw-Hill, New York.

Director S.W. (1971) A method for quick determination of the periodic steady-state in nonlinear networks, *Allerton Conf. Circuit Syst. Theory*, pp. 131–9.

Director S.W., Current K. W. (1976) Optimization of forced nonlinear periodic circuits, *IEEE Trans. Circuits Syst., vol. CAS-23, 6*, pp. 329–55.

Eesof Inc. (1985) Non-linear circuit analysis comes to microwave, *Microwave J., vol. 28, 8*, pp. 153–6.

El Rabaie S., Fusco V.F., Stewart C. (1988) Harmonic balance evaluation of nonlinear microwave circuits — A tutorial approach, *IEEE Trans. Education, vol. 31, 3*, pp. 181–92.

Filicori F., Naldi C. (1983) An algorithm for the periodic or quasi-periodic steady-state analysis of nonlinear circuits. *IEEE International Symposium on Circuits and Systems*, pp. 366–9.

Gilmore R. (1986) Nonlinear circuit design using the modified harmonic balance algorithm, *IEEE Trans. Microwave Theory Tech., vol. MTT-34, 12*, pp. 1294–307.

Gilmore R.G., Kiehne R., Rosenbaum F. J. (1985) Circuit design to reduce 3rd order intermodulation distortion in FET amplifiers, *IEEE MTS-S Digest*, pp. 413–16.

Gilmore R.J., Steer M.B. (1991) Nonlinear circuit analysis using the method of harmonic balance — A review of the art. Part I. Introductory concepts, *Int. J. Microwave and Millimeter-Wave Computer-Aided Eng., vol. 1, 1*, pp. 22–37. Part II. Advanced concepts, *Int. J. Microwave and Millimeter-Wave Computer-Aided Eng., vol. 2, 2*, pp. 159–80.

Gupta K.C., Garg R., Chadha R. (1981) *Computer-Aided Design of Microwave Circuits*, Artech House, Inc., Dedham.

Hicks R.G., Khan P.J. (1980) Numerical technique for determining pumped nonlinear device waveforms, *Electron. Lett., vol. 16*, 375–6.

Hicks R.G., Khan P.J. (1982a) Numerical analysis of nonlinear solid-state device excitation in microwave circuits, *IEEE Trans. Microwave Theory Tech., vol. MTT-30, 3*, pp. 251–9.

Hicks R.G., Khan P.J. (1982b) Numerical analysis of subharmonic mixers using accurate and approximate models, *IEEE Trans. Microwave Theory Tech., vol. MTT-30, 12*, pp. 2113–20.

Hu Y., Molier J.C., Obregon J. (1986) A new method of third-order intermodulation reduction in non-linear microwave systems, *IEEE Trans. Microwave Theory Tech., vol. MTT-34, 2*, pp. 245–50.

Hwang D., Itoh T. (1986) Large-signal modeling and analysis of GaAs MESFET, *Proc. 16th European Microwave Conf.*, pp. 189–94.

Jastrzebski A.K., Sobhy M.I. (1984) Analysis of non-linear microwave circuits using state-space approach, *Proc. IEEE Int. Symp. on Circuits and Systems, Montreal, vol. 2*, pp. 1119–22.

Javed A., Goud P.A., Syrett B.A. (1977) Analysis of microwave feedforward amplifier using Volterra series representation, *IEEE Trans. Commun., vol. 25*, pp. 355–60.

Kerns D.M., Beatty R.W. (1969) *Basic Theory of Waveguide Junctions and Introductory Microwave Network Analysis*, Pergamon Press, Oxford.

Kerr A.R. (1975) A technique for determining the local oscillator waveforms in a microwave mixer, *IEEE Trans. Microwave Theory Tech., vol. MTT-23, 10*, pp. 828–31.

Kerr A.R. (1979) Noise and loss in balanced and subharmonically pumped mixers: part 1-theory, *IEEE Trans. Microwave Theory Tech., vol. MTT-27, 12*, pp. 938–50.

Kneppo I., Fabian J. (1991) Inversion Scattering Matrix Method for Scattering to Transmission Matrix Transformation, *Elektrotechnický časopis, 42, 11–12*, pp. 626–8.

Lambranianou G.M., Aitchison C.S. (1977) Optimization of third-order amplifier using Volterra seriess analysis, *IEEE Trans. Microwave Theory Tech., vol. MTT-33, 12*, pp. 1395–403.

Minasian R.A. (1980) Intermodulation distortion analysis of MESFET amplifiers using the Volterra series representation, *IEEE Trans. Microwave Theory Tech., vol. MTT-28, 1*, pp. 1–8.

Minasian R.A. (1981) Volterra series analysis of MESFET mixers, *Int. J. Electron., vol. 50, 3*, pp. 215–19.

Monaco V.A., Tiberio P. (1970) On the transformation of a lumped element linear network into a circuit composed of multiports, *Alta Freq., vol. 39, 11*, pp. 1013–14.

Monaco V.A., Tiberio P. (1974) Computer-aided analysis of microwave circuits, *IEEE Trans. Microwave Theory Tech., vol. MTT-22, 3*, pp. 249–63.

Nagel L.W., Pederson D.O. (1973) SPICE (Simulation Program with Integrated Circuit Emphasis), Memorandum ERL M382, University of California, Electronics Research Laboratory.

Nakhla M.S., Branin F.H. (1977) Determining the periodic response of nonlinear systems by a gradient method, *Circuit Theory Appl., vol. 5*, pp. 255–73.

Nakhla M.S., Vlach J. (1976) A piecewise harmonic balance technique for determination of periodic response of nonlinear systems, *IEEE Trans. Circuits Syst., vol. CAS-23, 2*, pp. 85–91.

Norenkov I.P., Yevstifeyev Y.A., Manichev V.B. (1987) A steady-state analysis method for multiperiodic electronic circuit, *Radiotekhnika, 11*, pp. 86–9.

Obregon J. (1985) Non-linear analysis and design of microwave circuits, *European Microwave Conf., Paris*, pp. 1089–93.

Overfelt P.L., White D.J. (1989) Alternate forms of the generalized composite scattering matrix, *IEEE Trans. Microwave Theory Tech., vol. MTT-37, 8*, pp. 1267–8.

Pospíšil J. (1986) *Matrix Decompositions and some their Applications in Network Theory*, Publications of Technical and Scientific Papers of the Technical University in Brno, Brno.

Redheffer R. (1962) On the relation of transmission-line theory to scattering and transfer, *J. Math. Phys., vol. 41*, pp. 1–41.

Rizzoli V., Cecchetti C., Lipparini A., Mastri F. (1988) General-purpose harmonic balance analysis of nonlinear microwave circuits under multitone excitation, *IEEE Trans. Microwave Theory Tech., vol. MTT-36, 12*, pp. 1650–60.

Simonyi K. (1963) *Foundations of Electrical Engineering*, Pergamon Press,

Oxford, pp. 327–417.

Skelboe S. (1980) Computation of the periodic steady-state response of nonlinear networks by extrapolation methods, *IEEE Trans. Circuits and Systems, vol. CAS-27*, pp. 161–75.

Skelboe S. (1982) Time-domain steady-state analysis of nonlinear electrical systems, *Proc. IEEE. vol. 70., 10*, pp. 1210–28.

Sobhy M.I., Jastrzebski A.K. (1985) Direct integration methods of nonlinear microwave circuits, *Proc. 15th European Microwave Conf.*, pp. 1110–18.

Steer M.B., Chang C.-R., Rhyne G.W. (1991) Computer-aided analysis of nonlinear microwave circuits using frequency domain nonlinear analysis techniques; The state of the art, *Int. J. of Microwave and Millimetre-Wave Computer-Aided Eng., vol. 1, 2*, pp. 181–200.

Weinberg L. (1966) *Scattering Matrix and Transfer Scattering Matrix, in Amplifiers*, R. F. Shea, Ed., McGraw-Hill, New York.

Weiner D.D., Spina J.F. (1980) *Sinusoidal Analysis and Modeling of Weakly Nonlinear Circuits*, Van Nostrand Reinhold Co., New York.

REFERENCES 4.2

Alseyab S.A. (1982) A Novel Class of Generalized Chebyshev Low-Pass Prototype for Suspended Substrate Stripline Filters. *IEEE Trans. on MTT-S, Vol. 30, No. 9*, pp. 1341–7.

Alseyab S.A., Ashoor N. (1986) Element values of a generalized Chebyshev prototype filter for suspended subtrate stripline, *Int. J. Electronics, Vol. 60, No. 4*, pp. 439–49.

Atwater H.A. (1983) Microstrip reactive circuit elements, *IEEE Trans. on MTT-S, Vol. 31, No. 6*, pp. 488–91.

Bezoušek P. *et al.* (1977) Design of MIC — part B, Technical Note 103L2295, ÚVR Opočínek, CSFR.

Gupta C. (Dec. 1979) Design of Parallel Coupled Line Filter with Discontinuity Compensation in Microstrip, *Microwave Journal*, pp. 39–57.

Levy R., Lind L.F. (Feb. 1968) Synthesis of Symmetrical Branch-Guide Directional Coupler, *IEEE Trans. on MTT, Vol. 16, No. 2*, pp. 80–9.

Matthaei G.L., Young L., Jones E.M.T. (1964) *Microwave Filters, Impedance-Matching Networks and Coupling Structures*, McGraw-Hill Book Company.

Ou W.P. (1975) Design Equations for an Interdigitated Directional Coupler, *IEEE Trans. on MTT-S, Vol. 23, No. 2*, pp. 253–5.

Shelton J.P., Mosko J.A. (1966) Synthesis and design of wideband equal-ripple TEM directional couplers and fixed phase shifters, *IEEE Trans. on MTT, Vol. 14, No. 10*, pp. 462–73.

REFERENCES 4.3

Arden J. (June 1976) The Design, Performance and Application of the NEC V244 and V388 GaAs FETs, *Application Note, California Eastern Laboratories INC*, pp. 59.

Bezoušek P., Tubl R. (Aug. 1989) Microwave broadband integrated detector (in Czech), *Slaboproudy obzor, Vol. 50, No. 8*, pp. 388–93.

Carbin H. (Feb. 1978) A New Method of Broad-Band Equalization Applied to Microwave Amplifiers, *IEEE Trans. on MTT, Vol. 27*, pp. 93–8.

Euberg J. (Sept. 1974) Simultaneous Input Power and Noise Optimization Using Feedback, *4-th European Microwave Conference, Montreaux*, pp. 385–9.

Hartman K. (Oct. 1973) Changes of the Four Noise Parameters Due to General Changes of Linear Two Port Circuits, *IEEE Trans. on Elect. Dev., Vol. 20*, pp. 874–7.

Johnson K.M. (March 1979) Large Signal GaAs MESFET Oscilator Design, *IEEE Trans. on MTT, Vol. 27*, pp. 217–27.

Levy R., Lind L.F. (Feb. 1968) Synthesis of Symmetrical Branch-Guide Directional Coupler, *IEEE Trans. on MTT, Vol. 16, No. 2*, pp. 80–9.

Lusack J., Perlow S., Perlman B.S. (Dec. 1974) Automatic Load Contour Mapping for Microwave Power Transistors, *IEEE Trans. on MTT, Vol. 22*, pp. 1146–52.

Matthaei G.L., Young L., Jones E.M.T. (1964) Microwave filters, *McGraw-Hill Book Company, New York.*

Minisian R. (Jan. 1980) Intermodulation distortion analysis of MESFET amplifiers using the Voltera series representation, *IEEE Trans. on MTT, Vol. 28*, pp. 1–8.

Mitsui Y., Nakatami M., Mitsui S. (Dec. 1977) Design of GaAs MESFET Oscillator Using Large-Signal S-parameters, *IEEE Trans. on MTT, Vol. 25*, pp. 1981–4.

Niclas K. (June 1983) On Theory and Performance of Solid State Microwave Distributed Amplifiers, *IEEE Trans. on MTT, Vol. 31*, pp. 447–56.

Niclas K. (Aug. 1983a) On Noise in Distributed Amplifiers at Microwave Frequencies, *IEEE Trans. on MTT, Vol. 31*, pp. 661–8.

Shelton J.P., Mosko J.A. (Oct. 1966) Synthesis and design of wideband equal-ripple TEM directional couplers and fixed phase shifters, *IEEE Trans. on MTT, Vol. 14, No. 10*, pp. 462–73.

Vendelin G. (May 1975) Feedback Effects on the Noise Performance of GaAs MESFETs, *MTT-s Integrated Microwave Symposium Digest, Palo Alto*, pp. 324–6.

Wienert F. (1968) Scattering parameters speed design of high frequency transistor circuits, *HP Application Note 95*, pp. 2.1–11.

5

Measuring and testing

I. Kneppo and J. Fabian

The RF measurements are an inseparable part of the microwave integrated technology. By measurement, data are obtained concerning real characteristics of materials and structural elements, necessary in modelling, analysis and designing the future MIC, measurement serves for checking and classifying materials and parts prior to the production (measurements of the permittivity of substrates, choice of low-noise transistor chips), and last, measurements serve for complex testing of the final MICs. The set of high-frequency characteristics obtained by measurement is also an important basic material for the future application of the given MIC.

As far as the techniques of measurement itself are concerned, more or less conventional microwave measuring techniques are used (network analysers for S-coefficients measurements, noise figure measurement based on hot/cold techniques, power measurement, spectrum analysers, etc.). They are described in sufficient details in the literature (Bryant, 1988; Cappy, 1988; Gledhill, 1989; Griffin, 1989; Kneppo, 1988; Laverghetta, 1976; Laverghetta, 1991; Oldfield, 1989; Sinclair, 1989; Strid, 1981; Warner, 1989a; Warner, 1989b; Maury, 1991; Hewlett-Packard, 1989; Wiltron, 1989).

However, the complexity of the various MICs to be characterized by the measurement system requires the use of several different measuring instruments. Obviously, measurements of the device, performed separately on each instrument by manually connecting the device under test each time to each instrument, is a very inefficient way of testing. Not only is significant time wasted because of multiple connection cycles, but additional time is also lost through repetitive device set up. Using a fully integrated automatic measuring system is an ideal approach (Bahl *et al.*, 1991; Kuhn, 1983; Maury,

1990; Hewlett-Packard, 1985a; Hewlett-Packard, 1985b).

The specific feature of the current microwave measuring technique for the integrated technology is, however, the fact that it is implemented during the production still before encapsulating the devices. In the case of hybrid MICs this means that the measurement of basic electrical characteristics of circuit subsystems and devices is carried out before their completing into capsules and before equipping with coaxial high-frequency inlet lines. In the case of the MMICs this means the measurement of microwave characteristics of the circuit still before wafer dicing and subsequent operations. This approach exerts definite advantages given by the fact that the measured high-frequency characteristics of device are unaffected by non-defined parasitic factors of couplings and discontinuities of the capsule and the measuring precision is higher. Only direct measurements on wafers present a possibility to objectively estimate how much the device produced corresponds to the original theoretical design. On wafer microwave testing of MICs, at relatively early stages in the process, can dramatically reduce production costs by eliminating wafers having below standard devices. It is also possible to trim in the course of the production, particularly of hybrid MICs, certain elements in the circuit to achieve the required or also optimum parameters of the resulting circuit. After completion of the processing, on-wafer microwave measurement or testing of MICs will be a cost-effective method of sorting out substandard MICs on a wafer before packing. By the measurement and detailed analysis of data before and after capsulating of the given device, the valuable basic data are also obtained for the development of the optimum capsule.

These have led to the rapid acceptance of wafer-level measuring techniques for the two most important characteristics of the MMICs and of the components for MICs, namely for the scattering coefficients and noise figure (Reeve *et al.*, 1990; Pavio *et al.*, 1991). Accurate on wafer S-coefficients measurements are required for impedance matching and device modelling, and for determining the effects of load and source reflections on power output, noise figure correction, noise parameters measurement, etc. The successful determination of noise figure by on-wafer probing techniques is of a particular importance, since, as it follows from the principle of the noise measurement itself, this measurement is sensitive to the inserted losses between the noise source and input of the MIC measured. In the encapsulated condition, these losses would be undefinable and unmeasurable and thus, the measurement error originating in this way should not be corrected.

On the other hand, the key problem prior packing or on wafer-level microwave measurements is the proposal and design of the test fixture or probes, respectively and their calibration including the proposal and design of calibration standards. Today, several types of testing fixtures and probes are available, which have been tested and routinely used (Inter Continental Microwave, 1991; Argumens, 1992; Hewlett-Packard, 1992; Lang *et al.*, 1988; Cascade Microtech, 1991).

5.1 INCORPORATION OF THE MIC MEASURED INTO THE MEASURING SYSTEM: MICROWAVE TEST FIXTURES AND PROBES

The purpose of the test fixture is (1) to mechanically fix the device to be measured for the testing time and (2) to electrically match the inlet lines of the conventional microwave measuring technique (coaxial line or rectangular waveguide) and of the MIC to be measured (microstrip, coplanar waveguide), which are different by their shapes as well as impedances. Because of the wide variety of devices to be measured, there is no universal solution to the fixture design problem. However, some general considerations from which to start the design of a test fixture may be useful.

In principle, the test fixture consists of the following three parts: (1) end block with input launch connector(s), (2) centre section, and (3) end block with output launch connectors (Fig. 5.1). These parts can form one system which cannot be disassembled or they are independent mechanical parts, which are assembled for the measurement. The first method serves for solving a single-purpose test fixture, typically designed for one particular type of the MIC measured, whereas the second approach is typical for modular microcircuit packages.

The end blocks for the input and output are typically identical by their shape and design, the difference may be in the number of high-frequency inlet lines. From the electric standpoint, the most important node of the end block is the coaxial or waveguide to the microstrip launcher. Its purpose is properly to transform the

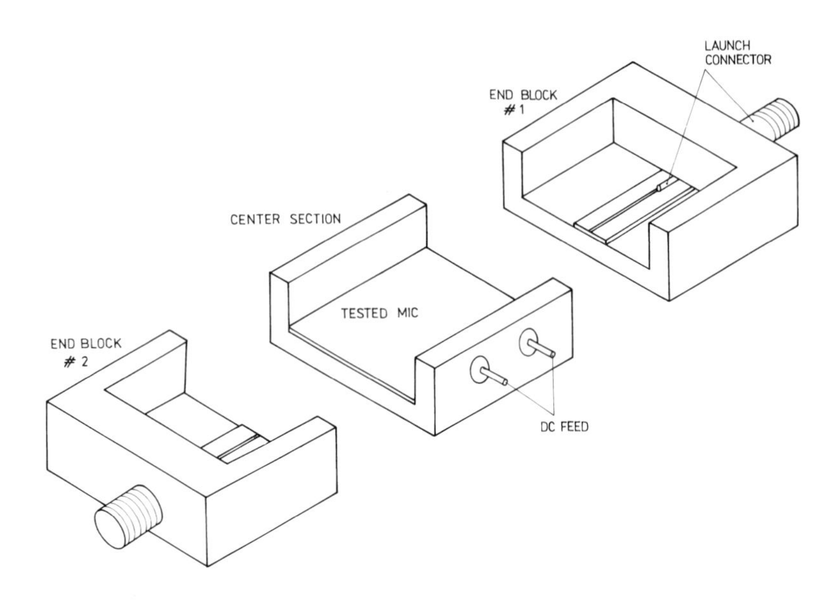

Fig. 5.1 Main parts of the test fixture.

impedance and to match the patterns of the electric field at the input and output. The type of the input coaxial connector depends on the requirement for the precision and reproducibility of the measurement, and for the width of the operation frequency band (see Table 5.1). For testing millimetre wave MICs (MMICs) above 50 GHz, at the input, there is usually a rectangular waveguide corresponding to the given frequency band. The centre section mechanically fixes the MIC measured itself and it is by its geometric dimensions matched to the substrate size. It usually also includes electric inlet line providing the supply and bias for the DUT.

A number of issues must be considered in designing and using a fixture for the testing of MIC. The basic requirement is the repeatability and mechanical stability of the test fixture, since otherwise, sources of measurement errors occur, which cannot be corrected. The fixture parameters must not be changed between the calibration and subsequent measurement. The proposal and design should be cautiously solved. Particular parts of the fixture and mainly fixing of the substrate or chip of the device measured should be designed with small tolerances. It is of importance that the fixture should not mechanically stress the substrate fixed and that it should provide

Table 5.1 Coaxial Connectors to Microstrip Transitions*

Connector type	Insertion loss [dB]	Return loss [dB]	Frequency [GHz]
APC 7	< 0.15	> 24	18
APC 3.5	< 0.2	> 18	26.5
K	< 0.2	> 18	26.5
OS 50	< 0.75	> 14	50

*(Inter Continental Microwave, 1991)

defined and reproducible substrate fixation and efficient screening of the DUT from the interference with the external electromagnetic radiation. It is of a great importance to locate the earthing connections in the vicinity of input and output inlet lines of the measuring RF signal. The coplanar MIC carrier satisfies to this condition very well since the RF and earthing contacts are situated on the same area. The microstrip carrier requires a much more complicated test fixture design and closer tolerance control of the substrate thickness. However, via holes allow the shortest possible ground path which minimizes the electrical discontinuity between the launcher and the circuit tested. The reproducibility of the testing fixture must be provided over a wide range of temperatures. The test fixture should further be electrically invariant with respect to the MIC measured. This means that all the electric discontinuities of the fixture, particularly of launchers must be as small as possible and they must not occur within the range of scattering and electromagnetic fields of the device measured. It is important to minimize leakage terms due to imperfect ground path. The amplitude of these terms is often dependent on the MIC topology.

The important electric requirement is the frequency-wide band applicability of the test fixture. It depends to a considerable extent on the coaxial inlet lines used. Waveguide launchers are acceptable only in the frequency band of the given waveguide dimension, when disregarding the use of higher modes. Low losses (below 0.5 dB), good return loss (up to 20 dB), and minimum crosstalk (below -30 dB) in the whole operation frequency band are ultimative conditions, particularly in noise measurements.

Besides this, as already mentioned a good testing fixture particularly when it should be suitable for testing a larger series of devices (control of the production, classing of the series) should be easy to

Table 5.2 Test fixtures*

Model	Connector type	Frequency range [GHz]	Insertion loss [dB]	Return loss [dB]	Repeatability of S_{21} [dB]
Inter Cont. Microwave					
TF-3001-A	APC 7	DC – 18	< 0.3	> 24	< 0.02
TF-3001-B	APC-3.5	DC – 26.5	< 0.3	> 22	< 0.02
TF-3001-L	K	DC – 40	< 0.5	> 18	< 0.05
TF-3001-G	OS 50	DC – 50	< 0.5	> 18	< 0.10
Hewlett-Packard					
HP 83040	APC 3.5	DC – 26.5	< 0.5	> 20	NA
Wiltron					
3680 K	K	DC – 40	NA	> 14	< 0.20
3680 V	V	DC – 60	NA	> 10	< 0.30

*(Inter Continental Microwave, 1991; Hewlett-Packard, 1992; Wiltron, 1990).

handle and reliable. The characteristics of certain types of commercial testing fixtures are summarized in Table 5.2.

The key node of the coplanar probe for high-frequency testing of MICs directly on the wafers is the coaxial or waveguide to the coplanar transition. At frequencies below 65 GHz a coax to CPW transition has been used (Jones, 1989; Gleason *et al.*, 1983; Carlton *et al.*, 1985; Jones *et al.*, 1987; Strid and Burchan, 1989). A rectangular waveguide to CPW transition was required to extend the upper frequency limit of wafer probes over this value (Godshalk, 1991). For a transition to be successful the proper impedance transformation must take place, and the electric field patterns are to be properly matched at the input and output. The CPW is formed by depositing a layer of gold on the bottom surface of the alumina substrate forming a structure of the probe. The coplanar waveguide is stepwise reduced in size from the original dimension corresponding to the span of the coaxial inlet line to the dimension of several micrometres (typically 25 μm). The point formed in this way is in a direct contact with metallic contact areas of the chip measured in the course of the measurement. To define the contact area precisely, small fingers of relatively hard metal are deposited on the gold CPW pattern at the probe tip. Probes are categorized by their pitch, which is the centre-to-centre spacing between adjacent fingers (Fig. 5.2).

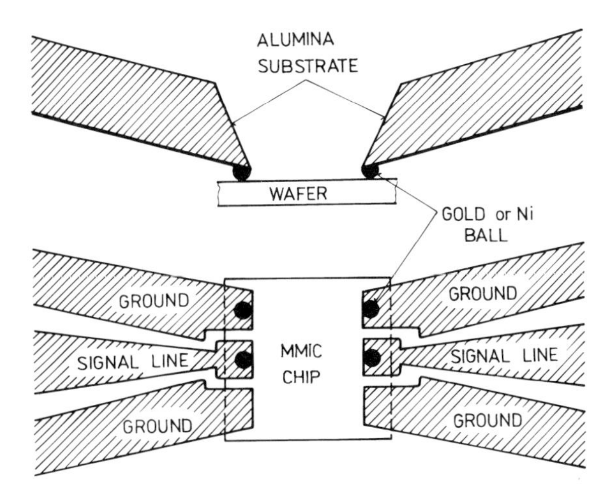

Fig. 5.2 Contact test probes.

Similarly as the testing fixture for the MIC, the RF probes for testing the MMIC should satisfy many mechanical and electrical conditions. In addition to the reproducible performance and mechanical stability the required characteristics are a good planarity for single and multiple contact probes, spatial resolution in the order 5 μm, good durability, medium (several watts) power handling capability, dc to millimetre band operation, good transition between the probe and the coaxial/CPW launcher, minimum electrical discontinuity between the probe and the circuit pad, single-mode propagation from the probe to the substrate, low loss, good return loss, and minimum crosstalk. In addition, MMIC substrates need RF ground on the top surface in order to connect the probe ground to the device tested. It is obvious that high-quality RF probes meeting to these high requirements can be produced only by relatively expensive thin-layer technology. Table 5.3 summarizes several types of RF probes for the MMIC on wafer testing. Thick-film wafer probes represent a less tedious alternative from the production as well as economic standpoint. They provide acceptable characteristics up to 18 GHz (Rodriquez-Tellez, 1991).

A special example of microwave probes for on wafer level testing of MMIC are non-contact probes. Non-contact probing uses optical techniques and is based on electro-optical effects in semiconductors or electro-optic crystals. Internal and external electro-optic measurement techniques, for the characterization of MICs and/or MMICs

Table 5.3 RF Probes for MMIC Testing*

PROBE TYPE	FREQUENCY RANGE [GHz]	CONNECTOR	PITCH MIN-MAX [μm]	RETURN LOSS [dB]	ATTEN [dB]
WPH-0##	DC – 18	SMA	50 – 250	> 10	< 2
WPH-1##	DC – 25.6	K	50 – 250	> 10	< 2
WPH-2##	DC – 50	2.4 mm	100 – 250	> 10	< 3
WPH-3##	DC – 40	2.4 mm	100 – 250	> 10	< 2.5
WPH-4##	DC – 65	1.85 mm	100 – 250	> 10	< 3
WPH-5##	50 – 75	Waveguide	100 – 250	> 9	< 5

*(Cascade Microtech, 1991)

have been developed (Baul *et al.*, 1991; Valdmanis *et al.*, 1983; Weingarten *et al.*, 1985; Freeman *et al.*, 1985; Kolner and Bloom, 1986; Rodwell *et al.*, 1986; Weingarten *et al.*, 1987; Weingarten *et al.*, 1988; Majidi-Ahy, 1989; Bierman, 1990).

In the internal electro-optical testing, a narrow beam of ps pulses of a sub-band gap laser is conducted into the GaAs substrate in such a way that it passes through the electric field produced by the conductor of the device measured. Voltage-induced perturbations occur in the polarization of the returning beam due to the electro-optic effect in the GaAs substrate. Polarization changes are measured in such a way that a perturbated beam strikes a photodetector after its passage through a polarizer.

External electro-optic probing technique takes the advantage of the electro-optic effect, which occurs in an extremely small volume of the electro-optic crystal situated in a close vicinity of the conductor of the IC tested. Thus, in principle this crystal acts as a sensor of the electric field. The conductor fringing field induces a change in the index of refraction in the electro-optic crystal, and a picosecond laser pulse reflecting on the bottom of the crystal resolves the speed of this index change.

A great advantage of the non-contact probing technique is its high spatial resolving power, which follows from the fact that the optical probe beam can be focused to a small diameter (below 5 μm). Further advantages are the long operating service time, very good reproducibility and large width of the operation frequency band exceeding 100 GHz. For the electro-optic probe, internal points of the IC tested are also admissible and this technique is in a certain sense of word universally acceptable for any set of microwave circuits on the wafer. On the other hand, the disadvantages of this testing technique include a rather complicated apparatus, its high cost and time consuming calibration procedure.

The optoelectronic technique also falls into the group of noncontact techniques of the MMICs characterization. In addition to the fact that it yields the possibility of sampling RF signals at the input and output of the IC measured, it also possesses means for generating testing signals in a wide frequency spectrum as well as in bands of millimetre waves (Arjavalingam *et al.*, 1989; Hung *et al.*, 1989).

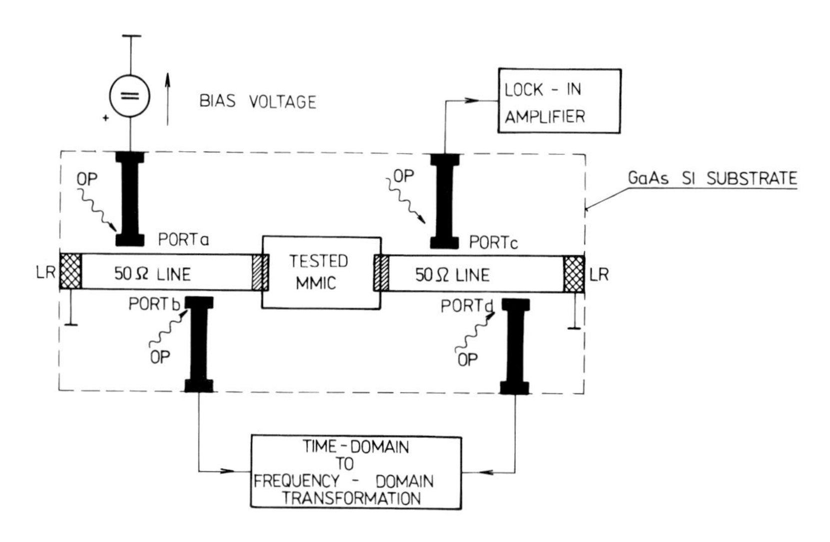

Fig. 5.3 Optoelectronic testing.

Fig. 5.3 shows a schematic diagram of the optoelectronic characterization system, including the MMIC tested. A short electric pulse is generated by a dc-biased photoconductive switch at port a and travels down the input transmission line towards the input port of the MMIC. The pulse travelling in the opposite direction is absorbed by the matched termination. The pulse approximates a delta function in the time domain, resulting in a very broad frequency spectrum. Since the input electrical pulse generated is in complete time synchronization with optical pulse, this electrical pulse can be precisely sampled at port b by a time-delayed laser pulse illuminating a second picosecond photoconductive switch on the input side of the MMIC under test. The output signal of the MMIC tested can be sampled using the same procedure, with a photoconductive switch at the output of the device (port c). By comparing the Fourier transforms of the transmitted and reflected waveforms to those of the incident waveform, the two-port scattering parameters can be determined.

5.2 MEASUREMENT TECHNIQUES

5.2.1 S-parameter measurement

In accordance with the definition, the network of S-parameters completely characterizes electric properties of any linear (for small signals also non-linear) microwave active or passive device. The scattering matrix is suitable for expressing the electrical properties of MICs because its elements can be measured by measuring the ratio of the emergent to incident waves, which can be done over the whole microwave range. As discussed in Section 4, from Equation (4.1) direct results are that

(I) if the MIC having n ports is connected to the measuring system so that the incident amplitude is zero in all ports exceed the k-th, then

$$S_{kk} = \frac{b_k}{a_k}, \qquad a_i = 0, \qquad \text{for } i \neq k, \qquad i = 1, 2, \ldots, n \tag{5.1}$$

(II) if the conditions in the measuring system are modified so that the incident wave is nonzero in the k-th port of the given MIC only, the other ports having no sources and all the ports with non-reflecting terminations, then

$$S_{jk} = \frac{b_j}{a_k}, \quad a_i = 0, \quad \text{for } i \neq k, \quad i = 1, 2, \ldots, n, \quad \text{and } j \neq k. \tag{5.2}$$

Equations (5.1) and (5.2) are important from the point of view of measurement of the S-parameters, since they make it possible to define the conditions to be fulfilled to enable the measurement of these S-parameters. They define the relationships between pairs composed of the measured and directly measurable quantities, namely

$(S_{k,k}, b_k/a_k)$ and $(S_{jk}, b_j/a_k)$. On the base of condition (I), the original multiport MIC is reduced to the microwave one-port of port k, while (II) reduces given MIC into the two-port, k and j being its input and output ports, respectively.

Principally there are two possible approaches to the measurement of scattering parameters of a MIC.

First, measuring reflection and transition coefficients of the given device under conditions specified by (5.1) and (5.2). Using this approach the S-parameter measurement is in fact a combination of the particular measuring techniques for reflection and transmission measurements in one arrangement of the measuring set. A microwave network analyser is an instrument measuring transfer and/or impedance functions of linear MICs through sine wave testing. A network analyser system accomplishes these measurements by configuring its various components around the MIC under test. The requirement of the measurement system is (1) a sine wave signal source to stimulate the MIC under test, (2) to separate the appropriate signals from the measurement ports of the device under test, and (3) to detect the separated signals, from the signal ratios desired in accordance with S-parameter definitions, and to display the results.

The second approach is characterized by the fact that the DUT is connected into the measuring system, enabling the measurement of the complex amplitude ratio of the emergent to incident waves in all the ports of the DUT. The unknown scattering parameters are determined by solving the corresponding scattering equations, which in number, correspond to the number of unknown S-parameters. The instrumentation for this kind of measurement is represented by network analysers based on multiple reflectometers (e.g., a dual reflectometer analysing microwave two-port MIC (Hoer, 1977)). Considering the large extent of computation, a computer ought to be an indispensable part of such measurement set.

The microwave network analyser is currently a conventional equipment for measuring S-parameters of MIC in a wide frequency range up to 1000 GHz (Dildine and Grace, 1988; AB Millimetre, 1992). With respect to the fact that it is employed for measuring complex scattering parameters (magnitude as well as phase) it is more precisely specified as a vector network analyzer in contrast to the scalar network analyser, which is used only for measuring the magnitude of scattering parameters (Bryant, 1988; Wiltron, 1989).

The measurement of scattering parameters of MIC analysers of circuits encounters errors of different type and origin. They can be divided into systematic and random errors. The systematic errors are repeated in each measurement and they are independent of time as well as MIC measured. With respect to the fact that they are repeated, they can be measured and their effect on the measurement result can be eliminated in contrast to random errors affecting the measurement result in a random manner, which does not allow for their correction. The errors of measuring microwave circuits by the MIC analyser result on the one hand from the analyser itself and, on the other hand from the testing fixture or possibly from probes.

The systematic errors most often encountered in network analyser measurement are frequency response tracking, directivity, source match, load match, and isolation errors. Frequency response tracking errors result from different frequency characteristics of the measuring and reference signal line. These differences result from using different electric components in signal ways and they affect the reflection as well as transmission measurement. The directivity error has its origin in the imperfectness of the directional coupler in the reflectomery. Due to a finite directivity, the incident and reflected waves are not ideally separated from each other. A portion of the energy of the propagating wave is thus added into a branch of the coupler sampling the reflected wave and it continues together with it further into the detector. The contribution of the directivity error to the uncertainty of the measurement depends on the device and it is mostly encountered in reflection measurements, especially for MICs with low S_{11} and S_{22}. The source match error is caused by a mismatch between the test port impedance and the measuring system impedance. If the test port impedance is not equal to the system impedance, reflections from the test port can be re-reflected by the device tested and introduce errors into the measurement. In contrast to the preceding errors, the present error is significant only when the device measured exerts a high reflection coefficient. The origin of the load match error is a mismatch between the impedance of the return port and system impedance. When the impedance of the return port does not equal that of the measuring system, then the energy reflected from the return port can be reflected again by the device measured back to the return port, thus introducing an undesirable error into the measurement. The contribution of the load match error to the uncertainty of measurement depends on the device measured and it is most considerably manifested when the two-port device measured exerts a high output reflection coefficient and low transmission losses. The isolation error is the leakage between the test and reference signals through the network analyser set and it is usually caused by crosstalk between the test and reference paths in the network analyser. The contribution from this source to the uncertainty of the measurement can depend on the device measured and it is particularly manifested when this device exerts high transmission losses.

The random errors in the measurement by the microwave network analyser have their origin in repeated connections, noise, compression and in the effect of spurious signals (Judish and Engen, 1987; Butter *et al.*, 1991).

The measuring fixture or probes included into the measuring system for the measurement of the microwave integrated circuit have their own effect on the uncertainty of the measurement of S-parameters of the microwave integrated circuit. This specific error source occurs due to the electrical imperfectness of the testing fixture with respect to probes. The systematic error models of a two-port test system include mismatch leakage signals, isolation characteristics between reference and test signal paths, and system frequency response. Serious sources of errors of the measuring fixture and/or probes of non-systematic nature are changes of characteristics of cables, connectors and fixture due to changes of the external temperature, non-reproducible contacts in repeated connecting of the device measured and interference with external electromagnetic fields.

Relevant calibration can serve as an approach to suppressing or even eliminating systematic sources of errors in measuring S-parameters of a microwave integrated circuits by a network analyser. From the nature of the problem it follows that the calibration should involve on the one hand the network analyser itself and, on the other hand, the testing fixture or probes. In an one-step calibration the complete measurement setup comprising the network analyser and the connected test fixture or probes is used for calibration. In such a case any calibration verification test would give only information on the total measurement uncertainty. A two-step calibration procedure of the measurement system, which means the network analyser calibration prior to the system calibration, is more suited to separate the residual calibration errors due to the network analyser used and the test fixture or probes error parameters extraction method employed. In spite of the fact that the calibration of the network analyser itself is not principally different from the calibration of the testing holder or probes, the differences are in the implementation (coaxial etalons for the calibration of the network analyser vs. microstrip or coplanar in the second case) and efficiency (the calibration makes an almost ideal measuring system from the network analyser, however, the error of the testing fixture or possibly probes cannot be completely eliminated).

A common feature of all the calibration procedure used is the attempt to represent the systematic errors of the measurement system by means of the scattering response of a virtual error network, EN assumed to interface the DUT to an ideal, error-free network-analyser system (INA), Fig. 5.4. The virtual error network has its n ports

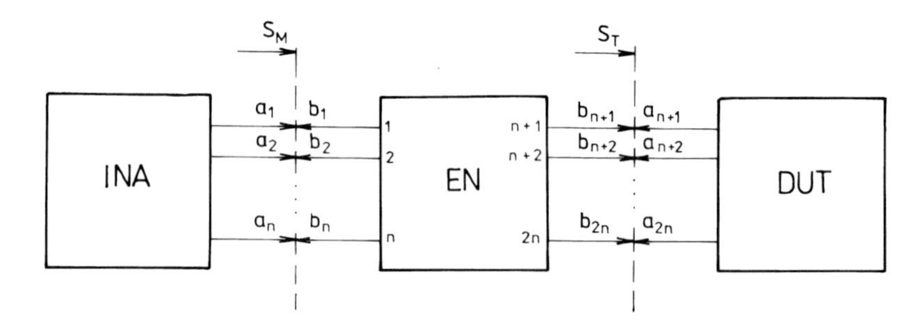

Fig. 5.4 Error network concept.

$1, 2, \ldots, n$ connected to the INA, and n ports $n+1, n+2, \ldots, 2n$ connected to the measured n-port device. The ports $n+1, n+2, \ldots, 2n$ represent the test leads of the real network analyser system, with defined terminal planes. Error network results in that the network analyser measures the improper scattering matrix $\mathbf{S}_\mathrm{M}$ and not the matrix $\mathbf{S}_\mathrm{T}$ of true scattering parameters of DUT. It can be proved (Speciale, 1977) that the relation between the erroneous measured scattering matrix $\mathbf{S}_\mathrm{M}$ and the corresponding $\mathbf{S}_\mathrm{T}$ matrix is given by

$$\mathbf{S}_\mathrm{M} = (\mathbf{T}_1\mathbf{S}_\mathrm{T} + \mathbf{T}_2)\,(\mathbf{T}_3\mathbf{S}_\mathrm{T} + \mathbf{T}_4)^{-1} \tag{5.3}$$

where $\mathbf{T}_1, \ldots, \mathbf{T}_4$ are the four $n \times n$ submatrices of the $2n \times 2n$

T-matrix,

$$\mathbf{T} = \begin{bmatrix} \mathbf{T}_1 & \mathbf{T}_2 \\ \mathbf{T}_3 & \mathbf{T}_4 \end{bmatrix}$$

relating the vector $[b_i, a_i]^{\mathrm{T}}$ where $b_i = [b_1, b_2, \ldots, b_n]$, $a_i = [a_1, a_2, \ldots, a_n]$ of the input interface waves to the vector $[a_j, b_j]^{\mathrm{T}}$ where $a_j = [a_{n+1}, a_{n+2}, \ldots, a_{2n}]$, $b_j = [b_{n+1}, b_{n+2}, \ldots, b_{2n}]$ of the output interface waves of the error network, according to the matrix expression

$$\begin{bmatrix} b_i \\ a_i \end{bmatrix} = \mathbf{T} \begin{bmatrix} a_j \\ b_j \end{bmatrix}$$

The calibration problem consists of (1) stepwise measurement of a set of matrices $\mathbf{S}_{\mathrm{M}i}$, $i = 1, 2, \ldots, k$, under a condition that the analyser is stepwise connected to k different calibration standards with known scattering matrices $\mathbf{S}_{\mathrm{S}i}$ and (2) calculation of entries of submatrices $\mathbf{T}_1, \ldots, \mathbf{T}_4$ on the basis of a corresponding number of matrix pairs $(\mathbf{S}_{\mathrm{M}i}, \mathbf{S}_{\mathrm{S}i})$. Once the T-matrix of the error network has been computed and assuming that it is non-singular, the removal of all systematic errors from the measured $\mathbf{S}_{\mathrm{M}}$ data may be performed computing the corrected $\mathbf{S}_{\mathrm{T}}$ matrix as follows

$$\mathbf{S}_{\mathrm{T}} = \left\{ \left(\mathbf{T}_1 - \mathbf{T}_2\mathbf{T}_4^{-1}\mathbf{T}_3\right)^{-1} \mathbf{S}_{\mathrm{M}} + \left(\mathbf{T}_3 - \mathbf{T}_4\mathbf{T}_2^{-1}\mathbf{T}_1\right)^{-1} \right\} \cdot \left\{ \left(\mathbf{T}_2 - \mathbf{T}_1\mathbf{T}_3^{-1}\mathbf{T}_4\right)^{-1} \mathbf{S}_{\mathrm{M}} + \left(\mathbf{T}_4 - \mathbf{T}_3\mathbf{T}_1^{-1}\mathbf{T}_2\right)^{-1} \right\}^{-1} \quad (5.4)$$

Basically, there are two important elements of calibrating network analyser system: calibration standards (including verification devices) and calibration method. The first step of the calibration process is the formation of calibration standards and calibration checking devices. The standards are used to establish an unchanging reference point on which all the measurements will be based. The verification devices are secondary standards that are used to check the accuracy of the calibration. The calibration is only as good as the calibration standards. Choosing the proper calibration method for MIC measurement is critical for obtaining accurate measurements.

Calibration standards used in the calibration of the microwave analyser of circuits itself are, dependent on the type of measuring inlets of the analyser, either coaxial or waveguide ones. These are microwave devices with electric characteristics almost ideally postulated on the basis of precise solving of the electromagnetic field equations or on the basis of a primary measurement by the absolute measuring method. These typical calibration standards are the air extension of given electrical length, short, open, and non-reflecting termination. It is believed that a large variety of calibration standards may be obtained by combining this lines, shorts and terminations in any predetermined way.

The calibration standards for determining the reference point of the test fixture or probes are not as precisely defined as in the case of coaxial or waveguide microwave standards. In their design the condition is considered to a certain extent of the shape compatibility of the standard with the devices measured and of the delimited space, where they should be situated for the time of the calibration. The calibration standards are typically in the microstrip or coplanar form. The current theory of these microwave devices presents no precise formulae for the calculation of parameters, and input assumptions for which these formulae were derived are also to a certain extent different from the true situation. The limited finite thickness of metallic strip conductors also results in discrepancies between the theoretical model and actual microwave device chosen as a calibration standard. It is also of importance that the current microwave measuring technique has no absolute measuring methods, which could make it possible to check the calibration standards of this form and which could introduce corrections to their theoretical models. A further important drawback of the microstrip and coplanar calibration standards can also be their time and possibly also temperature instability, inherently given by materials, which are these calibration etalons made of. The calibration standards of this type are frequently individually designed in laboratories in attempts to make adjustments for the testing fixture and substrate with the device tested (Standinger and Seely, 1987); commercially available calibration standards are also supplied with universal testing fixture (Argumens, 1992).

The calibration etalons for on wafer testing are fabricated on a semiconductor or sapphire substrate, including shorts, shifted shorts, transmissions lines, and loads. Open circuit is realized by lifting the probes over substrate. These on wafer standards are commercially available (known as impedance standard substrate (ISS) (Cascade Microtech, 1991), however, many laboratories designed and developed their own standards (Williams and Miers, 1989). The ISS also includes devices that are used to check the accuracy of the calibration, such as open circuits, short circuits, loads, inductors and capacitors.

It should be noted that the current technology of the MIC and MMIC testing uses commercially available standards, which are to a considerable extent conventional calibration standards and that the measurement result in different media with calibration etalons of different origin can be considerably different from each other. The result is a MIC and/or MMIC that, when compared to coaxial standards, does not perform as expected.

The calibration process includes the methods and the procedures that are implemented to remove systematic errors and to maximize the measurement repeatibility. A network analyser is typically used to vectorially remove measurement system errors. This is accomplished by measuring the calibration standards over the operating frequency range. After measuring the calibration standards, an error network model may be calculated and stored in the network analyser processor memory.

Depending on what set of k calibration standards with defined matrices $\mathbf{S}_{\mathrm{M}i}$, $i = 1, 2, \ldots, k$, serves as a basis of the calibration of the network analyser and/or test fixture (or probes), the use of several calibration techniques known in practice was developed named the Through-Short-Delay (TSD), Short-Open-Line-Through (SOLT), Through-Reflect-Line (TRL), Line-Reflect-Line (LRL), and Line-Reflect-Match (LRM) techniques.

TSD Calibration Technique (Franzen and Speciale, 1975; Speciale, 1977) is based on two-port devices as the primary standards. This two ports having no moving parts can be designed to physically fit in place to substitute the DUT directly at its defining terminal planes. This possibility includes any possible interconnecting network within the measurement system being calibrated. In particular, the TSD calibration procedure does not assume negligible measurement-port mismatch nor negligible response distortion by the external RF interfacing networks. As a consequence, for example, test fixture or wafer probes may be included in the measurement system while measuring MICs or MMICs. The three two-port calibration standards used are:

1) the Through, defined as a residual length l, of nominal impedance transmission line, with postulated scattering matrix expressed

as

$$\mathbf{S}_{\mathrm{ST}} = \begin{bmatrix} 0 & \mathrm{e}^{-\gamma l_1} \\ \mathrm{e}^{-\gamma l_1} & 0 \end{bmatrix} \tag{5.5}$$

where γ is the complex propagation constant,

2) the Short, defined as a pair of immediate shorts at both measurement leads, its scattering matrix postulated in the following manner

$$\mathbf{S}_{\mathrm{SS}} = \begin{bmatrix} -1 & 0 \\ 0 & -1 \end{bmatrix} \tag{5.6}$$

and

3) the Delay, defined as a substantial length l_2 nominal impedance transmission line, with S-matrix

$$\mathbf{S}_{\mathrm{SD}} = \begin{bmatrix} 0 & \mathrm{e}^{-\gamma l_2} \\ \mathrm{e}^{-\gamma l_2} & 0 \end{bmatrix} \tag{5.7}$$

The electrical lengths of l_1 and l_2 need not be accurately known in TSD calibration procedure, as they are computed in the process together with the respective insertion losses.

Stepwise measurement of the scattering parameters gives rise to corresponding set of matrices

$$\mathbf{S}_{\mathrm{MT}} = \begin{bmatrix} {}^{T}S_{11} & {}^{T}S_{12} \\ {}^{T}S_{21} & {}^{T}S_{22} \end{bmatrix} \tag{5.8}$$

$$\mathbf{S}_{\mathrm{MS}} = \begin{bmatrix} {}^{S}S_{11} & {}^{S}S_{12} \\ {}^{S}S_{21} & {}^{S}S_{22} \end{bmatrix} \tag{5.9}$$

and

$$\mathbf{S}_{\mathrm{MD}} = \begin{bmatrix} {}^{D}S_{11} & {}^{D}S_{12} \\ {}^{D}S_{21} & {}^{D}S_{22} \end{bmatrix} \tag{5.10}$$

On the basis of the measured matrices $\mathbf{S}_{\mathrm{MT}}$, $\mathbf{S}_{\mathrm{MS}}$, and $\mathbf{S}_{\mathrm{MD}}$, the unknown S-matrices

$$\mathbf{S}_{\mathrm{A}} = \begin{bmatrix} {}^{A}S_{11} & {}^{A}S_{12} \\ {}^{A}S_{21} & {}^{A}S_{22} \end{bmatrix} \tag{5.11}$$

and

$$\mathbf{S}_{\mathrm{B}} = \begin{bmatrix} {}^{B}S_{11} & {}^{B}S_{12} \\ {}^{B}S_{21} & {}^{B}S_{22} \end{bmatrix} \tag{5.12}$$

of the error two-ports A and B (Fig. 5.5) are computed using following formulae

$$
\begin{aligned}
{}^{A}S_{11} &= \frac{H_3}{H_1} = \frac{H_2}{H_4} \\
{}^{A}S_{22} &= \frac{\frac{H_3}{H_1} - {}^{S}S_{11}}{{}^{S}S_{11} - \frac{H_1}{H_4}} \\
{}^{A}S_{12}\,{}^{A}S_{21} &= \frac{-{}^{A}S_{22}}{H_4} \\
\det(\mathbf{S}_{\mathrm{A}}) &= {}^{A}S_{22}\frac{H_1}{H_4} = {}^{A}S_{22}\frac{H_3}{H_2} \\
{}^{B}S_{11} &= -\frac{{}^{S}S_{22} + \frac{K_4}{K_1}}{{}^{S}S_{22} + \frac{K_1}{K_3}} \\
{}^{B}S_{22} &= -\frac{K_4}{K_1} = -\frac{K_2}{K_3} \\
{}^{B}S_{12}\,{}^{B}S_{21} &= \frac{{}^{B}S_{11}}{K_3} \\
\det(\mathbf{S}_{\mathrm{B}}) &= -\frac{{}^{B}S_{11}K_1}{K_3} = -\frac{{}^{B}S_{11}K_4}{K_2}
\end{aligned}
\qquad (5.13)
$$

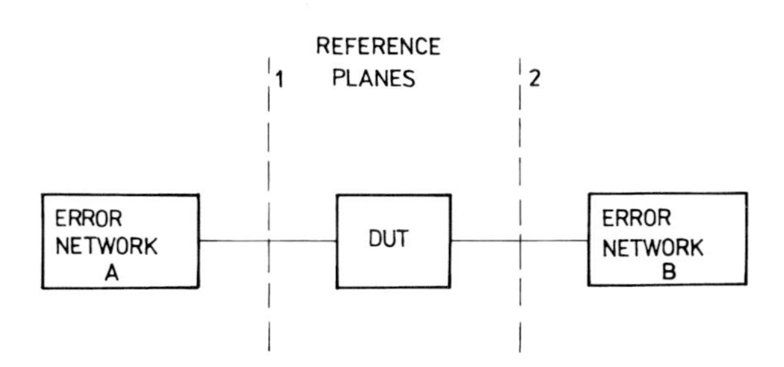

Fig. 5.5 Error networks of the test fixture.

Auxiliary functions H_1, K_1, $i = 1, 2, 3, 4$ are defined on the basis of T-matrix product

$$\mathbf{H} = \mathbf{T}_{\mathrm{T}}\mathbf{T}_{\mathrm{D}}^{-1}$$

and

$$\mathbf{K} = \mathbf{T}_\mathrm{D}^{-1}\mathbf{T}_\mathrm{T}$$

The measured T-matrices $\mathbf{T}_\mathrm{T}$ and $\mathbf{T}_\mathrm{D}$ correspond to the S-matrices $\mathbf{S}_\mathrm{T}$ and $\mathbf{S}_\mathrm{D}$, respectively, and are computed from these through the standard S- to T-parameter transformation. The auxiliary functions $H_1, \ldots, H_4$ and $K_1, \ldots, K_4$ are then as follows

$$\begin{aligned}
H_1 &= \frac{1}{2}\left[R\left(H_{22} - H_{11}\right) + 1\right] \\
H_2 &= \frac{1}{2}\left[R\left(H_{22} - H_{11}\right) - 1\right] \\
H_3 &= RH_{12} \\
H_4 &= -RH_{21} \\
K_1 &= \frac{1}{2}\left[R\left(K_{22} - K_{11}\right) + 1\right] \\
K_2 &= \frac{1}{2}\left[R\left(K_{22} - K_{11}\right) - 1\right] \\
K_3 &= -RK_{12} \\
K_4 &= RK_{21}
\end{aligned} \qquad (5.14)$$

where

$$R = \sqrt{\left(H_{11} + H_{22}\right)^2 - 4} = \sqrt{\left(K_{11} + K_{22}\right)^2 - 4}$$

As a consequence of the two elements $\exp(-\gamma l_1)$ and $\exp(-\gamma l_2)$ of the postulated S-matrices $\mathbf{S}_\mathrm{T}$ and $\mathbf{S}_\mathrm{D}$, two alternative solutions of the TSD calibration exists, corresponding to mutually reciprocal and opposite values for either of these matrix elements. However, if the electrical length of the line l_1 is assumed to be always $l_1 < \frac{\lambda}{4}$, then the imaginary part of $\exp(2\gamma l_1)$, must always be positive, signalling which of the two solutions is physically meaningful.

TRL Calibration Technique (Engen and Hoer, 1979; Standinger and Seely, 1987; Hewlett-Packard, 1987; Lan, 1989; Pantoza *et al.*, 1989; Marks and Phillips, 1989; Williams, 1989; Hoer, 1987) of calibration an microwave automatic network analyser (ANA) having sexless test-port coaxial connectors or rectangular waveguides, consists of the following steps:

(1) the measurement planes are connected together,

(2) one or more highly reflecting terminations are connected to both measurement planes of the ANA, and

(3) a length of precision transmission line is connected between the test ports.

As compared with TSD, this TRL calibration technique assumes that the reflection coefficient of the line, and the reflection coefficient of the highly reflecting terminations need not be known. The TRL calibration yields the parameters of the error network as well as the reflection coefficients of all the terminations used in the reflection measurement, and electrical length of the precision transmission line used in the line measurement. Thus, the only standards for this technique are equal impedance line lengths. When a length of uniform transmission line is used as a standard in calibrating an ANA all the measured values of S parameters of the DUT are related to the characteristic impedance of that line. Systematic errors are created by uncertainties in this impedance and by imperfections in connectors. The systematic error in S parameters S_{11} and S_{22} of a DUT are functions of all the S parameters of the DUT.

Generally, TRL calibration technique can be used for both a one-step (which calibrates the total measurement system to the test fixture or to the wafer probe heads directly) calibration or a two-step calibration (which, by using both the TRL standards provided by the ANA manufacturer and the microstrip or wafer standards, provides independent constants for the test fixture or for the probe heads, respectively). While only one length of line (other than the through) is referred to TRL calibration steps, at least one additional length of line will be used to ensure optimal values of delay at various frequencies. This is required since differential delays between line lengths that approach integral multiples of π radians do not provide proper solutions for the calibration procedure (Hoer and Engen, 1987). While additional line lengths are costly to provide in coaxial or waveguide standards sets, the cost is trivial in microstrip or coplanar waveguide devices. TRL calibration technique is easy to implement on wafers that can use coplanar probes or lower frequency microstrip launchers. The input and output probes can be moved apart thereby creating the proper line for TRL calibration procedure.

Mathematicaly, the calibration of ANA using a TRL techniques is described in the following way (Engen and Hoer, 1979; Rubin, 1990):

First, denoting the cascading matrices of the error two-ports A, B by $\mathbf{T}_{\mathrm{A}}$, $\mathbf{T}_{\mathrm{B}}$, respectively, the T-matrix $\mathbf{T}_{\mathrm{AB}}$ of the through connection is given by

$$\mathbf{T}_{\mathrm{AB}} = \mathbf{T}_{\mathrm{A}}\mathbf{T}_{\mathrm{B}} \tag{5.15}$$

Note that elements of matrix $\mathbf{T}_{\mathrm{AB}}$ are known since were found directly from the S-parameters of the through connection measurement followed by proper S-T transformation.

Second, inserting the line with cascading matrix $\mathbf{T}_{\mathrm{L}}$ another set of T-parameters (using S-parameter measurement and S-T transformation) namely

$$\mathbf{T}_{\mathrm{ALB}} = \mathbf{T}_{\mathrm{A}}\mathbf{T}_{\mathrm{L}}\mathbf{T}_{\mathrm{B}} \tag{5.16}$$

is obtained.

Find the matrix

$$\mathbf{T} = \mathbf{T}_{\mathrm{ALB}}\mathbf{T}_{\mathrm{AB}}^{-1} \tag{5.17}$$

Using (5.15) and (5.16) yields

$$\mathbf{T}\mathbf{T}_{\mathrm{A}} = \mathbf{T}_{\mathrm{A}}\mathbf{T}_{\mathrm{L}} \tag{5.18}$$

If γ and l represent respectively the propagation constant and length of the line, then assuming the line is nonreflecting

$$\mathbf{T}_{\mathrm{L}} = \begin{bmatrix} \mathrm{e}^{-\gamma l} & 0 \\ 0 & \mathrm{e}^{\gamma l} \end{bmatrix} \tag{5.19}$$

Using (5.18), expansion of (5.18) gives

$$\begin{bmatrix} T_{11} & T_{12} \\ T_{21} & T_{22} \end{bmatrix} \cdot \begin{bmatrix} {}^{A}T_{11} & {}^{A}T_{12} \\ {}^{A}T_{21} & {}^{A}T_{22} \end{bmatrix} = \begin{bmatrix} {}^{A}T_{11} & {}^{A}T_{12} \\ {}^{A}T_{21} & {}^{A}T_{22} \end{bmatrix} \cdot \begin{bmatrix} \mathrm{e}^{-\gamma l} & 0 \\ 0 & \mathrm{e}^{\gamma l} \end{bmatrix}$$

so

$$T_{11}\,{}^{A}T_{11} + T_{12}\,{}^{A}T_{21} = {}^{A}T_{11}\mathrm{e}^{-\gamma l} \tag{5.20}$$

$$T_{21}\,{}^{A}T_{11} + T_{22}\,{}^{A}T_{21} = {}^{A}T_{21}\mathrm{e}^{-\gamma l} \tag{5.21}$$

$$T_{11}\,{}^{A}T_{12} + T_{12}\,{}^{A}T_{22} = {}^{A}T_{12}\mathrm{e}^{\gamma l} \tag{5.22}$$

$$T_{21}\,{}^{A}T_{12} + T_{22}\,{}^{A}T_{22} = {}^{A}T_{22}\mathrm{e}^{\gamma l} \tag{5.23}$$

From (5.20) and (5.21) or (5.22) and (5.23) we have quadratic equation

$$\mathrm{e}^{-2\gamma l} - \mathrm{e}^{-\gamma l}\left(T_{11} + T_{22}\right) + \left(T_{11}T_{22} - T_{12}T_{21}\right) = 0 \tag{5.24}$$

or

$$\mathrm{e}^{2\gamma l} - \mathrm{e}^{\gamma l}\left(T_{11} + T_{22}\right) + \left(T_{11}T_{22} - T_{12}T_{21}\right) = 0 \tag{5.25}$$

respectively. The solutions for $\exp(-\gamma l)$ and $\exp(\gamma 1)$ are two solutions to the equation

$$G^2 - G\left(T_{11} + T_{22}\right) + \left(T_{11}T_{22} - T_{12}T_{21}\right) = 0$$

we find

$$G = B\left(1 + \sqrt{D}\right)$$

where

$$B = \frac{1}{2}\left(T_{11} + T_{22}\right)$$

$$C = T_{11}T_{22} - T_{12}T_{21}$$

$$D = 1 - \frac{C}{B^2}$$

The root assignment should be made such that $|G^2| < 1$, however, since the difference between $|G^2|$ and unity is usually small, it is possible for measurement error to mask this effect. If the line length differs in a known direction from the design centre of a quarter wavelength, this could also serve as a basis for root choice.

Third, using identical loads Γ_L at test port terminal planes, do the reflect measurements, finding Γ_{MA} and Γ_{MB}. The reflection coefficient Γ_{MA} which obtains at the fictitious input port for error two-port A is related to the reflection coefficient of the load Γ_L by

$$\Gamma_{MA} = \frac{A_1\Gamma_L + A_2}{A_3\Gamma + 1} \tag{5.26}$$

where

$$A_1 = \frac{{}^{A}T_{11}}{{}^{A}T_{22}}$$

$$A_2 = \frac{{}^{A}T_{12}}{{}^{A}T_{22}}$$

$$A_3 = \frac{{}^{A}T_{21}}{{}^{A}T_{22}}$$

Similarly, for error two-port B one obtain the relation

$$\Gamma_L^{-1} = \frac{B_1 + B_2\Gamma_{MB}}{B_3 + \Gamma_{MB}} \tag{5.27}$$

where

$$B_1 = \frac{{}^{B}T_{11}}{{}^{B}T_{22}}$$

$$B_2 = \frac{{}^{B}T_{12}}{{}^{B}T_{22}}$$

$$B_3 = \frac{{}^{B}T_{21}}{{}^{B}T_{22}}$$

Eliminating the unknown Γ_L from (5.26) and (5.17) gives

$$\frac{A_1}{B_1} = \frac{(\Gamma_{MA} - A_2)\left(1 + \frac{\Gamma_{MB}B_2}{B_1}\right)}{\left(1 - \frac{\Gamma_{MA}A_3}{A_1}\right)(B_3 + \Gamma_{MB})} \tag{5.28}$$

Unknown values B_1, B_2, B_3 are obtained on the basis of known matrix

$$\mathbf{T}_{\mathrm{AB}} = \begin{bmatrix} {}^{AB}T_{11} & {}^{AB}T_{12} \\ {}^{AB}T_{21} & {}^{AB}T_{22} \end{bmatrix}$$

remembering (5.15) in the form

$$\mathbf{T}_{\mathrm{B}} = \mathbf{T}_{\mathrm{A}}^{-1} \cdot \mathbf{T}_{\mathrm{AB}}$$

so that

$${}^{B}T_{22}\begin{bmatrix} B_1 & B_2 \\ B_3 & 1 \end{bmatrix} =$$

$$= \frac{1}{{}^{A}T_{22}\,(A_1 - A_2A_3)}\begin{bmatrix} 1 & -A_2 \\ -A_3 & 1 \end{bmatrix}\begin{bmatrix} {}^{AB}T_{11} & {}^{AB}T_{12} \\ {}^{AB}T_{21} & {}^{AB}T_{22} \end{bmatrix}$$

and finally

$$B_1 = \frac{C_1 - A_2C_3}{A_1\left(1 - \frac{C_2A_3}{A_1}\right)} \tag{5.29}$$

$$B_2 = \frac{C_2 - A_2}{A_1\left(1 - \frac{C_2 A_3}{A_1}\right)} \tag{5.30}$$

$$B_3 = \frac{\frac{C_3 C_1 A_3}{A_1}}{1 - \frac{C_2 A_3}{A_1}} \tag{5.31}$$

where

$$C_1 = \frac{{}^{AB}T_{11}}{{}^{AB}T_{22}}$$

$$C_2 = \frac{{}^{AB}T_{12}}{{}^{AB}T_{22}}$$

$$C_3 = \frac{{}^{AB}T_{21}}{{}^{AB}T_{22}}$$

Multiplying (5.28) by (5.29) yields two solutions for A_1, namely

$$A_1 = +\sqrt{\frac{(\Gamma_{MA} - A_2)\left(1 + \frac{\Gamma_{MB} B_2}{B_1}\right)(C_1 - A_2 C_3)}{(\Gamma_{MB} + B_3)\left(1 - \frac{\Gamma A_3}{A_1}\right)\left(1 - \frac{C_2 A_3}{A_1}\right)}} \tag{5.32}$$

Apart from the choice of sign in (5.32), the requirement that Γ_L be known has been eliminated. As a practical matter, a nominal short or open circuit continues to be a convenient choice for the unknown reflector. In either case, a nominal value for the argument of this reflection is available and with the help of (5.26) rearranged into form

$$A_1 = \frac{\Gamma_{MA} - A_2}{\Gamma_L\left(1 - \frac{\Gamma_{MA} A_3}{A_1}\right)}$$

this permits the proper sign choice to be made in (5.32).

As noted previously, the TRL technique can be applied only to ANA's having identical sexless connectors at the test ports. If the requirement No (1) for making a through connection is replaced by a measurement with a short length of line, then the calibration technique can be applied to an ANA having identical connectors of any type, not just sexless connectors. This modification gives rise to the Line-Reflect-Line (LRL) calibration procedure (Hoer and Engen, 1987). The LRL technique is useful having to use connector adapters or imperfect test ports in the measuring system in order to accept test fixtures with different connectors (Juroshek *et al.*, 1989). The LRL calibration technique can also be applied if one set of connectors is a waveguide, and the other set is a coax. It is convenient, that the computations and software used in the TRL solution can also be used in the LRL solution with only slight modifications based on using a nonzero length of the first line instead of a through.

In the TRL and/or LRL calibration techniques the optimum difference in the electrical lengths of the lines is $\pi/2$ or an odd multiple of this value. Differences near $k\pi$, $k = 1, 2, \ldots$ must be avoided, since otherwise the solution becomes ill conditioned. At higher frequencies the optimum length of line for the TRL calibration can become physically too short to be realized. However, if two lines are used, the first can be of some convenient length and the second one slightly longer so that the difference in electrical length is $\pi/2$ in the operating frequency band. For this reason, a LRL calibration procedure is a good solution.

The condition of optimum difference in the electrical length when applied to coaxial lines leads to second line length l_2 given as (Hoer, 1983)

$$l_2 = \frac{15}{f_1 + f_2} + l_1$$

where l_1 is the first line length in cm, and f_1, and f_2 lower and upper operating frequencies in GHz, respectively. If the ratio f_2/f_1 is large, the difference between lengths becomes to be too small and the operating frequency range may be broken into two (or more) ranges f_1 to f_i and f_i to f_2, where f_i is an intermediate frequency chosen such that

$$\frac{f_2}{f_i} = \frac{f_i}{f_1}$$

or explicitly

$$f_i = \sqrt{f_1 f_2}$$

Three line lengths l_1, l_2 and l_3 in cm, satisfying the relations

$$l_2 = \frac{15}{f_i + f_2} + l_1$$

and

$$l_3 = \frac{15}{f_1 + f_i} + l_1$$

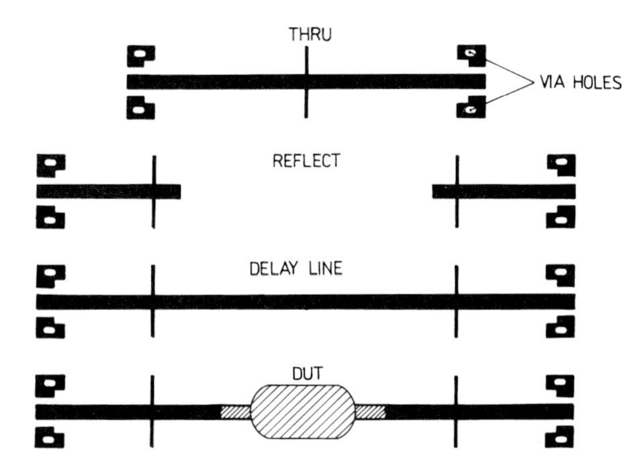

Fig. 5.6 TRL calibration standards.

are required to cover these two ranges.

Calibrating test fixture or probes a *microstrip calibration standards* (Fig. 5.6) for the TRL method are applicable. Reproducible and low parasitic via holes are required in this case. Thus, the TRL and/or LRL technique is useful for the MIC and MMIC measurement system calibration. This technique has simple standards which can be easily placed on the same substrate on which the measured devices are fabricated. The LRL calibration, however, has better flexibility in locating the reference planes and also minimizes radiative crosstalk between the input/output launchers or probes.

The SOLT Calibration Technique (Hewlett-Packard, 1980; Hewlett-Packard, 1985a, Hewlett-Packard, 1986) is applicable with coaxial as well as microstrip or coplanar standards. However, it has shown marginal utility in IC media especially above 15 GHz, because of the unavailability of well-defined open and short circuit standards.

5.2.2 Noise measurement

According to equation

$$F = \frac{kT_0BG_A + P_N}{kT_0BG_A} \tag{5.33}$$

the noise factor F, at a specified input frequency, is defined as the ratio of (1) the total noise power per unit bandwidth at a corresponding output frequency available at the output port when the noise temperature of the input termination is standard ($T_0 = 290\,\mathrm{K}$) to (2) that portion of (1) engendered at the input frequency by the input termination. In (5.33) k is Boltzmann's constant, B the bandwidth over which the noise power is measured, G_A the available power gain of the measured active circuit, and P_N the noise power added by active circuit itself (Haus *et al.*, 1960). The noise figure is a very important characteristic of the MIC to be used in input blocks of systems for processing weak signals (Reeve *et al.*, 1990).

The current automatic systems for measuring the noise figure almost exclusively operate on the principle of the Y-factor (Maury Microwave Corporation, 1991; Hewlett-Packard, 1987b), defined by a ratio of two levels of the noise power, at the MIC measured output, corresponding to two different power levels of the noise source (hot and cold state) connected to the input of the MIC, so that

$$Y = \frac{P_{N_2}}{P_{N_1}}$$

where

$$P_{N_1} = kBG_A T_h + P_N \tag{5.34}$$

and

$$P_{N_2} = kBG_A T_c + P_N \tag{5.35}$$

T_h and T_c are noise temperatures of the switched on and switched off noise source, respectively. By a subsequent solution of the system of two equations (5.34) and (5.35) with respect to unknown terms kBG_A and P_N (carried out in the computation block of an automatic noise measuring system) the resulting value of F is obtained.

As a matter of fact the noise figure itself cannot completely and unambiguously characterize the noise features of the given MIC, since it is also a function of parameters of the external network connected to the input port of the MIC measured as it follows from

$$F = F_{min} + 4R_N \frac{|\Gamma - \Gamma_{opt}|^2}{|1 + \Gamma_{opt}|^2 \left(1 - |\Gamma|^2\right)} \tag{5.36}$$

where Γ is the reflection coefficient of the load in the input port of the DUT. According to equation (5.36) two real parameters, minimum noise figure $F_{\min}$ and equivalent noise resistance R_N and one complex parameter, reflection coefficient Γ_{opt} corresponding to the minimum noise figure, i.e. four real parameters present a set of noise parameters completely and unambiguously characterizing noise features of any active microwave circuit. Further sets of noise parameters are also defined and used based on the impedance, admittance or wave model of the noise microwave circuit (Meys, 1978), which are equivalent to the above mentioned ones.

Finding the noise parameters of a MIC usually involves measuring the noise figure at two or more values of the source reflection coefficient. Figure 5.7 shows a block diagram of a measurement system suitable for noise parameters measurement. A noise figure meter displays noise figure, while a tuner in front of the measured MIC permits variation of the noise source reflection coefficient Γ. The measuring system is completed by calibrated noise source, a postamplifier to lower measurement system noise, and a filter/mixer combination for single-sideband downconversion.

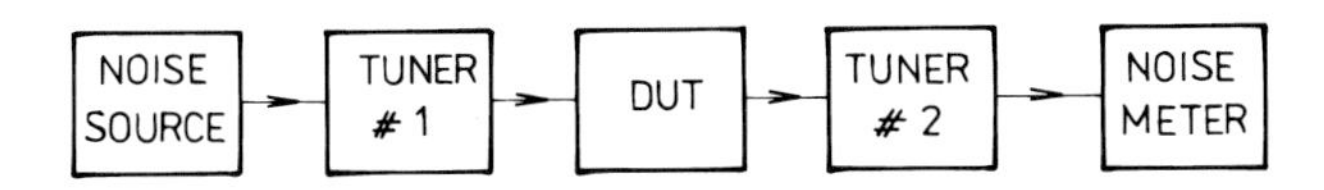

Fig. 5.7 Typical noise parameter measurement set up.

One way of using the system is to adjust the tuner until the noise figure measurement reaches its minimum. At this point F and Γ would presumably equal $F_{\min}$ and Γ_{opt}. Results of a second measurement at some other Γ would determine R_N. In spite of the fact that this procedure is simple, its implementation is connected with a number of disadvantages, particularly when it is necessary to automatize the measurement. These include (1) the necessity of tuning exactly on to Γ_{opt} in order to make an accurate measurement, (2) the possibility of missing the minimum for the MIC due to the variation in tuner loss, and (3) the difficulty of loading a shallow minimum using an automated search.

The second and better approach is to measure F at several distinct values of the noise source reflection coefficient. By inserting these data into equation (5.36), a system of equations is obtained whose solution results in determining the unknown noise parameters. Four measurements are principally sufficient for the determination of four scalar quantities in noise parameters. The redundant measurements are, however, more advantageous, since by fitting equation (5.36) to the data measured, for example by the least square method, the precision of the measurements can be considerably increased (Sannino, 1979).

The (Γ, F) data points which determine MIC noise parameters result from a series of reflection coefficient and noise figure measurements. The time and labour costs associated with these measurements are high, unless the measurements can be automatized. The key problem of the measurement automatizing, when disregarding the already reliably solved problems of the direct control of measuring instrument with a computer, for example by means of an interface bus, is the automatic operation of the tuner at the measured MIC input (Fraser *et al.*, 1988). The tuner should be fast and simple to use and its settings should be highly repeatable. Low and particularly stable inserted losses of the tuner are of importance, since they considerably contribute to the error of measuring the noise figure. An electronically controlled tuner is very suitable for the noise parameters measurement, consisting of segments of microstrip lines of different length overswitched by PIN diodes (Froelich, 1989). The other solution is the arrangement of the input as shown in Fig. 5.8, for which the noise power is injected between the input port DUT and programmable load (Hewlett-Packard, 1988a). This eliminates

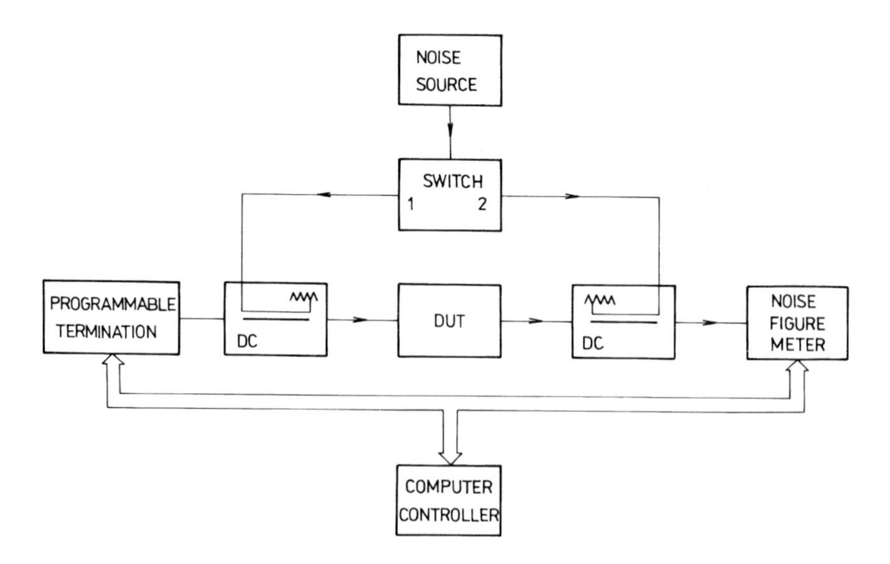

Fig. 5.8 Noise parameter measurement.

the problem of variable insertion loss through the tuner with each new tuner setting. All losses between input port of the MIC measured and the noise source are accounted for by a combination of the reflection data taken for input port and a calibration step which determines the amount of noise power injected out of input port from the noise source with different reflection coefficient settings. The switch in the arrangement is used to switch the noise source from input port to the output one, allowing for an S_{22} measurement of the MIC. The isolator improves the measurement system match, and the low noise amplifier reduces the measurement system's noise figure. The outlined technique of the measurement has a further advantage that in the measuring system no further tuner is necessary at the output of the MIC measured and thus also no further tuning to the maximum of the output power in each setting of the reflection coefficient of the noise source or during each change of the frequency. The value of the power gain available necessary for the calculation of the noise figure can be calculated for each measured value of the gain inserted. The noise figure for the corresponding coefficient of the noise source reflection and frequency is calculated from the equation

$$F_{meas} = F + \frac{(F_{sys} - 1)}{G}$$

where F is the MIC and tuner noise figure, G is the MIC and tuner available power gain, and F_{sys} is the measurement system noise figure.

The noise measurement is in principle a low-level measurement and thus, it is extremely sensitive to many factors mainly affecting the precision of the measurement. (Hewlett-Packard, 1988b). They can be summarized as follows:

(a) imperfect contacts in connectors, changes of the reflection coefficient and losses of connectors in repeated overswitching

(b) penetration of disturbing electromagnetic signals from the environment resulting in the measurement instability

(c) changes of the reflection coefficient of the noise source overswitched between the hot and cold state

(d) using insertion power gain instead of available power gain in the second stage correction equation

(e) discrepancies between calibrating and measuring frequency points

(f) differences in reading off the measured value in subsequent measurements resulting from random fluctuations of the system.

The disturbing effects are more considerable when the MIC measured has an extremely low noise figure and gain and further when it operates with a noise source, whose ENR is low (for example due to a reduction of the change of the reflection coefficient in overswitching between cold and hot states) and also when the measuring system itself has a considerable internal noise.

By the integration of the vector microwave network analyser into the measuring system, it is possible to efficiently increase the precision of the noise measurement, of course on account of the completeness of the whole equipment and measuring procedure (Davidson *et al.*, 1989).

REFERENCES

AB Millimetre (February 1992) 8–100 GHz vector network analyzer, *Microwave Eng. Europe, 18.*

Argumens (February 1992) Test-Fixture, *Microwave Eng. Europe, 46.*

Arjavalingam G., Pastol Y., Halbout J.M., Kopcsay G.V. (November 1989) Application of picosecond optoelectronics in broadband microwave material measurements, *Microwave J.*, pp. 133–40.

Bahl I., Lewis G., Jorgenson J. (1991) Automatic Testing of MMIC Wafers, *Int. J. Microwave Millimeter-Wave CAE, 1*, pp. 77–89.

Bierman H. (1990) Improved on-wafer techniques evolve for MMIC testing, *Microwave J., vol. 30, 3*, pp. 44–58.

Bryant G.H. (1988) *Principles of Microwave Measurements*, The Institution of Electrical Engineers and Peter Peregrinus Ltd., London.

Cappy A. (1988) Noise modeling and measurement techniques, *IEEE Trans. Microwave Theory Tech., vol. MTT-36, 1*, pp. 1–10.

Carlton D.E., Gleason K.R., Strid E.W. (1985) Microwave wafer probing achieves on-wafer measurements through 18 GHz, *Microwave System News, 15, 5*, pp. 99–115.

Cascade Microtech (1991) Beaverton: *Probe Head Selection Guide.*

Davidson A.C., Leake B.W., Strid E. (1989) Accuracy Improvements in Microwave Noise Parameter Measurements, *IEEE Trans. Microwave Theory Tech., vol. MTT-37, 12*, pp. 1973–8.

Dildine R.G., Grace J.D. (1988) Millimeter-wave vector network analysis, *Hewlett-Packard J., 39, 2*, pp. 12–8.

Engen G., Hoer C. (1979) Thru-reflect-line: An improved technique for calibrating the dual six-port automatic network analyzer, *IEEE Trans. Microwave Theory Tech., vol. MTT-27, 12*, pp. 987–93.

Franzen N.R., Speciale R.A. (1975) A new procedure for system calibration and error removal in automated S-parameter measurements, *Proc. 5th European Microwave Conf.*, pp. 69–73.

Fraser A., Strid E., Leake B., Burcham T. (November 1988) Repeatability and Verification of On-Wafer Noise Parameter Measurements, *Microwave J., 172–176.*

Freeman J.L. *et al.* (1985) Electro-optic sampling of planar digital integrated circuits, *Appl. Phys. Lett., vol. 47*, pp. 1083–4.

Froelich R. (1989) Automated Noise-Parameter Measurements Using A Microwave Probe, *Watkins-Jahnson Company Tech-Notes, 16, 1*, pp. 1–10.

Gleason K.R., Reeder T.M., Strid E.W. (1983) Precise MMIC parameters yielded by 18 GHz wafer probe, *Microwave System News, 13, 5*, pp. 55–65.

Gledhill S.J. (1989) Spectrum analysis, in *Microwave Measurements*, A. E. Bailey, ed., Peter Peregrinus Ltd., London.

Godshalk E. M. (1991) A V-band wafer probe using ridge-through waveguide, *IEEE Trans. Microwave Theory Tech., vol. MTT-39, 12*, pp. 2211–7.

Griffin E.J. (1989) An introduction to reflectometers, A. E. network analyzers, in *Microwave Measurements*, A. E. Bailey, ed., Peter Peregrinus Ltd., London.

Haus H.A. *et al.* (1960) IRE Standards on Methods of Measuring Noise in Linear Twoports, *1959, Proc. IRE, 48, 1*, pp. 60–9.

Hewlett-Packard (1980) Automating the HP 84108 Microwave Network Analyzer, *Application Note 221A, Palo Alto.*

Hewlett-Packard (1985a) *8510 Network Analyzer Operation and Programming Manual*, Palo Alto.

Hewlett-Packard (July 1985b) The New Standard in Microwave Network Analysis, *Technical Data.*

Hewlett-Packard (May 1986) On-wafer measurements using the HP 8510 network analyzer and Cascade Microtech wafer probes, Palo Alto.

Hewlett-Packard (October 1987a) Applying the HP 8410B TRL calibration for noncoaxial measurements, *Product Note 8510-8*, Palo Alto.

Hewlett-Packard (August 1987b) Application and Operation of the HP 8970B Noise Figure Meter and HP 8970S Microwave Noise Figure Measurement System, Product Note 8970B/S-2, Palo Alto.

Hewlett-Packard (1988a) Noise Parameter Measurement Using the HP 8970B Noise Figure Meter and the ATN NP4 Noise Parameter Test Set, *Product Note HP 8970B/S-3*, Palo Alto.

Hewlett-Packard (November 1988b) Noise Figure Measurement Accuracy, *Application Note 57-2*, Palo Alto.

Hewlett-Packard (1992) Palo Alto: *Test & Measurement Catalog.*

Hoer C.A. (1977) A network analyzer incorporating two six-port reflectometers, *IEEE Trans. Microwave Theory Tech., vol. MTT-25, 12*, pp. 1070–4.

Hoer C.A. (1983) Choosing line lengths for calibrating network analyzers,

IEEE Trans. Microwave Theory Tech., vol. MTT-31, 1, pp. 76–8.

Hoer C.A. (1987) On-line accuracy assessment for the dual six-port ANA: Treatment of systematic errors, *IEEE Trans. Instrum. Meas., vol. IM-36, 2*, pp. 514–9.

Hoer C.A., Engen G.F. (1987) On-line accuracy assessment for the dual six-port ANA: Extension to Nonmating connectors, *IEEE Trans. Instrum. Meas., vol. IM-36, 2*, pp. 524–9.

Hughes Microwave Product Division (1990) Torrance: *Millimeter-Wave Product Catalog.*

Hung H.A., Polak–Dingels P., Webb K.J., Smith T., Huang Ho C., Lee Chi H. (1989) Millimeter-wave monolithic integrated circuit characterization by a picosecond optoelectronic technique, *IEEE Trans. Microwave Theory Tech., vol. MTT-37, 8*, pp. 1223–31.

Inter Continental Microwave (1991) Santa Clara: *Test Fixtures and Calibration Standards.*

Jones K. (1989) Wafer probe captures test data at 65 GHz, *Microwave and RF, 28, 11*, pp. 146–9.

Jones K.E., Strid E.W., Gleason K.R. (1987) Mm-wave wafer probes span 0 to 50 GHz, *Microwave J., 4*.

Judish R.M., Engen G.F. (1987) On-line accuracy assessment for the dual six-port ANA: Statistical methods for random errors, *IEEE Trans. Instrum. Meas., vol. IM-36*, 2, pp. 507–13.

Juroshek J.R., Hoer C.A., Kaiser R.F. (1989) Calibrating Network Analyzers with Imperfect Test Ports, *IEEE Trans. Microwave Theory Tech., vol. MTT-38, 4*, pp. 898–901.

Kneppo I. (1988) *Microwave Measurement by Comparison Methods*, Elsevier, Amsterdam.

Kolner B.H., Bloom D.M. (1986) Electro-optic sampling in GaAs integrated circuits, *IEEE J. Quantum. Electron., vol. 22, 1*, pp. 97–3.

Kuhn N.J. (1983) Gettong Started with Automatic Noise Figure Measurements, *Microwave Systems News, 13, 1*, pp. 120–36.

Lang R.J., Jewett W.P., Merrill J.D. (May 1988) Test fixtures for frequencies from DC to 75 GHz, *Microwave J.*, pp. 365–71.

Lau W.Y. (1989) Network analysis verifies models in CAD packages, *Microwaves & RF, vol., 28, 11*, pp. 99–110.

Laverghetta T.S. (1976) Handbook of Microwave Testing, *Artech House Inc.*, Dedham.

Laverghetta T.S. (1991) *Modern Microwave Measurements and Techniques*, Artech House, Boston.

Majidi-Ahy R., Auld B.A., Bloom D.M. (1989) 100 GHz on-wafer S-parameter measurements by electrooptic sampling, *IEEE MTT Symp. Dig.*, pp. 299–302.

Marks R., Phillips K. (June 1989) Wafer-level ANA calibration at NIST, *34-t ARFTG Conf. Dig.*.

Maury Microwave Corporation (1990) Cucamonga: Automated Tuner System, *Technical Data 4T-050A.*

Maury Microwave Corporation (1991) Cucamonga: Noise Measurement

Instrumentation, *Technical Data 4N-001.*
Maury Microwave Corporation (1990/91) Precision Microwave Components & Instrumentation, *Short Form Catalog*, Cucamonga.
Meys R.P. (1978) A wave approach to the noise properties of linear microwave devices, *IEEE Trans. Microwave Theory Tech., vol. MTT-26, 1*, pp. 34–7.
Mitama M., Katoh H. (1979) An improved computational method for noise parameter measurement, *IEEE Trans. Microwave Theory Tech., vol. MTT-27*, pp. 612.
Oldfield L.C. (1989) Power measurement, in *Microwave Measurements*, A. E. Bailey, ed., Peter Peregrinus Ltd., London.
Pantoza R.R. *et al.* (1989) Improved calibration and measurement of the scattering parameters of microwave integrated circuits, *IEEE Trans. Microwave Theory Tech., vol. MTT-37, 11*, pp. 1675–80.
Pavio A.M., Pavio J.S., Chapman J.E., Jr., Boggan G. H. (1991) Computer-Aided Manufacturing of Advanced Microwave Modules, *Int. J. Microwave Millimeter-Wave CAE, 1, 1*, pp. 90–11.
Reeve G., Marks R., Blackburn D. (1990) Microwave Monolitic Integrated Circuit-Related Metrology at the National Institute of Standards and Technology, *IEEE Trans. Instrum. Meas., IM-36, 12*, pp. 958–61.
Rodriquez-Tellez J. (March 1991) A thick-film wafer probe for microwave measurements, *Microwave J.*, pp. 133–41.
Rodwell M.J.W. *et al.* (1986) Internal microwave propagation and distortion characteristics of traveling-wave amplifiers studied by electrooptic sampling, *IEEE Trans. Microwave Theory Tech., vol. MTT-34, 12*, pp. 1356–61.
Ross P.B., Geller B.D. (May 1987) A broadband microwave test fixture, *Microwave J.*, 233–48.
Sinclair M.W. (1989) Noise measurements, in *Microwave Measurements*, A. E. Bailey, ed., Peter Peregrinus Ltd., London.
Speciale A. (1977) A generalization of the TSD network-analyzer calibration procedure, covering n-port scattering parameter measurements, affected by leakage errors, *IEEE Trans. Microwave Theory Tech., vol. MTT-25, 12*, pp. 1100–15.
Staudinger J., Seely W. (February 1987) MMIC tests improved with standards on chip, *Microwave & RF, vol. 26*, pp. 107–14.
Strid E. (1981) Noise Measurements for Low-Noise GaAs FET Amplifiers, *Microwave Systems News, 11*, pp. 62–70.
Strid E.W., Burchan T. (August 1989) Wideband probing techniques for planar devices, *Solid State Technology*, pp. 49–50.
Valdmanis J.A., Mourou G.A., Gabel C.W. (1983) Subpicosecond electrical sampling, *IEEE J. Quantum Electron, vol., 19, 4*, pp. 664–7.
Warner F.L. (1989a) Attenuation measurement, in *Microwave Measurements*, A. E. Bailey, ed., Peter Peregrinus Ltd., London.
Warner F.L. (1989b) Microwave vector network analysers, in *Microwave Measurements*, A. E: Bailey, ed., Peter Peregrinus Ltd., London.

Weingarten K.J. *et al.* (1985) Direct electro-optic sampling of GaAs integrated circuits, *Electron. Lett., vol. 21*, pp. 765–6.

Weingarten K.J., Majidi-Ahy R., Bloom D.M. (1987) GaAs integrated circuit measurements using electrooptic sampling, *1987 IEEE GaAs Symp. Dig., 11–14.*

Weingarten K.J., Rodwell M.J.W, Bloom D.M. (1988) Picosecond optical sampling of GaAs integrated circuits, *IEEE J. Quantum Electron., vol. 24, 2*, pp. 198–220.

Williams D. (December 1989) On-wafer microwave standards at NIST, *34-th ARFTG Conf. Dig.*.

Williams D. F., Miers T. H. (1988) De-embeding coplanar probes with planar distributed standards, *IEEE Trans. Microwave Theory Tech., vol. MTT-36, 12*, pp. 1876–80.

Wiltron (1989) Morgan Hill: *Catalog.*

Index